# Detection and Management of New and Emerging Mystery Plant Virus Sources

# Detection and Management of New and Emerging Mystery Plant Virus Sources

*Edited By:*

**Alireza Golnaraghi, PhD**
**Rajarshi Kumar Gaur, PhD**
**Rakesh Kumar Verma, PhD**

AAP | APPLE ACADEMIC PRESS

First edition published 2025

**Apple Academic Press Inc.**
1265 Goldenrod Circle, NE,
Palm Bay, FL 32905 USA

760 Laurentian Drive, Unit 19,
Burlington, ON L7N 0A4, CANADA

**CRC Press**
2385 NW Executive Center Drive,
Suite 320, Boca Raton FL 33431

4 Park Square, Milton Park,
Abingdon, Oxon, OX14 4RN UK



**Library and Archives Canada Cataloguing in Publication**

CIP data on file with Canada Library and Archives

**Library of Congress Cataloging-in-Publication Data**

CIP data on file with US Library of Congress

ISBN: 978-1-77491-662-9 (hbk)
ISBN: 978-1-77491-663-6 (pbk)
ISBN: 978-1-00349-313-6 (ebk)

# About the Editors

**Alireza Golnaraghi, PhD**
*Assistant Professor, Department of Plant Protection, Science and Research Branch, Islamic Azad University, Tehran, Iran; Researcher, Department of Earth, Ocean & Atmospheric Sciences, University of British Columbia, Vancouver, British Columbia, Canada; Researcher, Department of Biodiversity, BoomZista Institute, Vancouver, British Columbia, Canada*

Alireza Golnaraghi earned his PhD in 2008, focusing on the biological, serological, and molecular characteristics of a Tospovirus isolated from a potato and its detection through modern techniques such as antibody microarrays. During his PhD thesis, he had the opportunity to collaborate with Prof. Kazusato Ohshima and his team at Saga University in Japan. Dr. Alireza's research primarily revolves around the detection and identification of DNA and RNA plant viruses, as well as their potential vectors. He has been involved in numerous national and international projects and has made significant contributions to the study of viruses and endophytic microbes, particularly in wild plants within natural ecosystems. He has published 53 national and international papers, authored two books, contributed to two book chapters, and presented over 65 papers at national and international conferences. In 2017, he joined a renowned environmental microbiology/virology group at the Department of Earth, Ocean, and Atmosphere Sciences at the University of British Columbia, Canada, where he has been conducting research on algal viruses in collaboration with the esteemed scientist, Prof. Curtis Suttle. In 2020, Alireza, along with international researchers and talented students, established a non-profit and nature-based organization in Canada called BoomZista Institute (boomzista.org), with a focus on various environmental issues. The institute aims to promote sciences, especially those relate to nature and our beautiful planet the Earth, to people worldwide, with an emphasis on children and young adults.

**Rajarshi Kumar Gaur, PhD**
*Professor, Department of Biotechnology, Deen Dayal Upadhyaya Gorakhpur University, Gorakhpur, Uttar Pradesh, India*

Rajarshi Kumar Gaur, PhD, is currently serving as a Professor and Head of the Department of Biotechnology at Deen Dayal Upadhyaya Gorakhpur University, Gorakhpur, Uttar Pradesh, India. He earned his PhD in 2005, focusing on the molecular characterization of sugarcane viruses such as mosaic, streak mosaic, and yellow luteovirus. He was awarded a MASHAV fellowship by the Government of Israel for his postdoctoral studies and conducted research at The Volcani Centre, Israel, and Ben-Gurion University, Negev, Israel. In 2007, he received a Visiting Scientist Fellowship from the Swedish Institute Fellowship, Sweden, and worked at Umea° University, Umea°, Sweden. Additionally, he was granted a postdoctoral fellowship from the International Center for Genetic Engineering and Biotechnology (ICGEB), Italy, in 2008. Dr. Gaur has made significant contributions to the field of sugarcane viruses and has published 130 national and international papers, authored 22 edited books, and presented approximately 50 papers at national and international conferences. He has been honored as a Fellow of the Linnaean Society, Fellow of the Royal Society of Biology, Fellow of the Society of Plant Research, Fellow of the Society of Applied Biology (FSAB), Fellow of the International Society of Biotechnology (FISBT), and Fellow of the Indian Virological Society. He has received numerous other awards, including the Prof. B.M. Johri Memorial Award from the Society of Plant Research (SPR), the Excellent Teaching Award from Astha Foundation, Meerut; the UGC Research Teacher Award and the Young Scientist Award 2012 in Biotechnology from the Society of Plant Research (SPR); and the Scientific and Applied Research Centre Gold Medal Award 2011 for Outstanding Contribution in the Field of Biotechnology. Dr. Gaur is also a member of the National Academy of Science India (NASI). He has visited several laboratories in the United States, Canada, New Zealand, United Kingdom, Thailand, Sweden, and Italy. Currently, he is involved in managing various national and international grants and international collaborative projects related to plant viruses and disease management.

**Rakesh Kumar Verma, PhD**

*Assistant Professor, Department of Biosciences, School of Liberal Arts and Sciences, Mody University of Science and Technology, Lakshmangarh, Sikar, Rajasthan, India*

Rakesh Kumar Verma, PhD, is an Assistant Professor in the Department of Biosciences, School of Liberal Arts and Sciences, Mody University of Science and Technology, Lakshmangarh, Sikar, Rajasthan, India. He is a life member of the Indian Science Congress Association (ISCA), the World Society of Virology (WSV), and The Association of Microbiologists of India. He received young scientist awards during several national conferences. During his PhD program, he worked on the molecular resolution of genetic variability in different potyviruses of India. He made a significant contribution to the diagnosis, management, and phylogeny of potyviruses and begomoviruses by using various molecular and bioinformatic tools. He is currently working on the molecular characterization of endophytes isolated from the Thar Desert and their plant growth promotion potential under abiotic and biotic stress conditions. He has published more than 25 research papers in journals of international repute and 10 book chapters and presented numerous papers at international and national conferences. Currently, he is handling two projects on endophytes and their involvement in biotic and abiotic stress management.

# Contents

# Contributors

**S. Ahmad**
Central Lab Sugarcane Research Institute, Shahjahanpur, Uttar Pradesh, India

**Sohail Akhtar**
University of Agriculture Faisalabad, Sub-Campus Burewala, Pakistan

**Md. Shamim Akhter**
Plant Pathology Division, Bangladesh Agricultural Research Institute (BARI), Joydebpur, Bangladesh

**Ritika Bandyopadhyay**
Department of Microbiology, Barrackpore Rastraguru Surendranath College, Kolkata, West Bengal, India

**Sachin Ashok Bhor**
Laboratory of Plant Molecular Biology and Virology, Faculty of Agriculture, Ehime University, Matsuyama, Ehime, Japan

**Irena Budzanivska**
Virology Department, ESC 'Institute of Biology and Medicine,' Taras Shevchenko National University of Kyiv, Kyiv, Ukraine

**Vanita Chandel**
Amity Institute of Virology and Immunology, Amity University Uttar Pradesh, Noida, Uttar Pradesh, India

**Ranjana Chauhan**
Insect Defense Laboratory, Molecular Biology and Biotechnology Division, CSIR–National Botanical Research Institute, Rana Pratap Marg, Lucknow, Uttar Pradesh, India

**Deepika Choudhary**
Department of Chemistry, University of Rajasthan, Jaipur, Rajasthan, India

**Sunny Dhir**
Department of Biotechnology, Maharishi Markandeshwar (Deemed to be University), Mullana, Ambala, Haryana, India

**Yashika Dhir**
Department of Biotechnology, Maharishi Markandeshwar (Deemed to be University), Mullana, Ambala, Haryana, India

**R. K. Gaur**
Department of Biotechnology, Deen Dayal Upadhyaya Gorakhpur University, Gorakhpur, Uttar Pradesh, India

**Gokul Uttamgir Gosavi**
Institute of Plant Protection, The Graduate School of Chinese Academy of Agricultural Sciences, Beijing, China

**Shaikhul Islam**
Plant Pathology Division, Bangladesh Wheat and Maize Research Institute, Nashipur, Dinajpur 5200, Bangladesh; The United Graduate School of Agricultural Science, Ehime University, Matsuyama, Ehime, Japan

**A. A. Khan**
Department of Botany, Gandhi Faiz-E-Aam College, Shahjahanpur (Affiliated with M.J.P.R.U. Bareilly), Uttar Pradesh, India

**Alok Kumar**
School of Plant Sciences, College of Agriculture and Environmental Sciences, Haramaya University, Dire Dawa, Ethiopia

**Rakesh Kumar**
Plant Pathology Department, Kalash Seeds Pvt. Ltd. Jalna, Maharashtra, India

**Anirban Mandal**
Department of Microbiology, Mrinalini Datta Mahavidyapith, Kolkata, West Bengal, India

**Avinash Marwal**
Department of Biotechnology, Mohanlal Sukhadia University, Udaipur, Rajasthan, India

**Manisha Mishra**
Insect Defense Laboratory, Molecular Biology and Biotechnology Division, CSIR–National Botanical Research Institute, Rana Pratap Marg, Lucknow, Uttar Pradesh, India

**Parneeta Mishra**
Insect Defense Laboratory, Molecular Biology and Biotechnology Division, CSIR–National Botanical Research Institute, Rana Pratap Marg, Lucknow, Uttar Pradesh, India

**Arindam Mukherjee**
Department of Botany, Mrinalini Datta Mahavidyapith, Kolkata, West Bengal, India

**Chitra Nehra**
Department of Biosciences, Mody University of Science and Technology, Lakshmangarh, Sikar, Rajasthan, India

**Rashmi Pandey**
Insect Defense Laboratory, Molecular Biology and Biotechnology Division, CSIR–National Botanical Research Institute, Rana Pratap Marg, Lucknow, Uttar Pradesh, India

**Anandrao Ashok Patil**
Department of Biology, University of Washington, Seattle, Washington, USA

**Asha Rani**
Amity Institute of Virology and Immunology, Amity University Uttar Pradesh, Noida, Uttar Pradesh, India

**Narayan Rishi**
Amity Institute of Virology and Immunology, Amity University Uttar Pradesh, Noida, Uttar Pradesh, India

**A. Sajjad**
Department of Botany, Gandhi Faiz-E-Aam College, Shahjahanpur (Affiliated with M.J.P.R.U. Bareilly), Uttar Pradesh, India

**Sharad Saurabh**
Insect Defense Laboratory, Molecular Biology and Biotechnology Division, CSIR–National Botanical Research Institute, Rana Pratap Marg, Lucknow, Uttar Pradesh, India

**Muhammad Shafiq Shahid**
Department of Plant Sciences, College of Agricultural and Marine Sciences, Sultan Qaboos University, Al-Khod, Muscat, Oman

**Oleksiy Shevchenko**
Virology Department, ESC 'Institute of Biology and Medicine,' Taras Shevchenko National University of Kyiv, Kyiv, Ukraine

**Tetiana Shevchenko**
Virology Department, ESC 'Institute of Biology and Medicine,' Taras Shevchenko National University of Kyiv, Kyiv, Ukraine

**Anoop K. Shukla**
Insect Defense Laboratory, Molecular Biology and Biotechnology Division, CSIR–National Botanical Research Institute, Rana Pratap Marg, Lucknow, Uttar Pradesh, India

**Pradhyumna K. Singh**
Insect Defense Laboratory, Molecular Biology and Biotechnology Division, CSIR–National Botanical Research Institute, Rana Pratap Marg, Lucknow, Uttar Pradesh, India

**Sneha Sinha**
Department of Biotechnology, St. Columbus College, Binoba Bhave University, Hazaribagh, Jharkhand, India

**Halyna Snihur**
Virology Department, ESC 'Institute of Biology and Medicine,' Taras Shevchenko National University of Kyiv, Kyiv, Ukraine

**Rashmi Suman**
Amity Institute of Virology and Immunology, Amity University Uttar Pradesh, Noida, Uttar Pradesh, India

**A. K. Tiwari**
Central Laboratory, Sugarcane Research Institute, Shahjahanpur, Uttar Pradesh, India

**Ashish Vishwakarama**
Insect Defense Laboratory, Molecular Biology and Biotechnology Division, CSIR–National Botanical Research Institute, Rana Pratap Marg, Lucknow, Uttar Pradesh, India

# Preface

Plant virus infections are estimated to cost billions of dollars in economic losses. Despite intensive efforts to manage and prevent plant viruses and their potential vectors in crop production processes, the threat posed by newly emerging or re-emerging viruses is becoming more common and poses potential threats even in previously uninfected areas. In all damages caused by these viruses, we can find virus diversifying events, appearance, and dissemination of efficient vectors, weather events, and most importantly, human activities. Plant viruses are widely distributed, found in many different sources, and considered a major threat to agriculture worldwide. Changes in land use, significant exchanges of plant materials, introduction of new plants to native ecosystems, and decrease of biodiversity have increasingly enhanced the possibility of emergence/re-emergence and outbreaks of new viruses worldwide.

Over the past decades, viral outbreaks have increased at an alarming rate. The outbreak of Coronavirus disease (COVID-19) and the Coronavirus period have alerted us to and emphasized the urgent need for efficient and cost-effective diagnostic techniques. To prevent the dispersal of plant viruses, disease management heavily relies on quarantine programs, certification of plant materials, and control of natural vectors. However, it is unlikely to have specific detection systems available at all desired locations, such as national borders. Therefore, diagnosing plant viruses at the border remains an ongoing challenge that requires a simple, rapid, sensitive, and accurate method.

Moreover, existing diagnostic methods may be expensive and ineffective in identifying many viruses. The extensive global trade of plant materials can lead to the rapid spread of infectious diseases, especially in developing/nondeveloped economies where specific detection of individual viruses is costly. Additionally, simultaneous infections of individual hosts with various viruses and viroids have been described in many studies. The use of multiplex detecting systems could be an efficient option for accurately detecting a greater number of viruses and viroids. Such diagnostic systems can simplify quarantine programs by reducing the number of tests, the amount of materials used, and the testing duration.

Alternatively, a broad-spectrum or universal detection of viruses and viroids may be necessary when no prior knowledge of the target pathogen exists, such as when testing wild plants, insect vectors, water, or soil as possible virus sources. However, conventional methods may not effectively control plant viruses in the future due to climate change and global warming, which are expected to expand the distribution of plant viruses to regions where no previous reports of their occurrence are available. This highlights the need for next-generation approaches and new technologies to overcome future outbreaks of plant viral diseases.

In this multi-authored book, the editors have undertaken the crucial task of reviewing the progress made and the challenges that lie ahead in the field of plant viruses. The book aims to uncover the hidden world of plant viruses, their impact on agriculture and food security, and the necessity for new detection systems and management strategies. Titled *Detection and Management of New and Emerging Mystery Plant Virus Sources*, the book covers various aspects of plant viruses in two parts: (I) Recent Approaches for Plant Virus Detection; and (II) Recent Trends in Viral Disease Control. Distinguished scientists from around the world have contributed their experiences and opinions, enriching the field of science.

This book illuminates the significance of undiscovered plant viruses and provides valuable information on their detection and management. It is expected to capture the attention of the general public, raising awareness about this area of study. We believe that readers will find this book instrumental in shaping their understanding of how to mitigate the potential threats posed by enigmatic plant viruses to plant production.

# PART I

# Recent Approaches for Plant Virus Detection

CHAPTER 1

# Advancement in Detection and Diagnostics of Plant Viruses

RAKESH KUMAR,[1] SNEHA SINHA,[2] and ALOK KUMAR[3]

[1]*Plant Pathology Department, Kalash Seeds Pvt. Ltd. Jalna, Maharashtra, India*

[2]*Department of Biotechnology, St. Columbus College, Binoba Bhave University, Hazaribagh, Jharkhand, India*

[3]*School of Plant Sciences, College of Agriculture and Environmental Sciences, Haramaya University, Dire Dawa, Ethiopia*

## ABSTRACT

Plant viruses are causing a number of economically important diseases in agricultural crops, resulting in an estimated loss of more than $60 billion in the global economy. The rapid evolution of viruses, driven by climate change and the transportation of planting materials to distant locations, has led to the spread of viruses to new areas and hosts. As the management strategy relies heavily on virus diagnosis and identification, it is crucial to implement a fast and accurate detection method to reduce losses caused by plant viral diseases.

Currently, more than 233 families and 2,606 genera of plant viruses have been reported in different crops, but many remain unknown, especially in weeds and wild plant species. Initially, viruses were identified by observing the symptoms developed on the host and indicator plants. However, this methodology does not always provide accurate information, as similar symptoms can be produced by other viruses or abiotic stress factors. Microscopic detection and diagnosis of plant viruses only provide information about their family but cannot identify the viruses at the species level. Similarly, antisera and genome-based techniques have their own pros and cons.

Detection and Management of New and Emerging Mystery Plant Virus Sources.
Alireza Golnaraghi, Rajarshi Kumar Gaur, and Rakesh Kumar Verma (Eds.)

This chapter will provide a summary of all available detection and diagnostic methods for plant viruses, along with the merits and limitations of each technique. This information will help in finalizing strategies to protect crops from virus infections and ultimately improve crop yield.

## 1.1 INTRODUCTION

Plant viruses are one of the major biotic constraints in world agriculture, causing qualitative and quantitative losses in crops. Viruses are known to cause up to 40% crop loss, with an estimated value of over 60 billion dollars per year in the global economy (Maksimov et al., 2019). This loss is gradually increasing due to the introduction of invasive pathogens into our ecosystem, resulting from increased global trade, transportation, and tourism. These invasive pathogens not only affect the health of agricultural and horticultural crops but also impact the biodiversity of these crops. Therefore, it is necessary to protect plants from these invasive pathogens. To achieve this, it is important to detect, diagnose, and identify the pathogens in the early stages of infection.

Plant viruses are ultra-microscopic in nature and depend on their host for survival. The virus can be transmitted to other plants/hosts either through vectors or by mechanical methods. Once infested, it utilizes the host machinery for multiplication and causes a reduction in plant growth, fruit quality, and lower yield, ultimately leading to a loss in crop productivity. The genome of plant viruses is small, encoding only a few proteins. The majority of plant viruses are known to have an RNA genome (Karen-Beth et al., 2011). Both DNA and RNA viruses evolve based on their host and environmental conditions (Elena et al., 2005). Evolution is the result of mutations in the viral genome, which are found to be higher in RNA viruses compared to DNA viruses, causing serious damage to crop and being considered more significant (Zhao et al., 2020). The rate of evolution in plant viruses is rapid due to an increase in mono-cultivation and international trading of agricultural materials and products. Due to this rapid evolution in the viral genome, the number of plant viruses is also increasing rapidly, and currently, a total of 10,434 species of viruses are known, which have been classified into 233 families and 2,606 genera in the current ICTV report (ICTV, 2022).

Management is the only option for controlling viral diseases as no direct method is available to control viruses, like other pathogens. Currently, viruses are managed through the eradication of virus sources, cultural practices, vector control, use of virus-free planting materials, use of resistant varieties, and quarantine practices (Kumar et al., 2022; Jones & Naidu, 2019). However, management practices entirely depend upon the accurate detection and diagnosis of viruses. Therefore, significant advancements have been made in the techniques used for the detection and precise identification of viruses (Figure 1.1). Identification of viruses based on the symptoms they cause on the host and the indicator plants were effective in the past. However, as the number of viruses increases, it becomes difficult to detect and differentiate closely related virus isolates based on symptoms alone. Additionally, symptoms developed in the case of mixed infection and sometimes due to abiotic stress can misguide in the identification of the causal agents. Therefore, it is not recommended to rely solely on symptoms. However, symptoms can provide initial information about virus infection in host plants and can be effective when combined with other advanced techniques. Other available techniques are based on microscopy, serology, and molecular methods. Each technique has its own benefits and limitations. Electron Microscopy (EM) is simple but not cost-effective; serological methods are cost-effective but less sensitive and time-consuming; molecular methods such as polymerase chain reactions (PCRs) are sensitive but fairly expensive; sequencing is precise but costly. Here, we are describing all the possible approaches that can be used for the detection, diagnosis, and identification of plant viruses in their chronological stage of development.

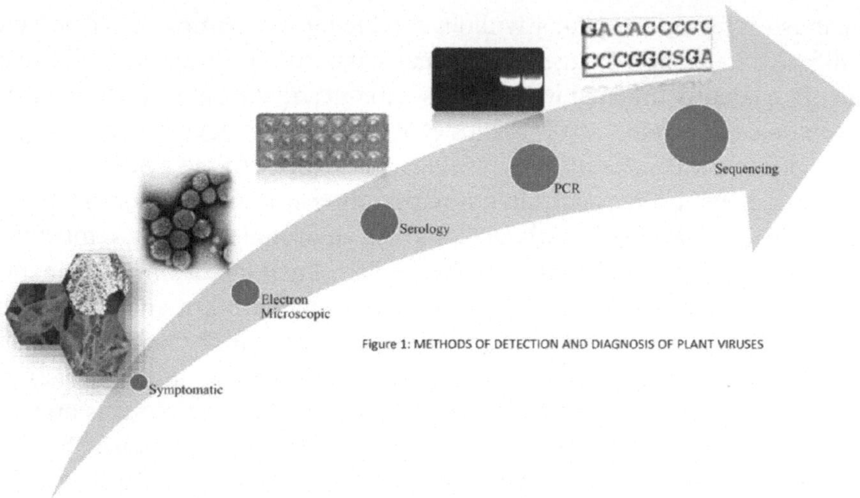


**FIGURE 1.1** Methods of detection and diagnosis of plant viruses.

## 1.2 SYMPTOM-BASED DETECTION

Generally, visual observation of infected plants and seeds is the most preferred method used for the detection and diagnosis of pathogens. In the beginning of the 20$^{th}$ century, symptoms of viral diseases affecting plants were proposed as a basis for the taxonomy of viruses (Scholthof et al., 2011). Detectable changes in the color, shape, size, and/or morphology of a plant and its parts are considered as symptoms that may develop in response to either biotic factors (pathogens such as fungus, bacteria, mycoplasma, virus, viroid, and nematodes) or abiotic factors (such as high or low temperature, excessive or insufficient water, pH, high salinity, chemical injury, heavy metal, ultraviolet radiation, nutritional deficiencies, etc.). It has also been observed that abiotic factors sometimes produce symptoms similar to those developed due to biotic factors, especially in the case of virus infection (Kennelly et al., 2012). However, diseases/symptoms caused by viruses are infectious and may spread within and between plants, whereas the same is not transmissible in the case of abiotic factors. Upon infection, viruses may develop local symptoms such as leaf spots or necrotic spots at their entry point. They may also exhibit systemic symptoms such as mosaic patterns, leaf curling, leaf rolling, or vein thickening in other parts of the plant away from the entry point. These distinct symptoms on the host and indicator plants aid in the detection and diagnosis of viruses based on their symptomatology. However, it has also been observed that sometimes viruses can establish infection in plants without producing any symptoms (Roossinck, 2005, 2010; Richert-Poggeler & Minarovits, 2014). In nature, wild plants are often latently infected with viruses without showing any apparent disease symptoms (Min et al., 2012; Shates et al., 2019). The development of symptoms due to virus infestation, as well as the severity of the symptoms due to diseases, have a direct correlation with a three-way interaction of the host (plant species/varieties), pathogen, and the environment (light, temperature, humidity, soil, pH, and fertility) (Figure 1.2). Each factor limits the expression of symptoms on the host plants. For example, okra enation leaf curl virus (OELCV) causes severe upward curling, vein thickening, and stunting on okra grown in coastal and humid areas but produces mild symptoms in other areas (Kumar et al., 2019). Similarly, cotton leaf curl virus is prevalent in cotton fields in northern India, but no incidence of this virus has been reported in southern India. Additionally, it has been observed that a virus can produce a variety of symptoms on different hosts, and many different viruses can develop similar symptoms on a single host (Holkar et al., 2017;

Takahashi et al., 2019). In both cases, identifying the exact causal virus based solely on symptomatic methods is difficult and needs to be supplemented with other detection techniques. Takahashi et al. (2019) reported the major viruses affecting crops worldwide and their peculiar symptoms on hosts.

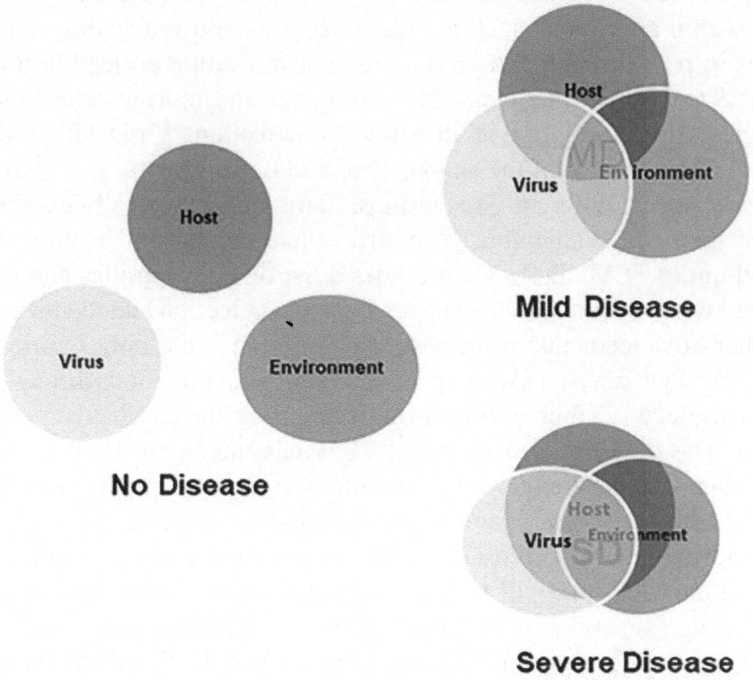


**FIGURE 1.2** Three-way interaction of factors affecting disease expression in plants.

## 1.3 MICROSCOPIC DETECTION

The size of plant viruses ranges from 20–500 nm in diameter and 700–1,000 nm in length (Britannica, 2021). These viruses cannot be seen alone in a light microscope. However, certain viruses such as tobacco mosaic virus (TMV), potato virus X (PVX), wheat mosaic virus (WMV), etc., produce clusters in the form of inclusions or Ivanovsky crystals after infection, which can be observed using light microscopes (Baker et al., 1985; Steere et al., 1953).

Electron microscopy (EM) has played an important role in plant virology and is considered a powerful tool for the detection, diagnosis, and classification of unknown viruses. It helps elucidate the shape, size, and surface details of the viruses. The technique is simple, as viruses can easily be identified using the sap of infected plant parts, without knowing their genetic composition. However, a high level of skill and expertise is required for virus identification using this technique.

The transmission electron microscope (TEM) is analogous to the compound light microscope and is capable of generating high-resolution images of both biological and non-biological specimens, including viruses. It utilizes thin specimens such as tissue sections and molecules, allowing electrons to pass through and producing high-resolution projection images. Compared to indirect methods like serological and molecular techniques, TEM offers the possibility of direct virus detection. Virus identification through TEM is based on the shape, size, and/or number of viral particles, which vary among different virus groups. Plant viruses typically exhibit one of four shapes (rod, filamentous, isometric or icosahedral, and bacillus) when observed under TEM. Table 1.1 provides a list of virus families and genera associated with specific particle shapes for virus detection and identification.

Further advancements in microscopic techniques not only improve the diagnostics of viruses, but also help in localizing viral infections through viral indicator molecules, thus aiding in understanding the mechanism of viral infection. The detection sensitivity of TEM has increased when combined with serology. The technique is named immune electron microscopy (IEM), where specific antiserum against a viral antigen is used to efficiently trap viral particles onto the grid used in EM, improving the efficiency of visualizing viral particles under EM. This technique has been efficiently used for the detection of several viruses, such as bean common mosaic virus (BCMV) and yellow mosaic virus (YMV) (Li et al., 2007; Kothandaraman et al., 2016). Solid phase immune electron microscopy (SPIEM) and immunosorbent electron microscopy (ISEM) are variants of IEM primarily used in viral diagnosis (Bhardwaj et al., 2020). Cryo-electron microscopy (cryo-EM), an advanced microscopy technique, provides nearly atomic-level structures of viruses, thus aiding in understanding the structural confirmations of viruses and investigating plant-virus interactions through the localization of viral indicator molecules (Zamora et al., 2017; Ertel et al., 2017). In total, EM is an efficient technique for studying plant viruses, and further improvements in the technique may play a significant role in plant virus detection, diagnosis, and classification.

## 1.4 SEROLOGICAL DETECTION

Serological methods are the most commonly used methods for routine diagnosis of plant viruses (Mehetre et al., 2021) and are based on the specific

**TABLE 1.1** Classification of Viruses Based on Shape, Structure, and Genome Under Microscope

| Shape of Virus | Genome (Number of Particle) | Family | Genus | Type Species | Structure of Viris Under Microscope |
|---|---|---|---|---|---|
| Rod-shaped | (+)ssRNA (1) | – | Tobamovirus | Tobacco mosaic virus |  |
| – | (+)ssRNAs (2) | – | Tobravirus | Tobacco rattle virus |  |
| – | (+)ssRNAs (2) | – | Furovirus | Soilborne wheat mosaic virus | |
| – | – | – | Pecluvirus | Peanut clump virus | |
| – | (+)ssRNAs (3) | – | Hordeivirus | Barley stripe mosaic virus |  |
| – | – | – | Pomovirus | Potato mop-top virus | |
| – | (+)ssRNAs (4) | – | Benyvirus | Beet necrotic yellow vein virus |  |
| Filamentous | (+)ssRNA (1) | – | Allexivirus | Shallot virus X |  |
| – | – | – | Carlavirus | Carnation latent virus | |
| – | – | – | Foveavirus | Apple stem pitting virus | |
| – | – | – | Potexvirus | Potato virus X | |
| – | – | – | Capillvirus | Apple stem grooving virus | |
| – | – | – | Trichovirus | Apple chlorotic leafspot virus | |
| – | – | – | Vitivirus | Grapevine virus A | |
| – | – | Potyviridae | Potyvirus | Potato virus Y |  |
| – | – | Potyviridae | Ipomovirus | Sweet pot mild mottle | |
| – | – | Potyviridae | Macluravirus | Maclura mosaic virus | |
| – | – | Potyviridae | Rymovirus | Ryegrass mosaic virus | |
| – | – | Potyviridae | Tritimovirus | Wheat streak mosaic virus | |
| – | – | Potyviridae | Bymovirus | Barley yellow mosaic virus | |

**TABLE 1.1** *(Continued)*

| Shape of Virus | Genome (Number of Particle) | Family | Genus | Type Species | Structure of Viris Under Microscope |
|---|---|---|---|---|---|
| – | (+)ssRNA (1 or 2) | Closteroviridae | Closterovirus | Beet yellows virus |  |
| – | – | Closteroviridae | Crinivirus | Lettuce inf. yellows virus | |
| Isometric | (+)ssRNA (1) | – | Sobemovirus | Southern bean mosaic virus |  |
| – | – | – | Marafivirus | Maize rayado fino virus | |
| – | – | – | Umbravirus | Carrot mottle virus | |
| – | – | – | Tymovirus | Turnip yellow mosaic virus | |
| – | (+)ssRNAs (2) | – | Idaeovirus | Raspberry bushy dwarf virus | – |
| – | ss(+)RNA (1) | Sequiviridae | Sequivirus | Parsnip yellow fleck virus |  |
| – | – | Sequiviridae | Waikavirus | Rice tungro spherical virus | |
| – | – | Tombusviridae | Tombusvirus | Tomato bushy stunt virus | |
| – | – | Tombusviridae | Aureusvirus | Pothos latent virus | |
| – | – | Tombusviridae | Avenavirus | Oat chlorotic stunt virus | |
| – | – | Tombusviridae | Carmovirus | Carnation mottle virus | |
| – | – | Tombusviridae | Dianthovirus | Carnation ring spot virus | |
| – | – | Tombusviridae | Machlomovirus | Maize chlorotic mottle virus | |
| – | – | Tombusviridae | Necrovirus | Tobacco necrosis virus | |
| – | – | Tombusviridae | Panicovirus | Panicum mosaic virus | |
| – | – | Luteoviridae | Luteovirus | Barley yellow dwarf virus | |
| – | – | Luteoviridae | Polerovirus | Potato leafroll virus | |
| – | – | Luteoviridae | Enamovirus | Pea enation mosaic virus | |

**TABLE 1.1** *(Continued)*

| Shape of Virus | Genome (Number of Particle) | Family | Genus | Type Species | Structure of Viris Under Microscope |
|---|---|---|---|---|---|
| – | ss(+)RNAs (2) | Comoviridae | Comovirus | Cowpea mosaic virus | 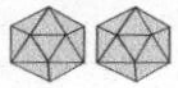 |
| – | – | Comoviridae | Fabavirus | Broad bean wilt virus | |
| – | – | Comoviridae | Nepovirus | Tobacco ring spot virus | |
| – | ss(+)RNAs (3) | Bromoviridae | Bromovirus | Brome mosaic virus |  |
| – | – | Bromoviridae | Cucumovirus | Cucumber mosaic virus | |
| – | – | Bromoviridae | Alfamovirus | Alfalfa mosaic virus | |
| – | – | Bromoviridae | Ilarvirus | Tobacco streak virus | |
| – | – | Bromoviridae | Oleavirus | Olive latent virus 2 | |
| – | dsRNA | Reoviridae | Phytoreovirus | Wound tumor virus |  |
| – | – | Reoviridae | Fijivirus | Fiji disease virus | |
| – | – | Reoviridae | Oryzavirus | Rice ragged stunt virus | |
| – | – | Partitiviridae | Alphacryptovirus | White clover crypto. virus 1 | |
| – | – | Partitiviridae | Betacryptovirus | White clover crypto. virus 2 | |
| – | – | Partitiviridae | Varicosavirus | Lettuce big-vein virus | |
| – | dsDNA | Caulimoviridae | Caulimovirus | Cauliflower mosaic virus |  |
| – | – | Caulimoviridae | – | Soybean chlorotic mottle virus-like | |
| – | – | Caulimoviridae | – | Cassava vein mosaic virus-like | |
| – | – | – | – | Petunia vein clearing virus-like | |
| – | – | Caulimoviridae | Badnavirus | Commelina yellow mottle virus | |
| – | – | Caulimoviridae | – | Rice tungro bacilliform virus-like | |

**TABLE 1.1** *(Continued)*

| Shape of Virus | Genome (Number of Particle) | Family | Genus | Type Species | Structure of Viris Under Microscope |
|---|---|---|---|---|---|
| – | (+)ssDNA | Geminiviridae | Mastrevirus | Maize streak virus | 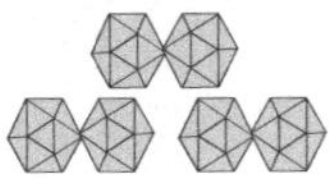 |
| – | – | Geminiviridae | Curtovirus | Beet curly top virus | |
| – | – | Geminiviridae | Begomovirus | Bean golden mosaic virus | |
| – | – | Geminiviridae | Topocuvirus | Tom. pseudocurly top vrius | |
| – | – | Circoviridae | Nanovirus | Subteranean clover stunt virus | |
| Membranous circular | – | Bunyaviridae | Tospovirus | Tomato spotted wilt virus | 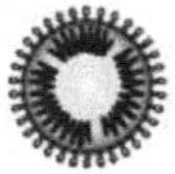 |
| Thin flexuous multipartite | – | – | Tenuivirus | Rice stripe virus | 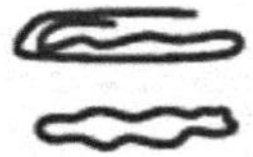 |
| – | – | – | Ophiovirus | Citrus psorosis virus | |
| Bacilliform | (+)ssRNAs (3) | – | Ourmiavirus | Ourmia melon virus | |
| Bacilliform | (−)ssRNA | Rhabdoviridae | Cytorhabdovirus | Lettuce necrosis yellows virus |  |
| – | – | Phabdoviridae | Nucleorhabdovirus | Potato yellow dwarf virus | |

binding of viral proteins with antibodies produced against the virus in animals (Clark & Adam, 1977). The method has been in use since the 1960s (Martin et al., 2000), but advancements have been made over time to increase sensitivity, specificity, and speed of detection. The use of recombinant polyclonal antiserum produced against individual viral genes reduces cross-reactivity compared to antiserum produced from purified virus preparations. Additionally, the use of monoclonal antibodies produced through hybridoma technology further enhances the specificity of the reaction (Holzloehner et al., 2013). Monoclonal antibodies bind to a single epitope, making them specific to a single strain or a group of closely related viral strains, while polyclonal antibodies recognize multiple epitopes on an individual antigen (Robison, 1995), making them a versatile tool for wide screening of plant viruses. Further advancements in the technical procedures of these tools have increased sensitivity and speed of detection. Currently, the commonly used serological tools by researchers include enzyme-linked immunosorbent assay (ELISA), dot immune binding assay (DIBA)/tissue blot immunoassay (TBIA), and lateral flow assay (LFA).

### *1.4.1 ENZYME-LINKED IMMUNOSORBENT ASSAY (ELISA)*

Enzyme-linked immunosorbent assay (ELISA) is a well-proven and widely used immunodiagnostic technique for plant virus detection, especially when a large number of samples need to be diagnosed in a short duration (Naidu & Hughes, 2001; Webster et al., 2004). The technique is performed in a 96-well polystyrene plate with the capacity to bind antibodies or antigens into the well. The presence of the virus in the samples is detected based on the yellow color development at the end of the process, which is a result of the reaction between the substrate and the enzyme attached to the anti-serum (Corning Life Science, 2001; Luminex, 2010). The technique has high sensitivity and can detect viral particles at very low concentrations. However, the sensitivity of the technique varies depending on the organism to be detected and the freshness of the samples used in the detection process (Martinelli et al., 2015). Additionally, the technique provides information about the level of infection in the samples, which is measured in terms of the optical density (the degree of coloration) of the ELISA reaction (Corning Life Science, 2001; Webster et al., 2004). Several modifications have been made to the technique to increase sensitivity, suitability, and speed of detection. Among the different versions of ELISA, direct antigen coating ELISA (DAC-ELISA), double antibody sandwich ELISA (DAS-ELISA), and triple antibody sandwich ELISA (TAS-ELISA) are most frequently used for plant virus diagnostics. In DAC-ELISA, the viral antigen directly binds to the microtiter plate and is detected using a virus-specific antiserum-enzyme conjugate. In DAS-ELISA, the virus-specific antiserum first binds to the microtiter plate, followed by trapping of the antigen with the antibody, and ultimately detection using a virus-specific antiserum-enzyme conjugate (Figure 1.3). TAS-ELISA is similar to DAS-ELISA, but it uses polyclonal antiserum to create a trapping layer in the microtiter plate, while detection of the antigen is performed using a virus-specific monoclonal antibody followed by a secondary antibody-enzyme conjugate (Figure 1.3) (Bhat & Roa, 2020). TAS-ELISA has higher sensitivity than DAC- and DAS-ELISA due to the use of monoclonal antibody in the reaction, while DAC-ELISA is the simplest, fastest, and cheapest among all the ELISAs.

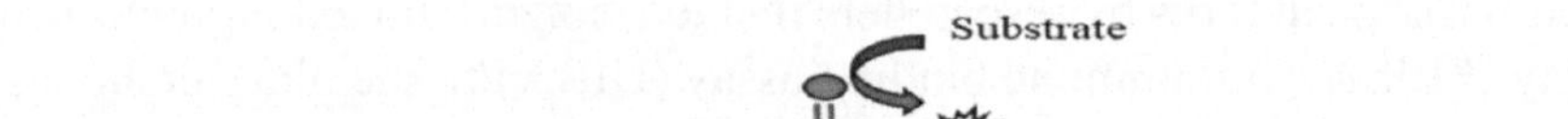


**FIGURE 1.3** Type of serological detection of plant viruses.

## *1.4.2 DOT IMMUNE BINDING ASSAY (DIBA)/TISSUE BLOT IMMUNOASSAY (TBIA)*

The principle of DIBA/TBIA is similar to ELISA, where the presence of the virus in the sample is detected based on the color development due to the reaction between the substrate and the enzyme conjugated with the virus-specific antiserum. However, in DIBA/TBIA, the antigen or antibody is bound to the nitrocellulose membrane (NCM) instead of a polystyrene plate, and the product developed through the enzymatic reaction is insoluble and purple in color (Almeida, 2001; Astier et al., 2007; Purcifull et al., 2001). There is a minor difference between DIBA and TBIA. In DIBA, a small amount (2–5 μl) of sample preparation (antigen) is blotted on NCM, whereas in TBIA, the freshly cut edge of a biological sample such as a leaf blade, leaf, stem, tuber, root, or an insect needs to be imprinted on NCM by firm and gentle force (Hancevic et al., 2012). However, both techniques have several advantages over ELISA in terms of detection time, cost, sensitivity, and convenience. Therefore, they are widely used for routine diagnosis of several viruses (Bove et al., 1988; Eid et al., 2008; Hančević et al., 2012; Lin et al., 1990; Makkouk & Kumari, 2006; Shang et al., 2011; Webster et al., 2004). The greatest advantage of DIBA/TBIA is that blotting can be done in the field or in the lab with minimum facilities. The blot can be stored for a longer duration and can be transported to distant locations before processing for virus detection (Naidu & Hughes, 2001). However, a major drawback of the technique is that weak positive reactions are sometimes not visible.

### *1.4.3 LATERAL FLOW ASSAY (LFAs)*

Lateral flow assay (LFAs) are prefabricated membrane-based strips coated with gold-conjugated virus-specific poly- or monoclonal antibodies for rapid in-field detection and monitoring of plant viruses (De Boer & Lopez, 2012; Varma & Singh, 2020). The technique was first used for cucumber mosaic virus (CMV) and tobacco mosaic virus (TMV) (Tsuda et al., 1992) and gradually adopted for several other viruses such as, *potato virus* M (PTM), Potato virus X (PVX), potato virus A (PVA), potato virus Y (PVY), potato virus S (PVS), impatiens necrotic virus, bean mild mosaic virus, spherical carnation mottle virus, tomato zonate spot virus, and tomato spotted wilt virus (TSWV) (Drygin et al., 2012; Byzova et al., 2009; Niu et al., 2018). The assay is simple and can be performed by a non-experienced person with high accuracy due to its high sensitivity (sensitivity range 0.08–0.5 μg/mL for 10 min), similar to ELISA and DIBA (Byzova et al., 2009). The sensitivity of the assay increases when monoclonal antibodies are used for detection (Mehetre et al., 2021). The increased popularity and high significance of this assay for in-field detection of plant viruses have promoted industries like

Agdia Inc. (https://orders.agdia.com/pathogen-tests/immunostrip-tests) and BIOREBA (https://www.bioreba.ch/catalog/agristrip-lateral-flow.aspx) for the commercial production of such ImmunoStrips/AgriStrips.

## 1.5 NUCLEIC ACID-BASED (NA-BASED) DETECTION

Advancements in molecular and genomic tools have greatly assisted in the detection and diagnosis of plant viruses using nucleic acid-based (NA-based) techniques. These techniques are highly efficient, robust, specific, and sensitive, making them the most suitable for accurate virus detection in routine diagnosis. One of the major advantages of NA-based techniques over biological and serodiagnostic tools is the ability to use any genomic region for virus detection, especially beneficial for viruses with poor immunogenicity (Zheng et al., 2010).

NA-based techniques are broadly classified into two categories: hybridization-based techniques and PCR-based techniques. While these techniques are highly sensitive, some of them can be time-consuming and require sophisticated instruments. However, combining NA-based techniques with serology can further enhance the efficiency of virus detection. Among the various NA-based tools available, PCR-based techniques are the most simple, rapid, specific, and highly sensitive (Edel et al., 1998; Henson et al., 1993; McCartney et al., 2003), and are commonly used for plant virus detection. However, hybridization-based techniques are equally important due to their unique advantages.

Both techniques generally involve three basic steps: extraction of nucleic acids (DNA/RNA), amplification/blotting of DNA/RNA, and product analysis. During the extraction step, the purity of the extracted DNA/RNA is crucial. In the amplification/blotting step, the design and synthesis of primers/probes play a crucial role. Product analysis is relatively simple for PCR on Agarose/sodium dodecyl sulfate (SDS) gel, but it can be more complicated and time-consuming for hybridization-based techniques. The pros and cons of each major NA-based technique are discussed in detail in this chapter.

### *1.5.1 HYBRIDIZATION-BASED TECHNIQUE*

The technique was first applied for the detection of viroids, where serology could not be used due to the lack of a protein component in their structure (Owens & Diener, 1981). It was subsequently used for plant virus detection due to its higher sensitivity compared to serology (Garger et al., 1983; Maule et al., 1983). The technique has intermediate sensitivity compared to serology and PCR-based methods (Pallias et al., 2018) and emerged as a promising tool for plant virus detection during the 1980s. The hybridization is based on the specific interaction of nucleic acid sequences of the target pathogen with a labeled complementary sequence, known as a probe, to produce hybrids. Probes can be made up of DNA (DNA probe) or RNA (riboprobe). However, as the majority of plant viruses have an RNA genome, riboprobes are most frequently used in plant virus diagnostics (Pallás et al., 2018). Additionally, the probes can be labeled either by a radioactive compound ($^{32}P$ or $^{12}C$) or a non-radioactive precursor (biotin or digoxigenin (DIG)-labeling). However, the use of radioactive precursors has been limited to specialized laboratories nowadays due to safety issues, and non-radioactive precursors are gaining popularity among users (Pallas et al., 2018). Among the various available non-radioactive precursors, DIG systems are the most widely used for nucleic acid hybridization due to their high sensitivity (Ruiz et al., 2002; Wu 2002; Du et al., 2004).

A number of hybridization techniques have been developed so far that can be used for plant virus detection, of which dot-blot hybridization is most commonly used due to its simplicity and rapidity. The major advantage of this technique is that it doesn't require the cumbersome nucleic acid extraction procedure, and crude extracts of the samples need to be immobilized onto the solid support (nylon or nitrocellulose membrane (NCM)) for hybridization. Regardless of the type of hybridization, the methodology includes three basic steps: (i) synthesis of the labeled probe; (ii) preparation of the samples and molecular hybridization with the appropriate probe; and (iii) detection using high-affinity antibody Fab-fragments coupled to alkaline phosphatase. The detection can be done either by autoradiography (for radioactive probes) or chemiluminescent/colorimetric methods (for non-radioactive probes). However, the chemiluminescent detection methods have been reported to be much more sensitive than the colorimetric method and are therefore popularly used (Bronstein et al., 1995). Although the molecular hybridization technique has several advantages, it is not exempt from weaknesses. The major drawback of this technique (Cañizares et al., 1999) is the presence of spurious hybridization signals caused by the availability of host RNAs in crude extract that have sequence similarity to plant viruses. However, this can be eliminated or reduced to some extent by incorporating additional steps, such as incubating membranes with RNase at high ionic strength after hybridization. Despite this limitation, the value of this technique has been enhanced by advancements that allow for the simultaneous detection of multiple viruses in a single sample. This has proven successful for phytosanitary certification of various crops (Saldarelli et al., 1996; Sánchez-Navarro et al., 1999, 2018; Saade et al., 2000; Ivars et al., 2004; Minutillo et al., 2012). In all these cases, probes (DNA or RNA) were mixed in the hybridization solution, resulting in reliable detection of all target viruses with high specificity and sensitivity comparable to that obtained using individual probes. However, this approach can lead to undesirable background signals that may render the results difficult to interpret. To address this issue, the membrane should be treated with RNase A after the washing steps, at least partially mitigating this drawback (Sánchez-Navarro et al., 1999).

### *1.5.2 AMPLIFICATION-BASED TECHNIQUES*

Amplification-based techniques offer several advantages over immunoassays and hybridization-based assays. These techniques are simple, rapid, specific, and cost-effective. They have the potential to detect unculturable pathogens such as viruses and can detect a single target (virus) in a sample infected with multiple pathogens. The sensitivity of these techniques is also higher than that of immunoassays and hybridization-based assays, making them commonly used for routine and accurate detection of plant viruses.

Another advantage of amplification-based techniques is the sequencing of the amplified products. Analysis of these sequences using bioinformatics tools helps in the identification, characterization, and determination of the diversity of the viruses. Amplification-based techniques can be classified into two types: polymerase chain reaction (PCR) and isothermal amplification, which has been discussed in subsections.

#### *1.5.2.1 POLYMERASE CHAIN REACTION (PCR)*

Polymerase chain reaction (PCR) is a widely accepted and frequently used technique for the detection of plant viruses. This technique is highly accurate and sensitive, relying on the presence of distinctive sequences in the viral genome. It allows for the amplification of millions of copies of a specific region of the viral genome within a few hours, which can be visualized through gel electrophoresis.

However, before conducting PCR, it is important to identify the suspected virus based on the symptoms observed in the plants. This information is crucial for selecting the appropriate primers for PCR. Primers play a key role in virus detection in PCR and can be designed and synthesized using known nucleotide sequences of viruses available in the NCBI database.

For unknown viruses and broad-spectrum detection of viruses in plant samples, group-specific degenerate primers are utilized (Martin et al., 2000; Zheng et al., 2010). Subsequently, specific primers can be designed based on the sequences of relatively similar virus species (Kumar et al., 2010). Degenerate primers target the conserved sites in the known sequences of viral genera and are successfully utilized for the detection of potyviruses, geminiviruses, closteroviruses, badnaviruses, etc., as well as for the initial screening and characterization of new variants or species infecting diverse crop species (Anderson et al., 2004; Deng et al., 1994). A typical PCR consists of three basic steps: (i) denaturation at above 94°C, resulting in the generation of single-stranded DNA (ssDNA) template for the synthesis of a new strand; (ii) annealing at 50–60°C, where oligonucleotide primers bind with the template DNA; and (iii) synthesis/extension at 72°C, in which a new strand is synthesized on the template strand. In a PCR reaction, these steps are repeated 20–40 times to make millions of copies of the target DNA. Several variants of PCR, such as reverse-transcriptase PCR (RT-PCR), nested PCR (nPCR), co-operational PCR (Co-PCR), quantitative PCR (qPCR), multiplex PCR (M-PCR), immunocapture PCR (IC-PCR), etc., have been developed based on the need and have advantages in one way or another over the traditional PCR. However, all the variants follow the same concept with little modification in the procedure, which have been discussed separately in this chapter.

#### *1.5.2.2 REVERSE TRANSCRIPTASE PCR (RT-PCR)*

In the case of viruses with an RNA genome, the use of RT-PCR is the most promising approach for their diagnosis. RT-PCR works on the principle of complementary DNA (cDNA) synthesis from the genomic RNA, followed by PCR. The process of RT-PCR requires an additional enzyme, reverse transcriptase (RT), and an additional step at 45–55°C for 30–45 minutes, which supports the synthesis of cDNA from the RNA genome (Lopez et al., 2008; Webster et al., 2004).

The reaction can be completed in a one-step or a two-step process. In the one-step process, cDNA synthesis and PCR take place together in one PCR tube, with an additional step of cDNA synthesis in the PCR program. On the other hand, in the two-step process, RNA is first subjected to cDNA synthesis, followed by PCR in a separate tube. However, the one-step RT-PCR method is found to be more sensitive and reliable compared to the two-step process, due to the chances of contamination during the setup of the second step of PCR using cDNA as a template. Therefore, the one-step method is most commonly used for the detection of a number of RNA viruses infecting crops (Li et al., 2007).

### *1.5.2.3 NESTED PCR (nPCR)*

This variant of PCR is mostly used in cases where viruses have low titer and/or an unstable target gene (Webster et al., 2004). The process consists of two rounds of PCR reactions using one or two new internal primers. In nPCR, the first PCR is carried out using a set of external primers to increase the concentration of the target sequence. This is followed by the second PCR using one or both new primers (internal primers), where the amplification product from the first PCR is used as a template for the second PCR reaction to amplify the target gene. The technique has been successfully utilized to detect several viruses that infect crops worldwide (Adkar-Purushothama et al., 2011; Helguera et al., 2001, 2002). However, contamination is a major issue when performing two rounds of amplification in different tubes, especially when performed for many samples.

### *1.5.2.4 CO-OPERATIONAL PCR (Co-PCR)*

The highly sensitive PCR technique developed for the amplification of plant RNA viruses and bacteria (Caruso et al., 2003) is known for its unique features. This technique, similar to nPCR but with minor differences, utilizes four primers – one external and three internal, as opposed to two external and two internal primers in nPCR (Olmos et al., 2002). It offers several advantages, including the ability to perform in a single tube, reducing time consumption and the risk of false positives while maintaining amplification sensitivity. Furthermore, it produces larger amplicons compared to nPCR, enables real-time virus detection, allows for coupling with dot blot hybridization, and can be applied to capillary air thermal cyclers (Olmos et al., 2002; Lopez et al., 2008).

The reaction process involves simultaneous reverse transcription of two different fragments from the same target, with one internal to the other. It also includes the production of four amplicons through the combination of two pairs of primers – one pair external and the other internal. The co-operational action of these amplicons results in the production of the largest fragment.

By using diluted samples in co-PCR, the issues of PCR inhibitors in conventional RT-PCR are overcome, leading to an increased sensitivity of at least 100 times compared to conventional RT-PCR (Olmos et al., 2002).

#### *1.5.2.5 REAL-TIME PCR (qPCR)*

The technique works on the principle of monitoring the amplification (detection and quantification) in real-time by labeling the primers or amplicon with fluorescent dyes. It is highly sensitive, accurate, and specific compared to conventional PCR, thus increasing its application in plant virus diagnostics (Schaad et al., 2002; Mumford et al., 2000). Additionally, the technique dramatically reduces the detection time due to the possibility of real-time monitoring of amplification, eliminating the need for gel electrophoresis to analyze the amplicon generated in qPCR (Lopez et al., 2008). Other advantages of this technique include the ability to detect both types (DNA and RNA) of viruses, reduced risk of sample contamination during PCR setup, and greater amenity to multiplexing. However, the technique has some drawbacks such as the high cost of equipment, the requirement of skilled personnel to set up the reaction and analyze the data, and the need to arrest the amplification process at a certain point during amplification (Gibson et al., 1996). Despite these limitations, real-time PCR has been increasingly used because it has shown valuable detection for plant viruses (McCartney et al., 2003).

#### *1.5.2.6 MULTIPLEX PCR/RT-PCR*

Multiplex PCR/RT-PCR enables the simultaneous detection of multiple viruses (DNA or RNA viruses) in a single reaction. This technique is valuable in plant pathology because many pathogens are known to infect a single host. By detecting all suspected pathogens in one reaction, it saves time and reduces the cost compared to individual detection using conventional PCR. The technique is simple and similar to conventional PCR, except for the use of multiple sets of primers (one set of specific primers for each suspected virus) in a single PCR reaction (Holkar et al., 2017). However, it has been reported that using more than six sets of primers in a single PCR reaction targeting different pathogens can affect the efficiency of detection (Sánchez-Navarro et al., 2005; Tao et al., 2012). Therefore, it is mostly used for the simultaneous detection of two to six plant viruses, with a few exceptions where up to nine viruses were detected simultaneously in one sample (Tao et al., 2012; Sánchez-Navarro et al., 2005). Despite these benefits, the technique has some limitations, which is why conventional PCR is still more commonly used. One major limitation of this technique is the difficulty in designing specific primers for each target virus that can differentiate viruses based on the amplicon size in agarose gel (Lopez et al., 2008; Webster et al., 2004). However, several approaches such as the use of species-specific biotinylated probes in streptavidin-coated microtiter wells (Charoenvilaisiri et al., 2014), the flow cytometric system developed by Luminex (Austin, TX, United States), multiplex real-time PCR (Zipper et al., 2004), etc., might be used to overcome this drawback.

#### *1.5.2.7 IMMUNOCAPTURE PCR (IC-PCR)*

Immunocapturing the viruses in a tube/microtiter plate followed by detection using PCR is found to be a versatile, sensitive, and robust tool for plant virus detection. The technique has particularly been useful for detecting viruses in species or tissues containing inhibitory substances such as polysaccharides and polyphenolic compounds that inhibit the DNA polymerase and reduce the sensitivity of the assay (Varveri, 2000). In principle, the process starts with the adsorption of the virus onto the antibody-coated PCR tube followed by heating in the presence of non-ionic surfactant Triton X-100, and PCR/RT-PCR to amplify the target gene. The major advantages of this technique are that it can detect the virus in a sample when present in low titer, and detection can be performed using crude plant extracts without the need for nucleic acid extraction. The technique is found to be more rapid, sensitive, and specific than either the RT-PCR or DAS-ELISA assays (Yang et al., 2012) therefore successfully utilized for detection of viruses such as, pepper mild mosaic virus (PMMoV), CMV, citrus tristeza virus (CTV), grapevine fanleaf virus (GFLV), and TSWV.

#### *1.5.2.8 ISOTHERMAL AMPLIFICATION*

As all PCR-based methods require specific equipment, such as a thermal cycler, they cannot be used for on-site/on-field detection of plant viruses.

Additionally, skilled personnel are required to perform the detection using these techniques. Therefore, research has been focused on developing tools that can detect viruses in plants in the field without the need for sophisticated instruments and skilled manpower. Nowadays, several methods are available, among which loop-mediated isothermal amplification (LAMP), recombinase polymerase amplification (RPA), and rolling circle amplification (RCA) are the most common methods used by researchers. These methods have been discussed in detail here under.

### *1.5.2.9 LOOP-MEDIATED ISOTHERMAL AMPLIFICATION (LAMP)*

Due to its feasibility for working under field conditions, LAMP has emerged as a robust, rapid, and sensitive method for in-field detection and diagnosis of plant viruses. The technique operates on the principle of strand displacement and auto-cycling, enabling amplification under isothermal conditions without the need for expensive devices like thermal cyclers. The technique is performed at a constant temperature of 60–65°C for approximately an hour using four primers (Notomi et al., 2000). The initial product formed in the reaction acts as a template, allowing for the multiplication of copies of DNA of different sizes through strand displacement and auto-cycling using Bst DNA polymerase. The reaction's result can be visualized either on an agarose gel, where a smear of multiple bands can be observed, or by adding dye to the reaction mixture, resulting in a color change. However, this technique also has limitations such as cross-contamination when handling a large number of samples and difficulties in designing primers specific to a virus. Despite these limitations, the technique has been well adopted by researchers and is extensively used for the detection of numerous plant viruses with DNA and RNA genomes.

### *1.5.2.10 ROLLING CIRCLE AMPLIFICATION (RCA)*

Rolling circle amplification (RCA) is an isothermal enzymatic process that uses unique DNA and RNA polymerases to produce long ssDNA and RNA. In RCR, the amplification of viral DNA occurs at a constant temperature of 37°C, where amplicon is generated in a sequence-independent manner using phi 29 DNA polymerase. The phi 29 DNA polymerase spontaneously and continuously adds nucleotides to the primers that bind to the circular template, generating long ssDNA with tandem repeats of tens to hundreds of orders of magnitude. The desired amplicon can be observed in agarose gel after digesting the RCA products using specific restriction enzymes. The technique has the specific advantage of amplifying novel sequence variants as it uses exo-resistant random hexamer primers instead of specific primers used in other PCRs (Johne et al., 2009; Jones, 2009). The method has been successfully used for the detection of many circular and single- or double-stranded DNA viruses (begomovirus, badnavirus) infecting crops worldwide and is also used for the characterization of begomoviruses (Pratap et al., 2011; Senanayake et al., 2013; Kumar et al., 2018; Hanamasagar et al., 2021) and badnaviruses (James et al., 2011) infecting different horticultural crops.

#### *1.5.2.11 RECOMBINASE POLYMERASE AMPLIFICATION (RPA)*

Recombinase polymerase amplification (RPA) amplifies the target DNA using recombinase, DNA polymerase, and single-strand-binding proteins (SSB proteins). A nucleoprotein complex is formed by the interaction of recombinase with primers, scanning the dsDNA molecules onto the target DNA, producing the primer template DNA hybrids. This is followed by the displacement of the non-template strand, which is stabilized by SSB proteins (Piepenburg et al., 2006; Zhao et al., 2015). The recombinase then dissociates from the 3′-end of the primer, which is used by DNA polymerase to generate a complete target amplicon. This amplicon serves as a template for continuous RPA, enabling exponential amplification of the target sequence through cyclic replication (Piepenburg et al., 2006; Zhao et al., 2015).

### *1.5.3 OTHERS*

All the above-mentioned techniques are common methods of detection used in plant virus diagnostics. However, several new methods such as DNA microarrays, digital-droplet PCR, new PCR variants, CRISPR/Cas12 technique, and biosensors have been developed and extensively used to meet the demand, increase the sensitivity, and simplify the detection process. These techniques are efficient, but due to their complexity, requirement of skilled personnel, and other limitations, they have not been widely adopted by the majority of researchers.

## 1.6 SEQUENCING PLATFORMS

Sanger sequencing (the first-generation sequencing platform) has widely been used for the detection and diagnosis of a number of plant viruses and is still considered a popular method due to the simplicity of the technique, low cost of sequencing, and high reliability of the generated sequences (Dumschott et al., 2020). However, advancements in sequencing technologies offer advantages over the traditional sequencing method (Lipkin & Firth, 2013). High-throughput sequencing (also called next-generation sequencing (NGS)) has changed the approaches to diagnosing plant viruses. The technique is considered a powerful tool not only for detecting the full spectrum of viruses, including novel viruses (Barzon et al., 2011; Roossinck, 2015; Villamor et al., 2016; Rott et al., 2017; Jo et al., 2018), but also for estimating genetic variations among populations, characterizing viruses at their taxonomic levels, studying synergistic interactions between viruses, and exploring genetic diversity and evolutionary mechanisms of virus populations (Kreuze et al., 2009; Wu et al., 2015; Pecman et al., 2017; Verdin et al., 2017; Kehoe et al., 2014; Hadidi et al., 2016; Roossinck, 2017; Tan et al., 2019; Xu et al., 2019). Additionally, the technique generates millions to billions of sequences reads ranging from 25 to 400 nt in parallel, compared to about 750 nt generated in Sanger sequencing from cloned DNA fragments. This increases the speed and reduces the time of detection. However, further advancements in NGS/HTS have improved the length of nucleotide reads to >750 nt (Prabha et al., 2013). Another major advantage of this technique is that it is sequence-independent, allowing for concurrent detection of RNA viruses, DNA viruses, viroids, phytoplasmas, and other plant pathogens, even if present in low titer (Adams et al., 2009; Kreuze et al., 2009; Baranwal et al., 2015).

The HTS can be divided into two types: second-generation and third-generation sequencing. In second-generation sequencing, total nucleic acid is extracted from the samples, random libraries of DNA/cDNA fragments are prepared, clonally amplified, ligated with adapters, sequenced using NGS platforms, and assembled using bioinformatic tools. Additionally, sequencing of double-stranded RNA (dsRNA) and small RNAs with pooling of RNAs from several plants/species increases the possibility of simultaneous detection of (+) and (–) ssRNA viruses, dsRNA viruses, dsDNA viruses, ssDNA viruses, and a viroid in a given sample, contributing to the reduction in time and costs of diagnostics (Rott et al., 2017; Verdin et al., 2017). The popular second-generation NGS platforms used in plant virus research are Roche 454 FLX, Illumina, Solid, and Ion Torrent (Bleidorn, 2016; Goodwin et al., 2016; Heather & Chain, 2016). Third-generation sequencing (also called single-molecule sequencing) has several advantages over second-generation sequencing and is gaining attention in plant virus diagnostics (Edwards &

Batley, 2010; Layezzo et al., 2016; Petersen et al., 2019; Tyler et al., 2018). The technique generates long reads of sequences from a single molecule in real-time without the need for clonal amplification, reducing the time of sequencing and sequence assembly using bioinformatics tools (Goodwin et al., 2016; van Dijk et al., 2018). Out of several platforms available for third-generation sequencing, the single-molecule real-time (SMRT) and MinION platforms are the most popular. The SMRT platform has been developed by Pacific Biosciences, USA. It continuously visualizes nucleotide sequences using a laser and camera system that detects the color and duration of the light emitted from labeled nucleotides. The system momentarily pauses during incorporation at the bottom of the picoliter wells immobilized with DNA polymerase. Another leading sequencing platform, the MinION nanopore, has been developed by Oxford Nanopore Technologies (ONT, UK). It utilizes base-specific fluctuations caused by the blockage of a nanopore and transforms these fluctuations into DNA sequence information (Lu et al., 2016). The major advantages of this technique over other HTS technologies are its pocket-sized device and the ability to process DNA/RNA for sequencing in real-time (Kiselev et al., 2020). This technique has successfully been utilized for the diagnosis of several known and unknown viruses (Bronzato et al., 2018; Chalupowicz et al., 2019; Naito et al., 2019) and has proven to be effective for real-time analysis with an average sequence length of more than 15 kb. However, the key limitation of this technique is its poor base accuracy, with an error rate of 5–15% (Lu et al., 2016). In general, contamination can occur at various steps of the HTS protocol. The requirement of validating NGS results using more familiar techniques such as PCR or ELISA, appropriate use of controls in the reactions performed, and the need for expertise in bioinformatics and plant virology have been identified as critical challenges in using this technique for routine diagnostics of plant viruses (Dickins et al., 2014; Massart et al., 2017).

## 1.7 CONCLUSIONS

Plant virus diagnostics are important for monitoring plant health, enabling quick detection and identification of known and unknown plant viruses of economic importance. Significant advancements have been made in diagnostic methods since the detection of the first plant virus (TMV) over a century ago. This chapter describes the major diagnostic tools developed for plant virus detection, along with their pros and cons.

Symptom-based diagnostics, although ancient and effective, are no longer feasible due to the significant increase in the number of closely related viruses. This method alone is not effective for accurate detection and identification of plant viruses. Serological methods such as ELISA and Dot-blot are still the most popular, especially for researchers dealing with a large number of samples or limited laboratory facilities.

PCR has become popular and widely used for plant virus detection due to its simplicity and higher rate of accuracy. Additionally, the development of isothermal amplification techniques like LAMP and RPA has provided an opportunity for on-site detection of viruses with limited resources and less skilled personnel.

The development of HTS/NGS technologies has opened up fast and precise detection and identification of even unknown viruses in a given sample. This has demonstrated that very little is known about plant viruses so far, and there is much more to be explored in the future.

However, every technique encounters challenges and limitations, and continuous improvement is required to overcome these challenges for fast and efficient detection of plant viruses, ultimately contributing to sustainable agriculture.

## KEYWORDS

- **immunocapture PCR**
- **isothermal amplification**
- **loop-mediated isothermal amplification**
- **polymerase chain reaction**
- **recombinase polymerase amplification**
- **reverse transcriptase PCR**
- **rolling circle amplification**

## REFERENCES

Adams, I. P., Glover, R. H., Monger, W. A., Mumford, R., Jackeviciene, E., Navalinskiene, M., et al., (2009). Next-generation sequencing and metagenomic analysis: A universal diagnostic tool in plant virology. *Mol. Plant Pathol., 10*, 537–545.

Adkar-Purushothama, C. J., Maheshwar, P. K., Sano, T., & Janardhana, G. R. A., (2011). Sensitive and reliable RT-nested PCR assay for detection of *Citrus tristeza virus* from naturally infected citrus plants. *Curr. Microbiol., 62*, 1455–1459.

Almeida, A., & Lima, J., (2001). Técnicas sorológicas aplicadas à fitovirologia. In: Almeida, A. M. R., & Lima, J. A. A., (eds.), *Princípios e Técnicas Aplicados em Fitovirologia* (pp. 33–62). Edições Sociedade Brasileira Fitopatologia, ISBN 901648, Fortaleza, Ceará, Brazil.

Anderson, P. K., Cunningham, A. A., Patel, N. G., Morales, F. J., Epstein, P. R., & Daszak, P., (2004). Emerging infectious diseases of plants: Pathogen pollution, climate change and agrotechnology drivers. *Trends Ecol. Evol., 19*, 535–544.

Astier, S., Albouy, J., Maury, Y., Robaglia, C., & Lecoq, H., (2007). *Genome, Pathogenicity, Virus Ecology.* Science, Principles of Plant Virology: Publishers, ISBN: 1578083168, New Hampshire, USA.

Baker, K. K., Ramsdell, D. C., & Gillett, J., (1985). Electron microscopy: Current applications to plant virology. *Plant Dis., 69*, 85–91.

Baranwal, V. K., Jain, P., Saritha, R. K., & Jain, R. K., (2015). Detection and partial characterization of cowpea mild mottle virus in mungbean and urdbean by deep sequencing and RT PCR. *Crop Prot. 75*, 77–79.

Barzon, L., Lavezzo, E., Militello, V., Toppo, S., & Palù, G., (2011). Applications of next-generation sequencing technologies to diagnostic virology. *Int. J. Mol. Sci., 12*, 7861–7884.

Bhardwaj, A., & Kulshrestha, S., (2020). Advance methods for the isolation and characterization of plant viruses infecting crops. In: *Applied Plant Virology* (pp. 39–53). Academic Press: Cambridge, MA, USA.

Bhat, A. I., & Roa, P. G., (2020). *Characterization of Plant Viruses: Methods and Protocols* (1st edn.). Springer US; Humana press, ISBN: 978-1-0716-0333-8,978-1-0716-0334-5.

Bleidorn, C., (2016). Third generation sequencing: Technology and its potential impact on evolutionary biodiversity research. *Syst. Biodivers, 14*, 1–8.

Bove, J. M., Vogel, R., Albertini, D., & Bove, J. M., (1988). Discovery of a strain of tristeza virus (K) inducing no symptoms in Mexican lime. *Proceedings of the 10th Conference of IOCV* (pp. 14–21). Spain. International Organization of Citrus Virologists Riverside, CA.

Bronstein, I., & Olesen, C. E. M., (1995). Detection methods using chemiluminescence. *Molecular Methods for Virus Detection.*, 147–174.

Bronzato, B. A., Sherman, D., Stone, A., Gopakumar, A., Wilson, V., Schneider, W., et al., (2018). Nanopore sequencing as a surveillance tool for plant pathogens in plant and insect tissues. *Plant Dis., 102*, 1648–1652.

Byzova, N. A., Safenkova, I. V., Chirkov, S. N., Zherdev, A. V., Blintsov, A. N., Dzantiev, B. B., & Atabekov, I. G., (2009). Development of immunochromatographic test systems for express detection of plant viruses. *Appl. Biochem. Microbiol., 45*, 204–209.

Cañizares, M. C., Marcos, J. F., & Pállas, V., (1999). Molecular characterization of an almond isolate of hop stunt viroid (HSVd) and conditions for eliminating spurious hybridization in its diagnosis in almond samples. *Eur. J. Plant Pathol., 105*, 553–558.

Caruso, P., Bertolini, E., Cambra, M., & Lopez, M. M., (2003). A new and sensitive co-operational polymerase chain reaction (Co- PCR) for a rapid detection of *Ralstonia solanacearum* in water. *J. Microbiol. Methods, 55*, 257–272.

Chalupowicz, L., Dombrovsky, A., Gaba, V., Luria, N., Reuven, M., Beerman, A., et al., (2019). Diagnosis of plant diseases using the nanopore sequencing platform. *Plant Pathol., 68*, 229–238.

Charoenvilaisiri, S., Seepiban, C., Bhunchoth, A., Warin, N., Luxananil, P., & Gajanandana, O., (2014). Development of a multiplex RT-PCR-ELISA to identify four distinct species of *tospovirus*. *J. Virol. Methods, 202*, 54–63.

Clark, M. F., & Adamas, A. N., (1977). Characteristic of the microplate of enzyme-linked immunosorbent assay for the detection of plant viruses. *Journal of General Virology*, *34*, 475–483.

Corning Life Science, (2001). Selecting the detection system: Colorimetric, fluorescent, luminescent methods. *ELISA Technical Bulletin 5*, 14. https://www.dutscher.com/data/pdf_guides/en/CLS-DD-AN-458.pdf (accessed on 28 February 2024).

Corning Life Science, (2014). *Selecting the Detection System: Colorimetric, Fluorescent, Luminescent Methods* (p. 14). ELISA Technical Bulletin 5.

De, Boer, S. H., Mar´ıa, M., & Lopez, M. M., (2012). New grower-friendly methods for plant pathogen monitoring. *Annual Review of Phytopathology*, *50*, 197–218.

Deng, D., McGrath, P. F., Robinson, D. J., & Harrison, B. D., (1994). Detection and differentiation of whitefly-transmitted *Geminivirus* in plants and vector insects by the polymerase chain reaction with degenerate primers. *Ann. Appl. Biol., 125*(2), 327–336.

Dickins, B., Rebolledo-Jaramillo, B., Su, M. S., Paul, I. M., Blankenberg, D., Stoler, N., et al., (2014). Controlling for contamination in re-sequencing studies with a reproducible web-based phylogenetic approach. *BioTechniques*, *56*, 134–141.

Drygin, Y. F., Blintsov, A. N., Grigorenko, V. G., Andreeva, I. P., Osipov, A. P., Varitzev, Y. A., Uskov, A. I., et al., (2012). Highly sensitive field test lateral flow immunodiagnostics of PVX infection. *Applied Microbiology and Biotechnology*, *93*, 179–189.

Du, G. Y., Wang, X. F., & Zhou, G. H., (2004). Digoxigenin-labelled cDNA probes for the detection of *Tobacco mosaic virus*, *Cucumber mosaic virus* and *Tobacco virus Y*. *Acta Phytopathologica Sinica*, *34*, 75–79.

Dumschott, K., Schmidt, M. H., Chawla, H. S., Snowdon, R., & Usadel, B., (2020). Oxford nanopore sequencing: New opportunities for plant genomics? *J. Exp. Bot., 71*, 5313–5322.

Edel, V., (1998). Polymerase chain reaction in mycology: An overview. In: Bridge, P. D., Arora, D. K., Reddy, C. A., & Elander, R. P., (eds.), *Applications of PCR in Mycology* (pp. 1–20). CAB International, New York.

Edwards, D., & Batley, J., (2010). Plant genome sequencing: Applications for crop improvement. *Plant Biotech. J., 8*, 2–9.

Eid, S., Atamian, H. S., Abou-Jawdah, Y., & Havey, M. J., (2008). Assessing the movement of cucurbit yellow stunting disorder virus in susceptible and tolerant cucumber germplasms using serological and nucleic acid based methods. *J. Phytopathol., 156*, 438–445.

Elena, S. F., & Sanjuán, R., (2005). Adaptive value of high mutation rates of RNA viruses: Separating causes from consequences. *J. Virol., 79*(18), 11555–11558.

Ertel, K. J., Benefield, D., Castano-Diez, D., Pennington, J. G., Horswill, M., Den, B. J. A., Otegui, M. S., & Ahlquist, P., (2017). Cryo-electron tomography reveals novel features of a viral RNA replication compartment. *eLife, 6*, 25940.

Garger, S. J., Turpen, T., Carrington, J. C., Morris, T. J., Dodds, J. A., Jordan, R. L., et al., (1983). Rapid detection of plant RNA viruses by dot blot hybridization. *Plant Mol. Biol. Rep., 1*, 21–25.

Gibson, U. E. M., Heid, C. A., & Williams, P. M., (1996). A novel method for real time quantitative RT-PCR. *Genome Res., 6*, 995–1001.

Goodwin, S., McPherson, J. D., & McCombie, W. R., (2016). Coming of age: Ten years of next–generation sequencing technologies. *Nat. Rev. Genet., 17*, 333–351.

Hadidi, A., Flores, R., Candresse, T., & Barba, M., (2016). Next-generation sequencing and genome editing in plant virology. *Front. Microbiol., 7*, 1325.

Hanamasagar, Y., Naganur, P., Shankarappa, K. S., et al., (2021). Characterization of *Tomato leaf curl Palampur virus* associated with leaf curl and yellowing disease of watermelon from India. *Indian Phytopathology*, *74*, 1075–1088.

Hancevic, K., Cerni, S., Radic, T., & Skoric, D., (2012). Comparison of different methods for *Citrus tristeza virus* detection in satsuma mandarins. *Journal of Plant Diseases and Protection*, *119*(1), 2–7.

Heather, J. M., & Chain, B., (2016). The sequence of sequencers: The history of sequencing DNA. *Genomics*, *107*, 1–8.

Helguera, P. R., Taborda, R., Docampo, D. M., & Ducasse, D. A., (2001). Immunocapture reverse transcription-polymerase chain reaction combined with nested PCR greatly increases the detection of *Prunus necrotic ring spot virus* in the peach. *J. Virol. Methods*, *95*, 93–100.

Henson, J. M., & French, R., (1993). The polymerase chain reaction and plant disease diagnosis. *Annu. Rev. Phytopathol., 31*, 81–109.

Holkar, S. K., Kumar, R., Yogita, M., et al., (2017). Diagnostic assays for two closely related *Tospovirus* species, Watermelon bud necrosis virus and groundnut bud necrosis virus and identification of new natural hosts. *J. Plant Biochem. Biotechno., 26*, 43–51.

Holzloehner, P., Schliebs, E., Maier, N., Füner, J., Micheel, B., & Heilmann, K., (2013). Production of monoclonal camelid antibodies by means of hybridoma technology. *J. Immunol., 190*:135.14.

Ivars, P., Alonso, M., Borja, M., & Hernandez, C., (2004). Development of a non-radioactive dot-blot hybridization assay for the detection of *Pelargonium* flower break virus and *Pelargonium* line pattern virus. *Eur. J. Plant Pathol., 110*, 275–283.

James, A. P., Geijskes, R. J., Dale, J. L., & Harding, R. M., (2011). Development of a novel rolling-circle amplification technique to detect *Banana streak virus* which also discriminates between integrated and episomal virus sequences. *Plant Dis., 95*, 57–62.

Jo, Y., Lian, S., Chu, H., Cho, J. K., Yoo, A. H., Choi, H., et al., (2018). Peach RNA viromes in six different peach cultivars. *Sci. Rep., 8*, 1844.

Jones, R. A. C., & Naidu, R. A., (2019). Global dimensions of plant virus diseases: Current status, and future perspectives. *Annu. Rev. Virol., 6*, 87–409.

Karen-Beth, G., S., Scott, A., Henryk, C., Peter, P., Emmanuel, J., et al., (2011). Top 10 plant viruses in molecular plant pathology. *Molecular Plant Pathology* (Vol. 12, No. 9, pp. 938–954). Wiley.

Kehoe, M. A., Coutts, B. A., Buirchell, B. J., & Jones, R. A. C., (2014). Plant virology and next generation sequencing: Experiences with a potyvirus. *PLoS One*, *9*, e104580.

Kennelly, M., O'Mara, J., Rivard, C., Miller, G. L., & Smith, D., (2012). Introduction to abiotic disorders in plants. *The Plant Health Instructor*.

Kiselev, D., Matsvay, A., Abramov, I., Dedkov, V., Shipulin, G., & Khafizov, K., (2020). Current trends in diagnostics of viral infections of unknown etiology. *Viruses*, *12*, 212.

Kothandaraman, S. V., Devadason, A., & Ganesan, M. V., (2016). Seed-borne nature of a *Begomovirus*, *Mung bean yellow mosaic virus* in black gram. *Appl. Microbiol. Biotechnol., 100*, 1925–1933.

Kreuze, J. F., Perez, A., Untiveros, M., Quispe, D., Fuentes, S., Barker, I., et al., (2009). Complete viral genome sequence and discovery of novel viruses by deep sequencing of small RNAs: A generic method for diagnosis, discovery, and sequencing of viruses. *Virology*, *388*, 1–7.

Kumar, A., Bag, M. K., Singh, R., Jailani, A. A. K., Mandal, B., & Roy, A., (2018). Natural infection of croton yellow vein mosaic virus and its cognate beta satellite in germplasm of different *Crambe* spp in India. *Virus Res., 243*, 60–64.

Kumar, R., Esakky, R., & Acharya, S., (2019). Molecular evidence of occurrence of *Tomato Leaf Curl New Delhi Virus* infecting cucurbits in several states in India. *Archives of Phytopathology and Plant Protection*, *52*(9, 10), 895–905.

Kumar, R., Kumar, A., & Sinha, S., (2022). Natural and engineered resistance: Implications for managing the cassava mosaic disease. In: *Detection, Diagnosis and Management* (pp. 531–548). Academic Press.

Kumar, R., Mandal, B., Geetanjali, A. S., Jain, R. K., & Jaiwal, P. K., (2010). Genome organization and sequence comparison suggest intraspecies incongruence in M RNA of Watermelon bud necrosis virus. *Arch Virol., 155*(8), 1361–1365.

Lavezzo, E., Barzon, L., Toppo, S., & Palu, G., (2016). Third generation sequencing technologies applied to diagnostic microbiology: Benefits and challenges in applications and data analysis. *Expert Rev. Mol. Diag., 16*, 1011–1023.

Li, R., & Hartung, J. S., (2007). Reverse transcription-polymerase chain reaction-based detection of plant viruses. *Curr. Protoc. Microbiol., 6*, 16C-1.

Lin, N. S., Hsu, Y. H., & Hsu, H. T., (1990). Immunological detection of plant viruses and a mycoplasma like organism by direct tissue blotting on nitrocellulose membranes. *Phytopathology*, *80*, 824–828.

Lipkin, W. I., & Firth, C., (2013). Viral surveillance and discovery. *Curr. Opin. Virol., 3*, 199–204.

Lopez, M. M., Llop, P., Olmos, A., Marco-Noales, E., Cambra, M., & Bertolini, E., (2008). Are molecular tools solving the challenges posed by detection of plant pathogenic bacteria and viruses? *Curr. Issues Mol. Biol., 11*, 13–45.

Lu, H., Giordano, F., & Ning, Z., (2016). Oxford nanopore MinION sequencing and genome assembly. *Genom. Proteom. Bioinf., 14*, 265–279.

Luminex, (2010). *xMAP Technology Technical Note: Overcoming the Cost and Performance Limitations of ELISA with xMAP Technology*. http://www.luminexcorp.com/prod/groups/public/documents/lmnxcorp/308-xmap-vs.-elisa-white-paper.pdf (accessed on 28 February 2024).

Makkouk, K. M., & Kumari, S. G., (2006). Molecular diagnosis of plant viruses. *Arab. J. Plant Protect., 24*, 135–138.

Maksimov, I. V., Sorokan, A. V., Burkhanova, G. F., Veselova, S. V., Alekseev, V. Y., Shein, M. Y., Avalbaev, A. M., Dhaware, P. D., Mehetre, G. T., Singh, B. P., et al., (2019). Mechanisms of plant tolerance to RNA viruses induced by plant-growth-promoting microorganisms. *Plants*, *8*, 575.

Martin, R. R., James, D., & Le´vesque, C. A., (2000). Impacts of molecular diagnostic technologies on plant disease management. *Annu Rev Phytopathol*, *38*, 207–239.

Martinelli, F., Riccardo, S., Salvatore, D., Stefano, P., Giuseppe, S., et al., (2015). *Advanced Methods of Plant Disease Detection: A Review* (Vol. 35, No. 1, pp. 1–25). Agronomy for Sustainable Development, Springer Verlag/EDP Sciences/INRA.

Massart, S., Candresse, T., Gil, J., Lacomme, C., Predajna, L., Ravnikar, M., et al., (2017). A Framework for the evaluation of biosecurity, commercial, regulatory, and scientific impacts of plant viruses and viroids identified by NGS technologies. *Front. Microbiol., 8*, 45.

Maule, A. J., Hull, R., & Donson, J., (1983). The application of spot hybridization to the detection of DNA and RNA viruses in plant tissues. *J. Virol. Methods, 6*, 215–224.

McCartney, A. H., Foster, S. J., Fraaige, B. A., & Ward, E., (2003). Molecular diagnostics for fungal plant pathogens. *Pest Manag. Sci., 59*, 129–142.

Mehetre, G. T., Leo, V. V., Singh, G., Sorokan, A., Maksimov, I., Yadav, M. K., Upadhyaya, K., Hashem, A., Alsaleh, A. N., Dawoud, T. M., et al., (2021). Current developments and challenges in plant viral diagnostics: A systematic review. *Viruses*, *13*, 412.

Min, B. E., Feldman, T. S., Ali, A., Wiley, G., Muthukumar, V., Roe, B. A., et al., (2012). Molecular characterization, ecology, and epidemiology of a novel *Tymovirus* in *Asclepias viridis* from Oklahoma. *Phytopathology*, *102*, 166–176.

Minutillo, S. A., Mascia, T., & Gallitelli, D., (2012). A DNA probe mix for the multiplex detection of ten artichoke viruses. *Eur. J. Plant Pathol., 134*, 459–465.

Mumford, R. A., Walsh, K., Barker, I., & Boonham, N., (2000). Detection of *Potato moptop virus* and *Tobacco rattle virus* using a multiplex real-time fluorescent reverse transcription polymerase chain reaction assay. *Phytopathology*, *90*, 448–453.

Naidu, R. A., & Hughes, J. A., (2001). Methods for the detection of plant virus diseases. In: *Proceedings of a Conference Organized by IITA, Plant Virology in Sub Saharan Africa International Institute of Tropical Agriculture* (pp. 233–260). Oyo State, Nigeria.

Naito, F. Y. B., Melo, F. L., Fonseca, M. E. N., Santos, C. A. F., Chanes, C. R., Ribeiro, B. M., et al., (2019). Nanopore sequencing of a novel bipartite new world *Begomovirus* infecting cowpea. *Arch. Virol., 164*, 1907–1910.

Niu, Y., Wang, D., Cui, L., Wang, B., Pang, X., & Yu, P., (2018). Monoclonal antibody-based colloid gold immunochromatographic strip for the rapid detection of tomato zonate spot *tospovirus*. *Virol. J., 15*, 1–9.

Notomi, T., Okayama, H., Masubuchi, H., Yonekawa, T., Watanabe, K., Amino, N., & Hase, T., (2000). Loop-mediated isothermal amplification of DNA. *Nucleic Acids Res., 28*, e63.

Olmos, A., Bertolini, E., & Cambra, M., (2002). Simultaneous and co-operational amplification (Co-PCR): A new concept for detection of plant viruses. *J. Virol. Methods*, *106*, 51–59.

Owens, R. A., & Diener, T. O., (1981). Sensitive and rapid diagnosis of potato spindle tuber viroid by nucleic acid spot hybridization. *Science, 213*, 67–672.

Pallás, V., Sánchez-Navarro, J. A., & James, D., (2018). Recent advances on the multiplex molecular detection of plant viruses and viroids. *Front. Microbiol., 9*, 2087.

Pecman, A., Kutnjak, D., Gutiérrez-Aguirre, I., Adams, I., Fox, A., Boonham, N., et al., (2017). Next generation sequencing for detection and discovery of plant viruses and viroids: Comparison of two approaches. *Front. Microbiol., 8*, 1998.

Petersen, L. M., Martin, I., Moschetti, W. E., Kershaw, C. M., & Tsongalis, G. J., (2019). Third-generation sequencing in the clinical laboratory: Exploring the advantages and challenges of nanopore sequencing. *J. Clin. Microbiol., 58*, e01315–19.

Piepenburg, O., Williams, C. H., Stemple, D. L., & Armes, N. A., (2006). DNA detection using recombination proteins. *PLoS Biol., 4*(7), e204.

Prabha, K., Baranwal, V. K., & Jain, R. K., (2013). Applications of next generation high throughput sequencing technologies in characterization, discovery, and molecular interaction of plant viruses. *Indian J. Virol., 24*, 157–165.

Pratap, D., Kashikar, A. R., & Mukherjee, S. K., (2011). Molecular characterization and infectivity of a *Tomato Leaf Curl New Delhi Virus* variant associated with newly emerging yellow mosaic disease of eggplant in India. *Virol. J., 16*(8), 305.

Purcifull, D. E., Hiebert, E., Petersen, M., & Webb, S., (2001). Virus detection – serology. In: Maloy, O. C., & Murray, T. D., (eds.), *Encyclopedia of Plant Pathology*. John Wiley & Sons, Inc. ISBN: 0-471-29817-4, 2, 1100–1109.

Richert-Pöggeler, K. R., & Minarovits, J., (2014). Diversity of latent plant-virus interactions and their impact on the virosphere. *Plant Virus Host Interact., 14*, 263–275.

Robison, B. J., (1995). Use of commercially available ELISA kits for detection of foodborne pathogens. In: Howard, J., & Whitcombe, D. M., (eds.), *Diagnostic Bacteriology Protocols* (pp. 123–132). Humana Press.

Roossinck, M. J., (2005). Symbiosis versus competition in plant virus evolution. *Nat. Rev. Microbiol., 3*, 917–924.

Roossinck, M. J., (2010). Lifestyles of plant viruses. *Philos. Trans. R. Soc. Lond. B Biol. Sci., 365*, 1899–1905.

Roossinck, M. J., (2015). Metagenomics of plant and fungal viruses reveals an abundance of persistent lifestyles. *Front. Microbiol., 5*, 767.

Roossinck, M. J., (2017). Deep sequencing for discovery and evolutionary analysis of plant viruses. *Virus Res., 239*, 82–86.

Rott, M., Xiang, Y., Boyes, I., Belton, M., Saeed, H., Kesanakurti, P., et al., (2017). Application of next generation sequencing for diagnostic testing of tree fruit viruses and viroids. *Plant Dis., 101*, 1489–1499.

Ruiz, L., Janssen, D., Velasco, L., Segund, E., & Cuadrado, I. M., (2002). Quantitation of cucurbit yellow stunting disorder virus in *Bemisia tabaci* (Genn.) using digoxigenin-labelled hybridization probes. *Journal of Virological Methods, 101*, 95–103.

Saade, M., Aparicio, F., Sánchez-Navarro, J. A., Herranz, M. C., Myrta, A., Di Terlizzi, B., et al., (2000). Simultaneous detection of three Ilarviruses affecting stone fruits by non-isotopic molecular hybridization and multiplex RT-PCR. *Phytopathology, 96*, 1330–1336.

Saldarelli, P., Barbarossa, L., Grieco, F., & Gallitelli, D., (1996). Digoxigenin-labeled riboprobes applied to phytosanitary certification of tomato in Italy. *Plant Dis., 80*, 1343–1346.

Sánchez-Navarro, J. A., Aparicio, F., Herranz, M. C., Minafra, A., Myrta, A., & Pállas, V., (2005). Simultaneous detection and identification of eight stone fruit viruses by one-step RT-PCR. *Eur. J. Plant Pathol., 111*, 77–84.

Sánchez-Navarro, J. A., Cañizares, M. C., Cano, E. A., & Pallás, V., (1999). Simultaneous detection of five carnation viruses by non-isotopic molecular hybridization. *J. Virol. Methods, 82*, 167–175.

Sánchez-Navarro, J. A., Fiore, N., Fajardo, T. V. M., & Pallás, V., (2018). Simultaneous detection of the 13 viruses and 5 viroids affecting grapevine by molecular hybridization using a unique probe or 'polyprobe.' In: *Proceedings of the 19th Conference of the International Council for the Study of Virus and Virus-like Diseases of the Grapevine, Santiago*.

Schaad, N. W., & Frederick, R. D., (2002). Real-time PCR and its application for rapid plant disease diagnostics. *Can. J. Plant Pathol., 24*, 250–258.

Scholthof, K. B., Adkins, S., Czosnek, H., Palukaitis, P., Jacquot, E., Hohn, T., Hohn, B., Saunders, K., Candresse, T., Ahlquist, P., et al., (2011). Top 10 plant viruses in molecular plant pathology. *Mol. Plant Pathol., 12*, 938–954.

Senanayake, D. M. J. B., Jayasinghe, J. E. A. R. M., Shilpi, S., et al., (2013). A new *Begomovirus*–betasatellite complex is associated with chilli leaf curl disease in Sri Lanka. *Virus Genes, 46*, 128–139.

Shang, H., Xie, Y., Zhou, X., Qian, Y., & Wu, J., (2011). Monoclonal antibody-based serological methods for detection of *Cucumber Green Mottle Mosaic Virus*. *Virol. J., 8*, 228–234.

Shates, T. M., Sun, P., Malmstrom, C. M., Dominguez, C., & Mauck, K. E., (2019). Addressing research needs in the field of plant virus ecology by defining knowledge gaps and developing wild dicot study systems. *Front. Microbiol., 9*, 3305.

Steere, R. L., & Williams, R. C., (1953). Identification of crystalline inclusion bodies extracted intact from plant cells infected with tobacco mosaic virus. *Am. J. Bot., 40*, 81–84.

Takahashi, H., Fukuhara, T., Kitazawa, H., & Kormelink, R., (2019). Virus latency and the impact on plants. *Frontiers in Microbiology, 10*, art. no. 2764.

Tan, S. H., Osman, F., Bodaghi, S., Dang, T., Greer, G., Huang, A., et al., (2019). Full genome characterization of 12 citrus tatter leaf virus isolates for the development of a detection assay. *PLoS One, 14*, e0223958.

Tao, Y., Man, J., & Wu, Y., (2012). Development of a multiplex polymerase chain reaction for simultaneous detection of wheat viruses and a phytoplasma in China. *Arch. Virol., 157*, 1261–1267.

Tsuda, S., Kameya-Iwaki, M., Hanada, K., Kouda, Y., Hikdtd, M., & Tomaru, K., (1992). A novel detection and identification technique for plant viruses: Rapid immunofilter paper assay (RIPA). *Plant Disease, 76*, 466–469.

Tyler, A. D., Mataseje, L., Urfano, C. J., Schmidt, L., Antonation, K. S., Mulvey, M. R., & Corbett, C. R., (2018). Evaluation of oxford nanopore's MinION sequencing device for microbial whole genome sequencing applications. *Sci. Rep., 8*, 10931.

Van Dijk, E. L., Jaszczyszyn, Y., Naquin, D., & Thermes, C., (2018). The third revolution in sequencing technology. *Trends Genet., 34*, 666–681.

Varma, A., & Singh, M. K. (2020). Diagnosis of plant virus diseases. *Applied Plant Virology*, 79–92.

Verdin, E., Wipf–Scheibel, C., Gognalons, P., Aller, F., Jacquemond, M., & Tepfer, M., (2017). Sequencing viral siRNAs to identify previously undescribed viruses and viroids in a panel of ornamental plant samples structured as a matrix of pools. *Virus Res., 241*, 19–28.

Villamor, D. E. V., Mekuria, T. A., Pillai, S. S., & Eastwell, K. C., (2016). High throughput sequencing identifies novel viruses in nectarine: Insights to the etiology of stem pitting disease. *Phytopathology, 106*, 519–527.

Webster, C. G., Wylie, J. S., & Jones, M. G. K., (2004). Diagnosis of plant viral pathogens. *Curr. Sci., 86*, 1604–1607.

Wu, Q., Ding, S. W., Zhang, Y., & Zhu, S., (2015). Identification of viruses and viroids by next-generation sequencing and homology-dependent and homology independent algorithms. *Annu. Rev. Phytopathol., 53*, 425–444.

Wu, X. Q., (2002). *The Molecular Identification and Detection Methods of Potato Viruses in Fujian.* Ph D dissertation, Fujian Agriculture and Forestry University.

Xu, Y., Li, S., Na, C., Yang, L., & Lu, M., (2019). Analyses of virus/viroid communities in nectarine trees by next-generation sequencing and insight into viral synergisms implication in host disease symptoms. *Sci. Rep., 9*, 12261.

Yang, J. G., Wang, F. L., Chen, D. X., Shen, L. L., Qian, Y. M., Liang, J. Y., Zhou, W. C., & Yan, T. H., (2012). Development of a one-step immunocapture real-time RT-PCR assay for detection of *Tobacco Mosaic Virus* in soil. *Sensors, 12*, 16685–16694.

Zamora, M., Méndez-López, E., Agirrezabala, X., Cuesta, R., Lavin, J. L., Sánchez-Pina, M. A., Aranda, M. A., & Valle, M., (2017). Potyvirus virion structure shows conserved protein fold and RNA binding site in ssRNA viruses. *Sci. Adv., 3*, 1–7.

Zhao, W., Vaithiyalingam, S., San, F. J., Maranon, D. G., Jimenez-Sainz, J., Fontenay, G. V., Kwon, Y., et al., (2015). Promotion of BRCA2-dependent homologous recombination by DSS1 via RPA targeting and DNA mimicry. *Mol. Cell., 59*(2), 176–187.

Zhao, Y., Yang, X., Zhou, G., et al., (2020). Engineering plant virus resistance: From RNA silencing to genome editing strategies. *Plant Biotechnol J.*, *18*, 328–336.

Zheng, L., Rodoni, B. C., Gibbs, M. J., & Gibbs, A. J., (2010). A novel pair of universal primers for the detection of potyviruses. *Plant Pathol., 59*, 211–220.

Zipper, H., Brunner, H., Bernhagen, J., & Vitzthum, F., (2004). Investigations on DNA intercalation and surface binding by SYBR green I, its structure determination, and methodological implications. *Nucleic Acids Res., 32*, e103.

## CHAPTER 2

# Recent Trends in Diagnostic Methods for Detection of Plant Viruses

RASHMI SUMAN,[1] ASHA RANI,[1] NARAYAN RISHI,[1] YASHIKA DHIR,[2] SUNNY DHIR,[2] and VANITA CHANDEL[1]

[1]*Amity Institute of Virology and Immunology, Amity University Uttar Pradesh, Noida, Uttar Pradesh, India*

[2]*Department of Biotechnology, Maharishi Markandeshwar (Deemed to be University), Mullana, Ambala, Haryana, India*

## ABSTRACT

Plant viruses pose a significant threat to the agricultural sector of developing countries. The transmission of viruses, along with their hosts and vectors, occurs between countries due to increased globalization of commerce and climate change. To ensure the production of healthy planting material, there is a need for rapid and specialized diagnostic tools for plant viruses. Traditional diagnostic procedures such as biological indexing and serological-based ELISA have been in use for many years. Molecular techniques like PCR and hybridization have provided more specific detection methods. Researchers have further developed nested and multiplex PCR (M-PCR) to detect multiple viruses. Other nucleic acid-based (NA-based) methods, including RT-LAMP, RPA, HAD, and NASBA, have reduced the time required for on-site testing. Real-time PCR offers integrated detection and quantification with excellent repeatability and accuracy. NGS technology is a more advanced method for detecting novel and multiple viruses simultaneously, requiring parallel sequencing and bioinformatic analysis. The CRISPR/Cas system is another modern technology that enables rapid and specific virus detection. It utilizes RPA or LAMP-based CRISPR/Cas12, 13, or 14 systems

Detection and Management of New and Emerging Mystery Plant Virus Sources.
Alireza Golnaraghi, Rajarshi Kumar Gaur, and Rakesh Kumar Verma (Eds.)

for fluorescent visualization of plant viruses. Additionally, artificial intelligence (AI) has gained attention for its potential in rapid virus detection. AI utilizes different sensors and neural networks like CNN (convolutional neural networks) for disease diagnosis. However, there are still various challenges in the diagnosis of plant viruses that require further research and innovation.

## 2.1 INTRODUCTION

Viruses are the most diverse organisms on Earth, and they possess the unique characteristic of being a parasitic entity with a small genome of RNA or DNA enclosed in a protein coat. Due to their rapid replication rate, which facilitates mutations and recombination, viruses have the ability to alter their genome organization, resulting in genetic variation and evolutionary changes. In agriculture, outbreaks of viral diseases in field crops can lead to significant financial losses, reaching millions of dollars. According to reports, viral infections can cause up to 40% crop losses, as well as a decline in both fruit quality and quantity (Maksimov et al., 2019; Hanˇcinský et al., 2020). The key factors contributing to viral emergence in any given region include: (i) virus evolution and adaptation to new environments; (ii) climate changes; (iii) agricultural fields with high plant density and low genetic diversity; and (iv) global trade of plant germplasm and seedlings across different geographical regions (Anderson et al., 2004; Jones, 2009; Elena et al., 2014). Once crops are extensively infected in the fields, it becomes challenging to cure viral diseases. Therefore, it is crucial to conduct early diagnosis of samples prior to planting, using certified crops obtained from nurseries.

There are multiple diagnostic methods that can be applied to detect viruses in plants. Some of them are diagrammatically represented in Figure 2.1 and described later in the chapter. The traditional approach for plant virus diagnosis was visual inspection based on symptoms. Then, people moved on to microscopic methods, which confirmed the presence of viral infection in plants. Biochemical techniques involve the purification of virus particles through ultracentrifugation, precipitation, and chromatography. After purifying virus particles, the virus needs to be visualized in a microscope or confirmed through serological techniques like ELISA using specific antibodies. Each of these methods was modulated as per the requirement based on certain virus properties. For example, visualization of the ultra-structure of the virus or their pathway inside plant cells was monitored using fluorescence microscope, transmission electron microscope (TEM), and scanning electron microscope (SEM). Similarly, molecular techniques involve PCR, RT-PCR, real-time PCR, multiplex PCR (M-PCR), microarray, immunocapture (IC)-PCR, and isothermal amplification methods such as loop-mediated isothermal amplification (LAMP), recombinase polymerase amplification (RPA), nucleic acid sequence-based amplification (NASBA), and cross priming amplification (CPA).

In the last decade, the discovery of CRISPR/Cas technology has brought another significant achievement in plant virus diagnostics. With the advent of Next Generation Sequencing (NGS), scientists can now discover novel viruses in any plant samples, along with known viruses, without any prior information. The metagenomics approach helps in the identification of novel

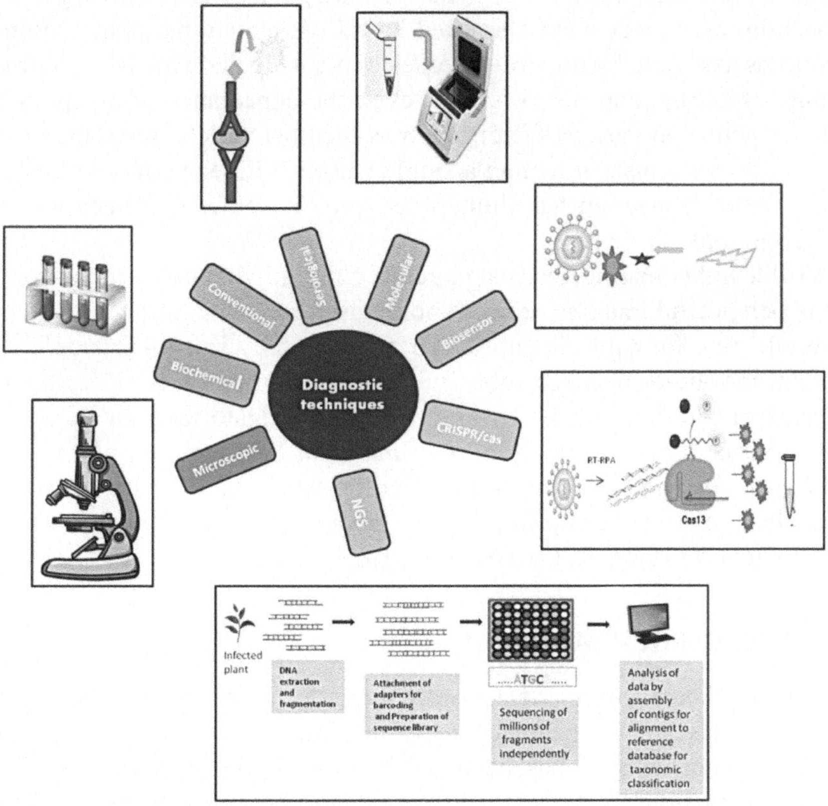


**FIGURE 2.1** Diagrammatic representation of different diagnostic techniques.

viruses and satellite molecules in plants, allowing for the defined taxonomic classification of new viral strains and their demarcation. NGS has revolutionized diagnostics in the past few years due to its capability to sequence millions of DNA/RNA molecules and analyze their data computationally. Artificial intelligence (AI) and deep learning (DL) techniques for image-based plant disease detection are the latest advancements being worked on for the diagnosis of plant viruses. In this chapter, we have discussed traditional methods for plant virus diagnosis, including serological, molecular, biochemical, and microscopic techniques. Additionally, we have explored recent techniques such as CRISPR/Cas, NGS, DL, and AI, which are highly sensitive, rapid, reliable, and cost-effective.

## 2.2 CONVENTIONAL METHODS

Centuries ago, the common way to detect plant pathogens was through visual inspection, as viruses were classified based on symptoms. Some common symptoms associated with viral diseases in plants include mosaic, chlorosis, curling, crinkling, and necrosis. However, the appearance of symptoms in plants depends on various factors such as the host variety, resistance to the virus, different viruses infecting a single variety, virulence of the virus, and more. Visual inspection has limitations when it comes to detecting latent viral infections.

Visible light imaging, thermal imaging, chlorophyll fluorescence imaging, and hyperspectral imaging are commonly used for visual inspection and are quite effective for viral diagnosis. To measure the virus titer, people used to count mosaic or necrotic spots on indicator plants and compare them to diseased plants. This method was used to investigate the titer of tobacco mosaic virus (TMV) using *Datura stramonium* L., *Nicotiana glutinosa* L., and *Phaseolus vulgaris* L.; tobacco necrotic virus (TNV) using *P. vulgaris* L.; and potato viruses using *Gomphrena globosa* L. and *Chenopodium album* L. (Yamaguchi, 1960; Tolkach et al., 2019).

## 2.3 MICROSCOPIC METHODS

Another method for virus detection is through a light microscope because certain viruses have a tendency to form Ivanovsky crystals or inclusion bodies in plant cells. It has been observed that each type of virus has a specific formation of crystals. For example, wheat mosaic virus (WMV) and potato virus X (PVX) form spherical crystals, while TMV forms needle-shaped and hexagonal crystals. With advancements in technology, more advanced microscopes have been invented for visualizing living particles, such as fluorescence microscopes, confocal microscopes (CM), and electron microscopes (EMs). These not only aid in virus identification but also help categorize them in different taxonomies based on their shape, size, and surface morphology.

The TEM has been instrumental in the morphological characterization and ultrastructural analysis of virus particles. It has allowed for the sub-cellular localization of virus particles in insect vectors. Several plant viruses, such as rice stripe mosaic virus (RSMV), tomato yellow leaf curl virus (TYLCV), Potato Viruses, and pepino mosaic virus (PMV), have been morphologically characterized using TEM. Ultrastructural alterations were observed in TMV in *N. tobacum* and zucchini yellow mosaic virus (ZYMV) in *Cucurbita pepo* upon microwave irradiation (Zechmann et al., 2009). TEM was also used to determine the ultrastructure of tomato spotted wilt virus (TSWV) and potato virus Y (PVY) in tomato and potato crops from Korea (Cho et al., 2009). In India, TEM examination of infected Gladiolus leaves revealed cytoplasmic inclusion bodies of laminated and scroll-shaped bean yellow mosaic virus (BYMV) (Kaur et al., 2015). The transmission mechanism of rice stripe virus (RSV) through plant hoppers was elucidated using TEM (Deng et al., 2013). TEM was employed to investigate the intracellular transport mechanism of prune dwarf virus (PDV), providing valuable insights into virus transport through plasmodesmata with the help of tubular structures generated by the movement protein (MP) (Kozieł et al., 2018). Furthermore, TEM observations of Plantago Asiatic Mosaic Virus (PlAMV) revealed that the N-terminal region of the coat protein (CP) is dispensable for virion assembly (Ozeki et al., 2009).

A CM was used to detect indicator molecules that co-localize with viruses in plants. For example, HSP70 and HSP90 were detected in tomato crops infected with TYLCV (Wieczorek et al., 2020). The CM also has another use in tracing viruses inside tissue by labeling them with either GFP or RFP for quantification purposes, as well as for observing the infectious process of the virus.

## 2.4 SEROLOGICAL TECHNIQUES

Antibodies are key components used in the serological detection of pathogens. The enzyme-linked immunosorbent assay (ELISA) was first used by

Adams and Clarks in 1977 for diagnosing plant diseases (Clark & Adams, 1977). The most common ELISA methods are the double antibody sandwich (DAS) and indirect ELISA, while modified versions include Dot-ELISA, tissue print ELISA (TP-ELISA), Protein A Sandwich ELISA (PAS-ELISA), Plate Trapped Enzyme ELISA (PTA-ELISA), antigen coated plate ELISA (ACP-ELISA), and triple antibody sandwich (TAS-ELISA) (Abd El-Aziz, 2019; Edwards & Cooper, 1985; Wu et al., 2013). These ELISA methods have been modified to enhance their specificity and sensitivity for virus diagnosis, considering factors such as virus titer and the presence of different viruses in the sample.

DAS-ELISA is highly strain-specific, as it sandwiches the antigen between capturing and detecting antibodies. TAS-ELISA is similar to DAS-ELISA but uses a monoclonal antibody specific to the antigen. Another approach is indirect ELISA, such as PTA-ELISA, which allows the virus or antigen to adsorb onto the plate surface by directly adding the test sample. In PAS-ELISA, protein A is pre-coated on a microtiter plate before adding the antibody, ensuring proper orientation of the antibody by binding with the Fc region, allowing the F(ab')2 portion to bind to the antigen or virus particles.

The principle of dot immunobinding assay (DIBA) is almost the same as ELISA, with the only difference being the antigen or antibody that binds to a nitrocellulose membrane (NCM) instead of a 96-well plate. Another difference is the formation of a precipitate on the membrane with a specific color due to the insoluble product of the enzyme reaction (Arakawa & Shimizu, 1985). TMV was detected from tobacco leaves using the DIBA technique. Similarly, Dot-ELISA is also a solid-phase immunoassay where a small amount of antigen is dotted onto a solid phase, such as PVDF or NCM or disc placed onto a microtiter plate. After incubation with an enzyme-labeled antibody, it forms a visible, blue-colored dot on the membrane (Pappas, 1988). Dot-ELISA was used for the detection of rice viruses, including Southern rice black-streaked dwarf virus (SRBSDV), rice dwarf virus, rice stipe mosaic virus, and potato leafroll virus (Smith & Banttari, 1987). Tissue immunoblotting assay (TIBA) is a modified version of DIBA, during which a finely and freshly cut edge of a leaf, leaf blade, tuber, stem, and root is blotted onto the NCM (Makkouk et al., 1993).

Initially, most plant viruses were serologically detected using one of the ELISA methods mentioned above due to their high specificity. There are numerous viruses that have been diagnosed through serological techniques, and a few of them are listed in Table 2.1.

**TABLE 2.1** Use of Different ELISA Techniques for Plant Virus Diagnosis

| Virus | Host | ELISA Techniques | References |
|---|---|---|---|
| Cucumber mosaic virus | – | TAS-ELISA | Yu et al. (2005) |
| Srilankan cassava mosaic virus | Cassava | TAS-ELISA | Charoenvilaisiri et al. (2021) |
| Southern rice black-streaked dwarf virus | Rice | Dot-ELISA | Wang et al. (2012) |
| Tomato spotted wilt virus | Crops | DAS-ELISA and Dot-ELISA | Gao & Wu (2022) |
| Zucchini yellow mosaic virus | Cucurbitaceous crops | DAS-ELISA and PTA ELISA | Chen et al. (2017) |
| Potato virus X | – | ELISA | Gallenberg & Jones (1985) |
| Potato virus Y | Potato | Tissue blot ELISA | Hamdy (1961) |
| Cowpea aphid-borne mosaic virus | Cowpea | PAS-ELISA | Ojuederie et al. (2010) |
| *Rubus canadensis* 1 | Blackberry | DAS-ELISA | Abou Ghanem et al. (2013) |
| Tomato yellow leaf curl Thailand virus | Tomato, pepper, okra, eggplant, and Cucurbitaceae | TAS-ELISA | Seepiban et al. (2017) |
| Grapevine leafroll associated closterovirus-3 | Grapevines | DAS-ELISA | Ling et al. (2000) |
| Citrus tristeza virus | Citrus plants | Dot-ELISA, tissue print-ELISA, and TAS-ELISA | Liu et al. (2016) |
| Rice stripe mosaic virus | Rice | Tissue prints ELISA, Dot-ELISA, and ACP-ELISA | Guo et al. (2020) |
| Rice dwarf virus | Rice | PTA-ELISA and Dot-ELISA | Wu et al. (2014) |
| Potato virus S | Potato | DAS-ELISA, tissue print ELISA and Dot-ELISA | Song et al. (2017) |
| Croton yellow vein mosaic virus | Croton | DAS-ELISA, TAS-ELISA | Pramesh et al. (2013) |
| Soyabean mosaic virus | Soyabean | Indirect ELISA, DAS-ELISA, Tissue prints ELISA, Dot-ELISA | Cui et al. (2011) |
| Tomato spotted wilt virus | Tomato | DAS-ELISA, Dot-ELISA | Gao & Wu (2022) |
| Potato leafroll virus | Potato | Dot-ELISA | Smith & Banttari (1987) |

Rapid detection of plant viruses on-site has become possible nowadays with the development of immunostrips. The immunochromatographic strips, based on colloidal gold, are specifically developed for each virus or related viral strain using antibodies specific to them, enabling rapid and specific detection. Another immunochromatography technique, known as lateral flow, was developed by Salomone et al. for the rapid detection of citrus tristeza virus (CTV) using polyclonal antibodies against it (Salomone et al., 2004). Gold nanoparticle-conjugated antibodies on an immunochromatography strip were used for the detection of grapevine leafroll-associated virus-3 (GLRaV-3) (Byzova et al., 2018). This method was also employed for the diagnosis of RSV in rice plants in China due to its quick, sensitive, and ready-to-use nature (Huang et al., 2019).

## 2.5 MOLECULAR TECHNIQUES

Looking back to prehistoric eras where the detection of viral infection in plants was limited to visual identification based on the symptoms caused by the viruses and biological local lesion assays, we now live in a modern world full of molecular and serological-based approaches for diagnosis. There has been a tremendous leap in this field. When comparing molecular and serological methods, molecular methods used for plant virus diagnosis are more sensitive and specific because they are based on nucleic acid. These methods are mainly used in disease surveys, plant quarantines, and certification programs.

Diagnosis of plant viruses involves the use of molecular techniques such as polymerase chain reaction (PCR), isothermal amplification, reverse transcriptase (RT) PCR, and quantitative or real-time (qPCR). These molecular techniques are more sensitive compared to serological methods. The process of molecular techniques includes isolating nucleic acid (DNA/RNA) from the virus, amplifying it through PCR, and analyzing its genome.

Conventional PCR is commonly used for identifying DNA viruses by using degenerate primers based on different strains or isolates of the same virus. However, for identifying RNA viruses, we need to convert RNA into cDNA and then amplify it using RT-PCR. Real-time PCR allows for the quantification of the virus and accurate discrimination of different virus strains and viroids.

M-PCR is another technique used for detecting multiple viruses in a single reaction, using a set of primers. This not only saves time but also reduces costs. For example, four cherry viruses (cherry necrotic rusty mottle virus (CNRMV), cherry virus A (CVA), little cherry virus 1 (LChV-1), and prunus necrotic ringspot virus (PNRSV)) were identified using M-PCR (Noorani et al., 2013).

Additionally, a multiplex nested RT-PCR has been developed for detecting viruses in olive plants, such as CMV, cherry leaf roll virus (CLRV), strawberry latent ringspot virus (SLRSV), and Arabis mosaic virus (ArMV). This technique is less time-consuming as it is performed in a single compartmentalized closed tube using 20 compatible primers (Bertolini et al., 2003).

Microarray, which is based on molecular hybridization, is another important method for plant virus detection. DNA microarrays consist of a surface including glass, nylon, or different polymers linked with numerous capture probes, where each probe is specific for a target RNA/DNA sequence. These probes are arrayed onto a single chip and amplified by PCR. The chip is then exposed to fluorescently labeled RNA or DNA sequences present in the sample (Lopez et al., 2003). Different fluorophores and laser technology are used by the detection system to detect the target DNA/RNA sequence present in the sample. Nanogen (San Diego, USA) used nanochip technology that was based on microelectronics combined with a microarray system on a solid support, which was exposed to streptavidin to increase the binding affinity with biotin-labeled sequences. DNA Microarray has the advantage that it can detect many viruses at one time if probes are available, but it can only be done for known viruses. It has been used for the identification of RNA viruses (Barba & Hadidi, 2011).

However, it has some disadvantages. It is very costly and generates a large amount of data that requires experts for interpretation.

Sensitivity of serological methods and specificity of molecular techniques are combined in immunocapture (IC) PCR/RT-PCR by some scientists for the detection of plant viruses. This method involves using antibodies to capture virus particles in a tube, followed by PCR amplification. An electrochemical-based immunosensor was developed for the highly sensitive, rapid, and easy-to-use detection of papaya ringspot virus (PRSV) in the field (Valekunja et al., 2016). Another study by Hamim et al. developed an ultra-sensitive diagnostic technique for PRSV detection using single-tube nPCR (Hamim et al., 2018).

With the need for faster diagnostic techniques, isothermal amplification methods such as LAMP, RPA, NASBA, and cross-priming amplification (CPA) have emerged. The advantage of these isothermal amplification methods is that they do not require a thermal cycler machine like traditional PCR, and they can be used for on-site detection within minutes. Additionally, they can be visualized directly in the tube through color changes, eliminating the need for a UV transilluminator. These methods have proven to be more rapid, sensitive, and robust in the field of virus diagnosis.

The LAMP or RT-LAMP method is based on amplification of target DNA or RNA at a constant temperature of 62–65°C without any denaturation step for double-stranded DNA prior to amplification and heavy equipment for measuring the amplicons. The RT-LAMP assay has been successfully applied for the detection of many viruses and viroids. Major apple viruses such as apple chlorotic leaf spot virus (ACLSV) and apple stem pitting virus (ASPV) were detected using the RT-LAMP technique from pear trees (Lu et al., 2018).

NASBA is a sensitive assay for the detection of RNA viruses as it is based on transcription of DNA into RNA at an isothermal temperature of 41°C using three enzymes: RT from avian myeloblastosis virus, T7 RNA polymerase, and RNase H, along with two specific oligonucleotide primers (Kievits et al., 1991). A real-time NASBA assay was developed for the identification of apple scar skin viroid (ASSVd) using electro-chemiluminescence and probe-capture hybridization in molecular beacons for the certification of disease-free apple cultivars (Heo et al., 2019).

CPA is another isothermal amplification technique that relies on strand displacing polymerase, namely Bst DNA polymerase (*Bacillus stearothermophilus* DNA Polymerase), and 5 primers: one cross primer, two displacement primers, and two amplification primers (Xu et al., 2012). It has been applied for the detection of many pathogens, including viruses like PNRSV (Huo et al., 2017) and cucumber green mottle mosaic virus (CGMMV) (Huo et al., 2017). Recently, Yang et al. have developed a highly sensitive assay based on RT-CPA along with a lateral flow dipstick for the identification of bean pod mottle virus (BPMV) along with four related viruses present in soybean seeds, such as tomato ringspot virus (ToRSV), tobacco ringspot virus (TRSV), southern bean mosaic virus (SBMV), and ArMV (Yang et al., 2022) (Table 2.2).

**TABLE 2.2** List of Plant Virus Detection Through Various Molecular Techniques

| Virus | Host | Molecular Techniques | References |
|---|---|---|---|
| Alfalfa mosaic virus | Cucurbits | IC-RT-PCR | Kılıç & Yardımcı (2015) |
| Ageratum enation virus | Ageratum | PCR | Srivastava et al. (2013) |
| Cucumber mosaic virus | Cucurbits | RT-PCR, multiplex RT-PCR | Chen et al. (2011) |
| Croton yellow vein mosaic virus | Croton and weeds | PCR | Sharma et al. (2019) |
| Potato virus X | Potato | RT-PCR, real time multiplex RT-PCR | Onozuka et al. (2021) |

**TABLE 2.2** *(Continued)*

| Virus | Host | Molecular Techniques | References |
|---|---|---|---|
| Potato virus Y | Potato | RT-PCR, real time multiplex RT-PCR | Onozuka et al. (2021) |
| Potato virus S | Potato | RT-PCR, multiplex RT-PCR, real time multiplex RT-PCR | Onozuka et al. (2021) |
| Tomato yellow leaf curl virus | Tomato | Duplex PCR, multiplex PCR | Yoon et al. (2015); Chen et al. (2015) |
| Tomato spotted wilt virus | Tomato, weeds, cucurbits | RT-PCR, real time RT-PCR | Roberts et al. (2000) |
| Tobacco mosaic virus | Tobacco, tomato, pepper | RT-PCR, multiplex RT-PCR, real time RT-PCR | Kumar et al. (2011) |
| Tomato mosaic virus | Tomato, pepper | RT-PCR, multiplex RT-PCR, real time RT-PCR | Kumar et al. (2011) |
| Papaya ringspot virus | Papaya | RT-PCR, multiplex RT-PCR, nested PCR, immunosensor | Huo et al. (2015); Hamim et al. (2018) |
| Papaya leaf distortion mosaic virus | Papaya | RT-PCR, RT-LAMP multiplex, RT-PCR | Shen et al. (2014); Huo et al. (2015) |
| Zucchini yellow mosaic virus | Pumpkin, cucumber, squash, melon | RT-PCR, multiplex RT-PCR, real time RT-PCR | Usher et al. (2012); Chinnadurai et al. (2018) |
| Papaya leaf curl virus | Papaya | PCR | Chang et al. (2003) |
| Melon necrotic spot virus | Melon | RT-PCR, RT-LAMP | Qiao et al. (2020) |
| Apple stem pitting virus | Apple | RT-PCR, multiplex RT-PCR, RT-LAMP | Menzel et al. (2002); Rana et al. (2009); Dhir et al. (2011) |
| Apple stem grooving virus | Apple | RT-PCR, multiplex RT-PCR | Menzel et al. (2002); Rani et al. (2021) |
| Apple chlorotic leaf spot virus | Apple, cherry | RT-PCR, multiplex RT-PCR, RT-LAMP | Menzel et al. (2002); Rana et al. (2008) |
| Grapevine fanleaf virus | Grapevines | RT-PCR, real time RT-PCT, IC RT-PCR | Koolivand et al. (2014) |
| Indian citrus tristeza | Citrus fruits | RT-duplex PCR, RT-LAMP | Kokane et al. (2020); Warghane et al. (2017) |
| Soyabean mosaic virus | Soyabean | RT-PCR, real time PCR, multiplex RT-PCR, IC-RT-PCR | Wu et al. (2014); Naghavi (2008) |
| Rice stripe mosaic virus | Rice | RT-PCR, multiplex RT-PCR | Yang et al. (2017) |
| Prunus necrotic ring spot virus | Apple, peach | RT-PCR, ELISA | Chandel et al. (2007, 2008) |

## 2.6 NEXT GENERATION SEQUENCING (NGS)

In the last decade, high-throughput sequencing technology (HTS) has emerged as the most accessible and advanced technique for diagnosing plant viruses. HTS can sequence millions of DNA/RNA fragments in parallel, without prior knowledge of the viral genome, enabling the diagnosis of all viruses present in each plant sample. Another significant advantage of next-generation sequencing (NGS) is its ability to detect unknown viruses in plant cells (Roossnick, 2015). HTS has revolutionized the field of diagnostics, helping to elucidate the mysterious etiology of certain diseases. However, it does not establish a direct correlation between the disease and the virus present in the plant, which can only be achieved by satisfying Koch's postulate (Vives et al., 2013; Tomasechová et al., 2020; Mumo et al., 2020). HTS offers multiple advantages beyond diagnostics, including the study of virus epidemiology, genetic diversity, evolutionary changes, and synergistic or antagonistic interactions between viruses in plants (Hadidi et al., 2016; Roossnick, 2017; Tan et al., 2019).

Previously, Sanger sequencing dominated the diagnostic field due to its high specificity, output, and cost-effectiveness. It was known as the first-generation sequencing method. However, with the advent of second and third-generation sequencing technologies, it has become outdated, and researchers have shifted towards omics fields such as genomics, proteomics, metagenomics, transcriptomics, metabolomics, etc., due to the ability of NGS to explore millions of molecules present in plant or animal cells and analyze their data.

Diagnosis of emerging viruses through second-generation sequencing utilizes millions of DNA fragments in the case of DNA viruses or cDNA in the case of RNA viruses after retrotranscription, for the preparation of a random DNA library using oligo-dT or random primers. These libraries are amplified and sequenced in parallel after attaching synthetic DNA adaptors. This generates short sequence reads of around 100–500 nt, which are assembled in an overlapping manner and analyzed through nucleotide identity using software like the Geneious package. Some of the commercial platforms used for second-generation sequencing are HiSeq, NextSeq, MiSeq by Illumina, Inc; GS FLX by Roche Diagnostics and 454 Life Sciences; SOLiD by ABI; Ion Torrent PGM, S5, Proton, and BGISeq-500 (Ambardar et al., 2016). These platforms are being used for plant virus diagnosis more than traditional ones due to their high sensitivity and efficiency.

Nanopore sequencing and single molecule real-time sequencing (SMRT) are third-generation sequencing techniques developed by some companies based on sequencing of single molecules in real time, giving long reads of kilobases without clonal amplification. Some of the advantages associated with nanopore sequencing are that it is rapid, has high mobility due to the small size of the sequencer, and there is no need for reverse transcription for RNA viruses (Kiselev et al., 2020). This technology was used for the detection of pathogens, including bacterial and viral, from tomato plants, and their study reported the presence of DNA viruses TYLCV and watermelon chlorosis stunt virus (WmCSV) (Chalupowicz et al., 2019). A UK-based company used Oxford Nanopore Technology MinION sequencer for the surveillance of plum pox virus (PPV) infected plant tissues of stone fruits and in the insect vector, and 95% of the reads generated were from PPV infected tissues (Badial et al., 2018). Another report has mentioned the use of Nanopore sequencing technology in the detection of a novel Begomovirus in a cowpea plant with bright golden mosaic symptoms and tentatively named it as cowpea bright yellow mosaic virus (CoBYMV) (Naito et al., 2019).

NGS provides a robust platform for the diagnosis of viruses, as it does not require prior information. It has become the leading tool of choice for detecting viruses with low titers. The use of NGS has greatly improved the identification, epidemiology, and ecology of viruses. For example, Grapevine Clearing Virus was detected in 6 cultivars across Indiana, Missouri, and Illinois using deep and whole genome sequencing (Zhang et al., 2011). NGS with metagenomic analysis also detected the presence of PMV and Gayfeather Mild Mottle Virus in *Solanum lycopersicum* and *Gomphrena globose* (Adams et al., 2009). Additionally, siRNA sequencing aids in the identification of quasi-species of viruses and helps us understand the mechanisms involved in the formation of these siRNAs and the defense mechanisms (Seguin et al., 2014).

NGS was used to identify 22 isolates of PVY, TVBMV, CMV, TMV, and BYV infecting tobacco plants in China (Akinyemi et al., 2016). Furthermore, NGS has facilitated the identification of ssRNA viruses, closteroviruses, and cytorhabdoviruses (through ribo-depleted RNA-based datasets). When combined with automated viral surveillance and diagnosis, NGS makes the identification of viruses much easier (Barrero et al., 2017). NGS outperforms traditional bioassays, especially for novel viruses and those with poor characterization. For instance, NGS was used to detect novel isolates of badnavirus and potyvirus in *Dioscorea spp*. Additionally, 7 sunflower varieties were identified using NGS, and small RNA data confirmed the presence of endornavirus in 2 of the varieties imported from the U.S. to the Netherlands. Finally, different algorithms in high-throughput sequencing were employed to detect 12 plant viruses from three different infected plants (Massart et al., 2019).

HTS helped in the identification of PTM (potato virus M) isolates (3) when testing tomato plants in Slovakia. Around 10 variants of PSTV (potato spindle tuber viroid) in tomato plants were identified and analyzed using an *in-silico* approach, utilizing small RNAs of the viroid (Adkar-Purushothama et al., 2020). NGS helped in the identification of ageratum yellow vein Malaysia virus. NGS and RT-PCR aided in the detection of Broad Bean Wilt Virus 2, Gaillardia Latent Virus, Asparagus Virus 2, and Tomato Spotted Wilt Orthotospovirus in *Valriana fauriei* (Uehara-Ichiki et al., 2020). NGS has also assisted in the identification of new viruses, contributing to the validation of new genera. The discovery of new viruses is made possible through NGS, as seen in the case of *Camellia japonica* plants infected with Camellia Yellow Ringspot Virus, Camellia Associated Marafivirus, and Camellia Associated Badnavirus. NGS aided in the identification of variants exhibiting different symptoms, such as a sequence variant of ACLSV displaying ring-shaped rust on apple trees. Another variant of ASPV (apple stem pitting virus) exhibited green crinkles on apple fruits (Li et al., 2020). HTS technology has proven helpful in reporting 70 new viruses (Bejerman et al., 2020). NGS, along with bioinformatics algorithms and RT-PCR techniques, has had a significant impact on the discovery and identification of new viruses. It is also unraveling more strategies for crop improvement, particularly with the assistance of CRISPR-based techniques for the development of resistant varieties.

## 2.7 CRISPR/CAS

Clustered regularly interspaced short palindromic repeats (CRISPR) were discovered as part of the adaptive immune system in prokaryotes. Its locus was first discovered in *E. coli*. This technique targets DNA or RNA to provide protection against viruses. In CRISPR, short repeated sequences are interspaced by spacers, which are unique sequences derived from viral genomes or plasmids. This observation led to its recognition as a part of the defense mechanism. Spacers are sequences that are 26 to 72 base pairs long and are present between the repeats. The spacer sequence is similar to different viral genome sequences to which the organism was previously exposed. Thus, spacers are derived from the viral genome and are added to the host genome during viral infection. These sequences can be later identified by the host immune system when exposed to the same virus again. In the technique, these sequences are added to the repeats, which aids in the identification and destruction of more viral particles through interference mechanisms. After manipulating the genome by adding more such sequences, the changes are passed on to the next generations as well. CRISPR does not work alone. It is also associated with proteins (Cas) that help in the immune response (Devashish et al., 2015).

The process involves the exogenous engineering of nucleases to cleave the target region from DNA or RNA, using a domain attached to the nuclease for target site recognition. These two components are combined exogenously. After the target sequence is cleaved from the DNA or RNA, the broken fragments are repaired using two methods. One method is non-homologous end joining (NHEJ), while the other method is homology directed repair (HDR). These repair methods are involved in repairing double strand breaks in DNA.

NHEJ is error-prone and can cause mutations after the double strand break is joined, while HDR is more reliable. The nucleases play a crucial role in this technique, with the specificity and flexibility of the nucleases being key factors. Nucleases such as zinc finger nuclease, transcription activator-like effector nuclease, and the more flexible Cas9 nuclease have been used in this technique. The two nucleases, ZFN and TALEN, are suitable for closely spaced DNA sequences as they require dimerization of the nuclease domain Fok1 for activation and specificity. On the other hand, Cas9 is easy to design, highly efficient, and inexpensive compared to the other two nucleases (Wada et al., 2020). This system is widely used to date (Dong et al., 2021).

Cas9 nuclease is accompanied by guide RNA (gRNA). The gRNA is 20 nucleotides long and complementary to the target DNA sequence. It utilizes DNA-RNA interaction-based gene editing. In addition to Cas9 and gRNA, the third component required for the system to function is the PAM sequence at the 3′-end. The PAM sequence is a short nucleotide sequence known as the protospacer adjacent motif (PAM). For example, *Streptococcus pyogenes* (Sp) Cas9 nuclease uses the NGG sequence as the PAM (Wada et al., 2020).

The genetic manipulation using this method has been proven successful in enhancing biotic and abiotic stress resistance, manipulating metabolic pathways to increase yield, and more. In an independent experiment, six genes of wild tomatoes were manipulated to bring about changes in fruit number, size, nutrient content, shape, and plant organization. Two rounds of genome editing were required to manipulate all six genes. Many other crops, including maize, sorghum, and wheat, have also been used in manipulation (Zsögön et al., 2018).

Viral infections in plants have a significant impact on yield, nutritional security, and the economy. While plants have defense mechanisms to fight against viruses, they can be infected by multiple viruses. The defense mechanisms against virus infections can be specific to a particular virus or act against all viruses infecting the host. Specific defense mechanisms may not provide complete resistance and may result in partial immunity. Plants exhibit two types of defenses against viruses, particularly if the virus is a non-host. Plant viruses face RNA interference (RNAi) antiviral silencing mechanisms, which are the first response of plants against infection. To effectively manage diseases, accurate and complete diagnosis is crucial. Plant breeding practices aimed at developing disease-resistant plants have limitations, including non-specificity, time-consuming processes, and labor-intensive efforts.

Diagnostics have shown significant growth in the use of CRISPR-Cas based tools. The initial diagnosis involved the recognition of dsDNA using CRISPR/Cas9. Other nucleases such as Cas12a, 14, and 13a have also been reported for nuclcic acid detection. Gootenberg developed a rapid and sensitive detection method by combining CRISPR/Cas13a with isothermal amplification, known as Specific High Sensitivity Enzymatic Reporter UnLOCKing (SHERLOCK). This method is highly sensitive and can detect nucleic acid concentrations in attomoles. It can even detect single base mismatches (Gootenberg et al., 2017).

Another method called DETECTR, the first Cas12a-based detection method, was used for dsDNA. As the name suggests (DNA endoribonuclease targeted CRISPR trans reporter), this method utilizes crRNA to target dsDNA, resulting in collateral cleavage of ssDNA carrying quenchers and fluorophores. This cleavage confirms the recognition of the target by generating fluorescent signals. One-hour low-cost multipurpose highly efficient system (HOLMES) combines the Cas12a system with LAMP to detect DNA and RNA targets. Another method involves using RPA or RT-PCR to increase the concentration of target DNA and RNA, respectively. Cas12/13 are then used after transcription of the RPA product to RNA. Although CRISPR-based virus detection technology shows promise, it has not been fully implemented and utilized. A paper strip method has also been developed for on-spot virus detection. In another method, detection can be done in a one-step RT-RPA-Cas12a assay (Gootenberg et al., 2018).

A report has been published regarding the detection of nucleic acid of PVX and TMV. The method used was a one-step *in vitro* specific CRISPR-based assay for nucleic acid detection called iSCAN-OP. This method involved RT-RPA amplification followed by collateral activity using Cas12a.

Similarly, CRISPR/Cas12a has been reported to aid in the detection of ASGV, ApNMV, ACLSV, ASPV, and ASSVd, both viruses and viroids. This method was found to be much more sensitive than RT-PCR. Other viruses, such as Begomoviruses-TYLCV, were also detected using the same technique. However, there are limitations, particularly in the case of femtomolar (fM) concentration of viruses. Lower concentration detection still relies on the amplification step. Additionally, the use of heating devices for the detection steps and sample preparation poses another set of limitations. To achieve broader application, these limitations need to be overcome (Sharma et al., 2021).

## 2.8 ARTIFICIAL INTELLIGENCE (AI)

The traditional approach for virus identification relied on symptom identification, which proved to be unsuccessful in the case of latent viruses and was also prone to inaccuracy and inefficiency. Other molecular methods, such as PCR and ELISA, are mostly used. However, these techniques require expertise and are time-consuming. After a pathogen attack, many changes occur in the plant structure and other parts as well. Initially, remote sensors were used to detect these changes. However, they were unable to differentiate between changes caused by biotic or abiotic stress, or determine the specific pathogen involved in the case of biotic stress. Data analysis using hyperspectral imaging (HSI) with a spectrum of large wavelengths makes diagnosing specific plant diseases easier. This concept of phenotyping stress allows for the detection of minor changes that are beyond the range of human vision. Machine learning (ML) models are another useful tool in this regard. To train an ML model, a dataset with images of diseased plants from various channels is required. One available database is 'Plantvillage.' A testing dataset, such as HSI, can be used to provide high-dimensional data and additional information. Feature extraction is then performed to reduce the number of features involved and ensure a smooth process. However, expertise is needed to carry out this process. DL is another technique that enables automatic feature extraction and can handle a large number of samples. The limitation in this case is obtaining field images of plants, as the conditions of temperature, humidity, and other factors are not controlled as they are in a lab setting (Ghosh et al., 2022).

Early detection of the cassava brown streak virus (CBSV), which causes cassava brown streak disease (CBSD) and reduces cassava productivity in Africa, was made possible through active multispectral imaging (A-MSI) (Peng et al., 2022). Scientists utilized an A-MSI sensor system and ML methods for early screening of CBSV in Africa. This system effectively distinguished between healthy and CBSV-infected plants based on spectral and spatial imaging data. It proved to be a reliable approach for the early and rapid detection of CBSD compared to PCR.

Identification and classification of TYLCV and tomato spotted wilt virus (TSWV) from infected tomato leaves were documented by Mokhtar and colleagues (Mokhtar et al., 2015). They utilized the support vector machine (SVM) algorithm with four different kernel functions. This approach achieved a general classification accuracy of 90%, which increased to 92% when using quadratic kernel functions after N-fold cross-validation.

In a similar study, Zhu et al. conducted research on the pre-symptomatic identification and classification of TMV using hyperspectral imaging techniques (Zhu et al., 2016). They employed various algorithms including back-propagation neural networks (BPNNs), Support Vector Machine (SVM), Least Squares Support Vector Machine (LS-SVM), extreme learning machine (ELM), latent Dirichlet allocation (LDA), random forest (RF), and Partial Least Squares Discrimination Analysis (PLS-DA) (Zhu et al., 2017). The BPNN models achieved an accuracy of 95%, while the chemometric models achieved 80% accuracy.

## 2.9 FUTURE PERSPECTIVE

The challenges of food security and the emergence of new viral diseases in plants, causing significant economic losses, have prompted the development of various diagnostic techniques. Recent advancements in diagnostic techniques, such as NGS, artificial intelligence (AI), Hyperspectral Imaging, and DL, have enabled researchers to detect viral infections in plants in epidemic areas within minutes. These techniques are highly reliable, cost-effective, and less time-consuming compared to traditional approaches.

When comparing molecular and serological diagnostic tests, HTS (High-Throughput Sequencing) stands out as a universal and highly accurate method for profiling plant pathogens, particularly viruses. However, its widespread acceptance is hindered by the high cost of using complex machines, the complexity of sample preparation, and the need for skilled professionals to analyze the large amount of data generated through HTS.

Similarly, diagnostic methods based on DL and AI are user-friendly and provide rapid detection of plant viruses in the field using portable devices.

## KEYWORDS


- **artificial intelligence**
- **CRISPR/Cas**
- **deep learning**
- **diagnostic methods**
- **food security**
- **microarray**
- **next-generation sequencing**
- **polymerase chain reaction**
- **serological diagnostic tests**


## REFERENCES


Abd El-Aziz, M., (2019). Three modern serological methods to detect plant viruses. *Journal of Plant Science and Phytopathology*, *3*(3), 101–106. https://doi.org/10.29328/journal.jpsp.1001039.

Abou Ghanem-Sabanadzovic, N., Tzanetakis, I. E., & Sabanadzovic, S., (2013). Rubus canadensis virus 1, a novel betaflexivirus identified in blackberry. *Archives of Virology*, *158*(2), 445–449. https://doi.org/10.1007/s00705-012-1484-7.

Adams, I. P., Glover, R. H., Monger, W. A., Mumford, R., Jackeviciene, E., Navalinskiene, M., Samuitiene, M., & Boonham, N., (2009). Next-generation sequencing and metagenomic analysis: A universal diagnostic tool in plant virology. *Molecular Plant Pathology*, *10*(4), 537–545. https://doi.org/10.1111/j.1364-3703.2009.00545.x.

Adkar-Purushothama, C. R., Bolduc, F., Bru, P., & Perreault, J. P., (2020). Insights into potato spindle tuber viroid quasi-species from infection to disease. *Frontiers in Microbiology*, *11*. https://doi.org/10.3389/fmicb.2020.01235.

Akinyemi, I. A., Wang, F., Zhou, B., Qi, S., & Wu, Q., (2016). Ecogenomic survey of plant viruses infecting tobacco by next generation sequencing. *Virology Journal*, *13*(1), 1–12. https://doi.org/10.1186/s12985-016-0639-7.

Ambardar, S., Gupta, R., Trakroo, D., Lal, R., & Vakhlu, J., (2016). High throughput sequencing: An overview of sequencing chemistry. *Ind. J. Microbiol.*, *56*, 394–404.

Anderson, P. K., Cunningham, A. A., Patel, N. G., Morales, F. J., Epstein, P. R., & Daszak, P., (2004). Emerging infectious diseases of plants: Pathogen pollution, climate change and agrotechnology drivers. *Trends Ecol. Evol., 19*, 535–544. doi: 10.1016/j.tree.2004.07.021.

Arakawa, A., & Shimizu, S., (1985). A dot immunobinding assay for the detection of densonucleosis virus of the silkworm, Bombyx mori. *The Journal of Sericultural Science of Japan*, *54*(6), 500–503. https://doi.org/10.11416/kontyushigen1930.54.500.

Badial, A. B., Sherman, D., Stone, A., Gopakumar, A., Wilson, V., Schneider, W., & King, J., (2018). Nanopore sequencing as a surveillance tool for plant pathogens in plant and insect tissues. *Plant Disease*, *102*(8), 1648–1652. https://doi.org/10.1094/PDIS-04-17-0488-RE.

Barba, M., & Hadidi, A., (2011). DNA microarrays and other future trends in detection and typing of viruses, viroids, and phytoplasmas. In: Hadidi, A., Barba, M., Candresse, T., & Jelkmann, W., (eds.), *Virus and Virus-Like Diseases of Pome and Stone Fruits* (pp. 361–372). T. Paul MN: APS Press.

Barrero, R. A., Napier, K. R., Cunnington, J., Liefting, L., Keenan, S., Frampton, R. A., Szabo, T., et al., (2017). An internet-based bioinformatics toolkit for plant biosecurity diagnosis and surveillance of viruses and viroids. *BMC Bioinformatics*, *18*(1). https://doi.org/10.1186/s12859-016-1428-4.

Bejerman, N., Debat, H., & Dietzgen, R. G., (2020). The plant negative-sense RNA virosphere: Virus discovery through new eyes. In: *Frontiers in Microbiology* (Vol. 11). Frontiers Media S.A. https://doi.org/10.3389/fmicb.2020.588427.

Bertolini, E., Olmos, A., Lo´pez, M. M., & Cambra, M., (2003). Multiplex nested reverse-transcription polymerase chain reaction in a single tube for sensitive and simultaneous detection of four RNA viruses and *Pseudomonas savastanoi* pv. savastanoi in olive trees. *Phytopathology, 93*, 286–292.

Byzova, N. A., Vinogradova, S. V., Porotikova, E. V., Terekhova, U. D., Zherdev, A. V., & Dzantiev, B. B., (2018). Lateral flow immunoassay for rapid detection of grapevine leafroll-associated virus. *Biosensors*, *8*(4), 1–14. https://doi.org/10.3390/bios8040111.

Chalupowicz, L., Dombrovsky, A., Gaba, V., Luria, N., Reuven, M., Beerman, A., Lachman, O., et al., (2019). Diagnosis of plant diseases using the nanopore sequencing platform. *Plant Pathology*, *68*(2), 229–238. https://doi.org/10.1111/ppa.12957.

Chandel, V., Rana, T., Hallan, V., & Zaidi, A. A., (2007). Evidence for the occurrence of *Prunus necrotic ringspot virus* on peach in India by serological and molecular methods. *Canadian Journal of Plant Pathology*, *29*(3), 311–316.

Chandel, V., Rana, T., Handa, A., Thakur, P. D., Hallan, V., & Zaidi, A. A., (2008). Incidence of *Prunus necrotic ring spot virus* on malus domestica in India. *Journal of Phytopathology*, *156*(6), 382–384.

Chang, L. S., Lee, Y. S., Su, H. J., & Hung, T. H., (2003). First report of papaya leaf curl virus infecting papaya plants in Taiwan. *Plant Disease*, *87*(2), 204–204. https://doi.org/10.1094/PDIS.2003.87.2.204A.

Charoenvilaisiri, S., Seepiban, C., Kumpoosiri, M., Rukpratanporn, S., Warin, N., Phuangrat, B., Chitchuea, P., et al., (2021). Development of a triple antibody sandwich enzyme-linked immunosorbent assay for cassava mosaic disease detection using a monoclonal antibody to *Sri Lankan cassava mosaic virus*. *Virology Journal*, *18*(1), 100. https://doi.org/10.1186/s12985-021-01572-6.

Chen, H. M., Lin, C. Y., Yoshida, M., Hanson, P., & Schafleitn, R., (2015). Multiplex PCR for detection of tomato yellow leaf curl disease and root-knot nematode resistance genes in tomato (*Solanum lycopersicum* L.). *International Journal of Plant Breeding and Genetics*, *9*(2), 44–56. https://doi.org/10.3923/ijpbg.2015.44.56.

Chen, S., Gu, H., Wang, X., Chen, J., & Zhu, W., (2011). Multiplex RT-PCR detection of *Cucumber mosaic virus* subgroups and *Tobamoviruses* infecting tomato using 18S rRNA as an internal control. *Acta Biochimica et BiophysicaSinica*, *43*(6), 465–471. https://doi.org/10.1093/abbs/gmr031.

Chen, Z., Zhang, M. H., Zhou, X. P., & Wu, J. X., (2017). Development and detection application of monoclonal antibodies against *Zucchini yellow mosaic virus*. *Journal of Integrative Agriculture*, *16*(1), 115–124. https://doi.org/10.1016/S2095-3119(16)61416-8.

Chinnadurai, C., Ramsubhag, A., & Jayaraman, J., (2018). Optimization of multiplex PCR technique for the detection of viruses causing severe mosaic disease in pumpkin. *Tropical Agriculture*, *95*, 50–56.

Cho, J. D., Kim, J. S., Kim, J. Y., Choi, G. S., & Chung, B. N., (2009). Biological characteristics and nucleotide relationships in Korean tomato spotted wilt virus isolates. *Plant Pathology J.*, *25*, 26–37.

Clark, M. F., & Adams, A. N., (1977). Characteristics of the microplate method of enzyme-linked immunosorbent assay for the detection of plant viruses. *The Journal of General Virology*, *34*(3), 475–483. https://doi.org/10.1099/0022-1317-34-3-475.

Cui, X., Chen, X., & Wang, A., (2011). Detection, understanding and control of *Soyabean mosaic virus*. In: *Soyabean: Molecular Aspects of Breeding* (No. 2014, pp. 335–354). https://doi.org/10.5772/14298.

Deng, J., Li, S., Hong, J., Ji, Y., & Zhou, Y., (2013). Investigation on subcellular localization of *Rice stripe virus* in its vector small brown planthopper by electron microscopy. *Virol. J.*, *10*, 310. https://doi.org/10.1186/1743-422X-10-310.

Devashish, R., Lina, A., Aechana, R., & Magnus, L., (2015). The CRISPR-Cas immune system: Biology, mechanisms, and applications. *Biochimie*, *117*, 119–128.

Dhir, S., Ram, R., Hallan, V., & Zaidi, A. A., (2011). Molecular characterization of an Indian variant of apple stem pitting virus: Evidence of recombination. *Journal of Plant Pathology*, *93*(2), 471–478. http://www.jstor.org/stable/41999020.

Dong, H., Cui, Y., & Zhang, D., (2021). CRISPR/Cas technologies and their applications in *Escherichia coli*. In: *Frontiers in Bioengineering and Biotechnology* (Vol. 9). Frontiers Media S.A. https://doi.org/10.3389/fbioe.2021.762676.

Edwards, M. L., & Cooper, J. I., (1985). Plant virus detection using a new form of indirect ELISA. *Journal of Virological Methods*, *11*(4), 309–319. https://doi.org/10.1016/0166-0934(85)90024-2.

Elena, S. F., Fraile, A., & García–Arenal, F., (2014). Evolution and emergence of plant viruses. *Adv. Virus Res.*, *88*, 161–191. doi: 10.1016/B978-0-12-800098- 4.00003-9.

Gallenberg, D. J., & Jones, E. D., (1985). Detection of potato viruses X and S in tissue culture plantlets. *American Potato Journal*, *62*(3), 111–118. https://doi.org/10.1007/BF02871339.

Gao, S., & Wu, J., (2022). Detection of *Tomato spotted wilt virus* (TSWV) infection in plants using DAS-ELISA and dot-ELISA. *Methods in Molecular Biology (Clifton, N.J.)*, *2400*, 253–261. https://doi.org/10.1007/978-1-0716-1835-6_24.

Ghosh, D., Chakraborty, S., Kodamana, H., & Chakraborty, S., (2022). Application of machine learning in understanding plant virus pathogenesis: Trends and perspectives on emergence, diagnosis, host-virus interplay, and management. In: *Virology Journal* (Vol. 19, No. 1). BioMed Central Ltd. https://doi.org/10.1186/s12985-022-01767-5.

Gootenberg, J. S., Abudayyeh, O. O., Kellner, M. J., Joung, J., Collins, J. J., & Zhang, F., (2018). Multiplexed and portable nucleic acid detection platform with Cas13, Cas12a and Csm6. *Science*, *360*(6387), 439–444. https://doi.org/10.1126/science.aaq0179.

Gootenberg, J. S., Abudayyeh, O. O., Lee, J. W., Essletzbichler, P., Dy, A. J., Joung, J., Verdine, V., et al., (2017). Nucleic acid detection with CRISPR-Cas13a/C2c2. *Science, 356*(6336), 438–442. https://doi.org/10.1126/science.aam9321.

Guo, L., Wu, J., Chen, R., Hong, J., Zhou, X., & Wu, J., (2020). Monoclonal antibody-based serological detection of rice stripe mosaic virus infection in rice plants or leafhoppers. *Virologica Sinica, 35*(2), 227–234. https://doi.org/10.1007/s12250-019-00186-1.

Hadidi, A., Flores, R., Candresse, T., & Barba, M., (2016). Next-generation sequencing and genome editing in plant virology. *Front. Microbiol. 7*, 1325. doi:10.3389/fmicb.2016.01325.

Hamdy, M. A. E. A., (1961). The importance of potato virus X. *American Potato Journal, 38*(12), 440–447. https://doi.org/10.1007/BF02861108.

Hamim, I., Borth, W., Melzer, M. J., & Hu, J., (2018). Ultra-sensitive detection of papaya ringspot virus using single-tube nested PCR Abbreviations: Dpi = day(s)-post-inoculation; LAMP = loop-mediated isothermal amplification; PRSV = papaya ringspot virus; STNP = single-tube nested PCR. *Acta Virologica, 62*, 379–385. https://doi.org/10.4149/av.

Hančinský, R., Mihálik, D., Mrkvová, M., Candresse, T., & Glasa, M., (2020). Plant viruses infecting *Solanaceae* family members in the cultivated and wild environments: A review. *Plants, 9*(5), 667. https://doi.org/10.3390/plants9050667.

Heo, S., Kim, H. R., & Lee, H. J., (2019). Development of a quantitative real-time nucleic acid sequence based amplification (NASBA) assay for early detection of apple scar skin viroid. *The Plant Pathology Journal, 35*(2), 164–171. https://doi.org/10.5423/PPJ.OA.10.2018.0206.

Huang, D. Q., Chen, R., Wang, Y. Q., Hong, J., Zhou, X. P., & Wu, J. X., (2019). Development of a colloidal gold-based immunochromatographic strip for rapid detection of *Rice stripe virus. Journal of Zhejiang University. Science. B, 20*(4), 343–354. https://doi.org/10.1631/jzus.b1800563.

Huo, P., Shen, W. T., Yan, P., Tuo, D. C., Li, X. Y., & Zhou, P., (2015). Simultaneous detection of papaya ringspot virus, papaya leaf distortion mosaic virus, and papaya mosaic virus by multiplex real-time reverse transcription PCR. *Acta Virologica, 59*(4), 380–388. https://doi.org/10.4149/av_2015_04_380.

Huo, Y. Y., et al., (2017). Development of nucleic acid test strip for detection of *Cucumber green mottle mosaic virus*. *Nor. Hortic., 10*, 100–104.

Huo, Y. Y., Li, G. F., Qiu, Y. H., Li, W. M., & Zhang, Y. J., (2017). Rapid detection of *Prunus necrotic ringspot virus* by reverse transcription-cross-priming amplification coupled with nucleic acid test strip cassette. *Sci. Rep., 7*, 16175. https://doi.org/10.1038/s41598-017-16536-6.

Jones, R. A. C., (2009). Plant virus emergence and evolution: Origins, new encounter scenarios, factors driving emergence, effects of changing world conditions, and prospects for control. *Virus Res., 141*, 113–130. doi: 10.1016/j.virusres.2008.07.028.

Kaur, C., Kumar, S., Raj, S. K., Chauhan, P. S., & Sharma, N. (2015). Characterization of a new isolate of Bean yellow mosaic virus Group-IV Associated with Mosaic disease of Gladiolus in India. *J. Plant Pathol. Microb., 6*, 309.

Kievits, T., Van, G. B., Van, S. D., Schukkink, R., Dircks, M., Adriaanse, H., & Lens, P., (1991). NASBA isothermal enzymatic *in vitro* nucleic acid amplification optimized for the diagnosis of HIV-1 infection. *Journal of Virological Methods, 35*(3), 273–286. https://doi.org/10.1016/0166-0934(91)90069-c.

Kılıç, H. Ç., & Yardımcı, N., (2015). Occurrence of *Alfalfa mosaic virus* (AMV) infecting bean crop in Burdur Province, Turkey. *Asian Journal of Agriculture and Food Science, 3*(2), 173–177.

Kiselev, D., Matsvay, A., Abramov, I., Dedkov, V., Shipulin, G., & Khafizov, K., (2020). Current trends in diagnostics of viral infections of unknown etiology. *Viruses, 12*(2). https://doi.org/10.3390/v12020211.

Kokane, A., Lawrence, K., Surwase, D., Misra, P., Warghane, A., & Ghosh, D. K., (2020). Development of reverse transcription duplex PCR (RT-d-PCR) for simultaneous detection of the *Citrus tristeza virus* and Indian *Citrus Ringspot Virus*. *International Journal of Innovative Horticulture, 9*(2), 131–138. https://doi.org/10.5958/2582-2527.2020.00016.0.

Koolivand, D., Sokhandan-Bashir, N., Behjatnia, S. A. A., & Jafari, J. R. A., (2014). Detection of *Grapevine fanleaf virus* by immunocapture reverse transcription-polymerase chain reaction (IC-RT-PCR) with recombinant antibody. *Archives of Phytopathology and Plant Protection, 47*(17), 2070–2077. https://doi.org/10.1080/03235408.2013.868697.

Kozieł, E., Otulak-Kozieł, K., & Bujarski, J. J., (2018). Ultrastructural analysis of *Prune Dwarf Virus* intercellular transport and pathogenesis. *International Journal of Molecular Sciences, 19*(9), 2570. https://doi.org/10.3390/ijms19092570.

Kumar, S., Udaya, S. A. C., Nayaka, S. C., Lund, O. S., & Prakash, H. S., (2011). Detection of *Tobacco mosaic virus* and *Tomato mosaic virus* in pepper and tomato by multiplex RT-PCR. *Letters in Applied Microbiology, 53*(3), 359–363. https://doi.org/10.1111/j.1472-765X.2011.03117.x.

Li, B., Yan, J., Zhang, Y., Li, W., Zeng, C., Zhao, W., Hou, X., et al., (2020). CRISPR-Cas12a possesses unconventional DNase activity that can be inactivated by synthetic oligonucleotides. *Molecular Therapy – Nucleic Acids, 19*, 1043–1052. https://doi.org/10.1016/j.omtn.2019.12.038.

Ling, K. S., Zhu, H. Y., Jiang, Z. Y., & Gonsalves, D., (2000). Effective application of DAS-ELISA for detection of grapevine leafroll associated closterovirus-3 using a polyclonal antiserum developed from recombinant coat protein. *European Journal of Plant Pathology, 106*(4), 301–309. https://doi.org/10.1023/A:1008713319294.

Liu, Z., Chen, Z., Hong, J., Wang, X., Zhou, C., Zhou, X., & Wu, J., (2016). Monoclonal antibody-based serological methods for detecting *Citrus tristeza virus* in citrus groves. *Virologica Sinica, 31*(4), 324–330. https://doi.org/10.1007/s12250-016-3718-4.

Lo´Pez, M. M., Bertolini, E., Olmos, A., Caruso, P., Gorris, M. T., Llop, P., Penyalver, R., & Cambra, M., (2003). Innovative tools for detection of plant pathogenic viruses and bacteria. *Int. Microbiol., 6*, 233–243.

Lu, Y., Yao, B., Wang, G., & Hong, N., (2018). The detection of ACLSV and ASPV in pear plants by RT-LAMP assays. *Journal of Virological Methods, 252*, 80–85. https://doi.org/10.1016/j.jviromet.2017.11.010.

Makkouk, K. M., Hsu, H. T., & Kumari, S. G., (1993). Detection of three plant viruses by dot-blot and tissue-blot immunoassays using chemiluminescent and chromogenic substrates. *Journal of Phytopathology, 139*, 97–102.

Maksimov, I. V., Sorokan, A. V., Burkhanova, G. F., Veselova, S. V., Alekseev, V. Y., Shein, M. Y., Avalbaev, A. M., et al., (2019). Mechanisms of plant tolerance to RNA viruses induced by plant-growth-promoting microorganisms. *Plants (Basel, Switzerland), 8*(12), 575. https://doi.org/10.3390/plants8120575.

Massart, S., Chiumenti, M., De Jonghe, K., Glover, R., Haegeman, A., Koloniuk, I., Komínek, P., et al., (2019). Virus detection by high-throughput sequencing of small RNAs: Large-scale performance testing of sequence analysis strategies. *Phytopathology*, *109*(3), 488–497. https://doi.org/10.1094/PHYTO-02-18-0067-R.

Menzel, W., Jelkmann, W., & Maiss, E., (2002). Detection of four apple viruses by multiplex RT-PCR assays with coamplification of plant mRNA as internal control. *Journal of Virological Methods*, *99*(1, 2), 81–92. https://doi.org/10.1016/s0166-0934(01)00381-0.

Mokhtar, U., Ali, M. A. S., Hassanien, A. E., & Hefny, H., (2015). Identifying two of *Tomatoes leaf viruses* using support vector machine. In: Mand, J. K., (ed.), *Information Systems Design and Intelligent Applications, Advances in Intelligent Systems and Computing* (pp. 771–782). Springer India. https://doi.org/10.1007/978-81-322-2250-7_77.

Mumo, N. N., Mamati, G. E., Ateka, E. M., Rimberia, F. K., Asudi, G. O., Boykin, L. M., et al., (2020). Metagenomic analysis of plant viruses associated with papaya ringspot disease in *Carica papaya* L. @ in Kenya. Front. *Microbiol., 11*, 205. doi: 10.3389/fmicb.2020.00205.

Naghavi, A., Habibi, M. K., & Firouzabadi, N. F., (2008). Detection and identification of some *Soybean viral mosaic viruses*, using molecular techniques in Lorestan Province, South West of Iran. *Asian Journal of Plant Science*, *7*, 557–562.

Naito, F. Y. B., Melo, F. L., Fonseca, M. E. N., Santos, C. A. F., Chanes, C. R., Ribeiro, B. M., Gilbertson, R. L., et al., (2019). Nanopore sequencing of a novel bipartite new world *Begomovirus* infecting cowpea. *Archives of Virology*, *164*(7), 1907–1910. https://doi.org/10.1007/s00705-019-04254-5.

Ojuederie, O., Odu, B., & Ilori, C., (2010). Serological detection of seed borne viruses in cowpea regenerated germplasm using protein a sandwich enzyme linked immunorsorbent assay. *African Crop Science Journal*, *17*(3). https://doi.org/10.4314/acsj.v17i3.54212.

Onozuka, N., Ohki, T., Oka, N., & Maoka, T., (2021). One-step real-time multiplex reverse transcription-polymerase chain reaction assay with melt curve analysis for detection of *Potato leafroll virus, Potato virus S, Potato virus X*, and *Potato virus Y. Virology Journal*, *18*(1), 1–10. https://doi.org/10.1186/s12985-021-01591-3.

Ozeki, J., Hashimoto, M., Komatsu, K., Maejima, K., Himeno, M., Senshu, H., Kawanishi, T., et al., (2009). The N-terminal region of the *Plantago asiatica mosaic virus* coat protein is required for cell-to-cell movement but is dispensable for virion assembly. *Molecular Plant-Microbe Interactions: MPMI*, *22*(6), 677–685.

Pappas, M. G., (1988). Recent applications of the Dot-ELISA in immunoparasitology. *Veterinary Parasitology*, *29*(2, 3), 105–129. https://doi.org/10.1016/0304-4017(88)90120-3.

Peng, Y., Dallas, M. M., Ascencio-Ibáñez, J. T., Hoyer, J. S., Legg, J., Hanley-Bowdoin, L., Grieve, B., & Yin, H., (2022). Early detection of plant virus infection using multispectral imaging and spatial–spectral machine learning. *Scientific Reports*, *12*(1), 1–14. https://doi.org/10.1038/s41598-022-06372-8.

Qiao, N., Dai, H., Liu, J., Zhu, X., Li, J., Zhang, D., & Liu, Y., (2020). Detection of melon necrotic spot virus by one-step reverse transcription loop-mediated isothermal amplification assay. *PLoS One*, *15*(3), 1–10. https://doi.org/10.1371/journal.pone.0230023.

Rana, T., Chandel, V., Hallan, V., & Zaidi, A. A., (2009). Molecular evidence for the presence of *Apple chlorotic leaf spot virus* in infected peach trees in India. *Scientia Horticulturae*, *120*(2), 296–299.

Rani, A., Suman, R., Chandel, V., Srivastava, A., Noorani, M. S., Rishi, N., & Dhir, S., (2021). Biological and molecular characterization of two isolates of apple stem grooving virus revealed distinct properties and indicates virus emergence driven by recombination. *Journal of Phytopathology*, *169*(11, 12), 740–751.

Roberts, C. A., Dietzgen, R. G., Heelan, L. A., & Maclean, D. J., (2000). Real-time RT-PCR fluorescent detection of tomato spotted wilt virus. *Journal of Virological Methods*, *88*(1), 1–8. https://doi.org/10.1016/s0166-0934(00)00156-7.

Roossinck, M. J., (2015). Metagenomics of plant and fungal viruses reveals an abundance of persistent lifestyles. *Front. Microbiol., 5*, 767. doi: 10.3389/fmicb.2014.00767.

Roossinck, M. J., (2017). Deep sequencing for discovery and evolutionary analysis of plant viruses. *Virus Res., 239*, 82–86. doi: 10.1016/j.virusres.2016.11.019.

Salomone, A., Mongelli, M., Roggero, P., & Boscia, D., (2004). Reliability of detection of citrus tristeza virus by an immunochromatographic lateral flow assay in comparison with Elisa. *Journal of Plant Pathology*, *86*(1), 43–48.

Seepiban, C., Charoenvilaisiri, S., Warin, N., Bhunchoth, A., Phironrit, N., Phuangrat, B., Chatchawankanphanich, O., et al., (2017). Development and application of triple antibody sandwich enzyme-linked immunosorbent assays for *Begomovirus* detection using monoclonal antibodies against *Tomato yellow leaf curl Thailand virus*. *Virology Journal*, *14*(1), 99. https://doi.org/10.1186/s12985-017-0763-z.

Seguin, J., Rajeswaran, R., Malpica-López, N., Martin, R. R., Kasschau, K., Dolja, V. V., Otten, P., et al., (2014). De novo reconstruction of consensus master genomes of plant RNA and DNA viruses from siRNAs. *PLoS One*, *9*(2). https://doi.org/10.1371/journal.pone.0088513.

Sharma, R., Chandel, V., & Rishi, N., (2019). Croton yellow vein mosaic virus found infecting naturally growing weeds in India. *Annals of Biology*, *35*(2), 316–319.

Sharma, S. K., Gupta, O. P., Pathaw, N., Sharma, D., Maibam, A., Sharma, P., Sanasam, J., et al., (2021). CRISPR-cas-led revolution in diagnosis and management of emerging plant viruses: New avenues toward food and nutritional security. In: *Frontiers in Nutrition* (Vol. 8). Frontiers Media S.A. https://doi.org/10.3389/fnut.2021.751512.

Shen, W., Tuo, D., Yan, P., Li, X., & Zhou, P., (2014). Detection of Papaya leaf distortion mosaic virus by reverse-transcription loop-mediated isothermal amplification. *Journal of Virological Methods*, *195*, 174–179. https://doi.org/10.1016/j.jviromet.2013.09.011.

Smith, F. D., & Banttari, E. E., (1987). Dot-ELISA on nitrocellulose membrane for detection of *Potato leafroll virus*. *Plant Disease*, *71*, 795–799.

Song, G., Wu, J. Y., Xie, Y., Liu, Y., Qian, Y. J., Zhou, X. P., & Wu, J. X., (2017). Monoclonal antibody-based serological assays for detection of *Potato virus* S in potato plants. *Journal of Zhejiang University. Science. B*, *18*(12), 1075–1082. https://doi.org/10.1631/jzus.B1600561.

Srivastava, A., Raj, S. K., Kumar, S., Snehi, S. K., Kulshreshtha, A., Hallan, V., & Pande, S. S., (2013). Molecular identification of *Ageratum enation virus*, betasatellite and alphasatellite molecules isolated from yellow vein diseased *Amaranthus cruentus* in India molecular identification of *Ageratum enation virus*, betasatellite and alphasatellite molecules. *Virus Genes*. https://doi.org/10.1007/s11262-013-0971-6.

Tan, S. H., Osman, F., Bodaghi, S., Dang, T., Greer, G., Huang, A., et al., (2019). Full genome characterization of 12 citrus tatter leaf virus isolates for the development of a detection assay. *PLoS One*, *14*, e0223958. doi: 10.1371/journal.pone.0223958.

Tolkach, V. F., Kakareka, N. N., Volkov, Yu. G., Kozlovskaya, Z. N., Sapotskiy, M. V., Pleshakova, T. I., D'yakonov, K. P., & Shchelkanov, M. Yu., (2019). Virus diseases of vegetable and melon crops in the South of the Russian Far East. *South of Russia: Ecology, Development*, *14*(4), 121–133. In Russ. https://doi.org/10.18470/1992-1098-2019-4-121-133.

Tomašechová, J., Hančinský, R., Predajňa, L., Kraic, J., Mihálik, D., Šoltys, K., et al., (2020). High-throughput sequencing reveals bell pepper endornavirus infection in pepper (*Capsicum annum*) in Slovakia and enables its further molecular characterization. *Plants*, *9*, 41. doi: 10.3390/plants9010041.

Uehara-Ichiki, T., Ohashi, M., Hanada, K., Igarashi, M., & Hishida, A., (2020). Identification and pathogenicity of plant viruses infecting *Valeriana fauriei* in Japan. *Journal of General Plant Pathology*, *86*(4), 310–315. https://doi.org/10.1007/s10327-020-00925-3.

Usher, L., Sivparsad, B., & Gubba, A., (2012). Isolation, identification, and molecular characterization of an isolate of *Zucchini yellow mosaic virus* occurring in KwaZulu-Natal, South Africa. *South African Journal of Plant and Soil*, *29*(2), 65–71. https://doi.org/10.1080/02571862.2012.701669.

Valekunja, R. B., Kamakoti, V., Peter, A., Phadnis, S., Prasad, S., & Nagaraj, V. J., (2016). The detection of papaya ringspot virus coat protein using an electrochemical immunosensor. *Anal. Methods*, *8*(48), 8534–8541. https://doi.org/10.1039/C6AY02201D.

Vives, M. C., Velazquez, K., Pina, J. A., Moreno, P., Guerri, J., & Navarro, L., (2013). Identification of a new enamovirus associated with citrus vein enation disease by deep sequencing of small RNAs. *Phytopathology*, *103*, 1077–1086. doi: 10.1094/PHYTO-03-13-0068-R.

Wada, N., Ueta, R., Osakabe, Y., & Osakabe, K., (2020a). Precision genome editing in plants: State-of-the-art in CRISPR/Cas9-based genome engineering. In: *BMC Plant Biology* (Vol. 20, No. 1). BioMed Central Ltd. https://doi.org/10.1186/s12870-020-02385-5.

Wada, N., Ueta, R., Osakabe, Y., & Osakabe, K., (2020b). Precision genome editing in plants: State-of-the-art in CRISPR/Cas9-based genome engineering. In: *BMC Plant Biology* (Vol. 20, No. 1). BioMed Central Ltd. https://doi.org/10.1186/s12870-020-02385-5.

Wang, Z., Yu, D., Li, X., Zeng, M., Chen, Z., Bi, L., Liu, J., et al., (2012). The development and application of a dot-ELISA assay for diagnosis of southern rice black-streaked dwarf disease in the field. *Viruses*, *4*(1), 167–183. https://doi.org/10.3390/v4010167.

Warghane, A., Misra, P., Bhose, S., Biswas, K. K., Sharma, A. K., Reddy, M. K., & Ghosh, D. K., (2017). Development of a simple and rapid reverse transcription-loop mediated isothermal amplification (RT-LAMP) assay for sensitive detection of *Citrus tristeza virus*. *Journal of Virological Methods*, *250*, 6–10. https://doi.org/10.1016/j.jviromet.2017.09.018.

Wieczorek, P., Budziszewska, M., Frąckowiak, P., & Obrępalska-Stęplowska, A., (2020). Development of a new *Tomato Torrado virus*-based vector tagged with GFP for monitoring virus movement in plants. *Viruses*, *12*(10), 1195. https://doi.org/10.3390/v12101195.

Wu, J., Ni, Y., Liu, H., Ding, M., & Zhou, X., (2014). Monoclonal antibody-based serological assays and immunocapture-RT-PCR for detecting *Rice dwarf virus* in field rice plants and leafhopper vectors. *Journal of Virological Methods*, *195*, 134–140. https://doi.org/https://doi.org/10.1016/j.jviromet.2013.09.013.

Wu, J., Ni, Y., Liu, H., Rao, L., Zhou, Y., & Zhou, X., (2013). Development and use of three monoclonal antibodies for the detection of rice black-streaked dwarf virus in field plants and planthopper vectors. *Virology Journal*, *10*(1), 114. https://doi.org/10.1186/1743-422X-10-114.

Xu, G., Hu, L., Zhong, H., Wang, H., Yusa, S. I., Weiss, T. C., & You, Q., (2012). Cross priming amplification: Mechanism and optimization for isothermal DNA amplification. *Scientific Reports*, *2*, 246. https://doi.org/10.1038/srep00246.

Yamaguchi, A., (1960). Increased respiration of leaves bearing necrotic local virus lesions. *Virology. 10*, 287–293.

Yang, Q. Q., Zhao, X. X., Wang, D., Zhang, P. J., Hu, X. N., Wei, S., & Yu, X. P., (2022). A reverse transcription-cross-priming amplification method with lateral flow dipstick assay for the rapid detection of *Bean pod mottle virus*. *Scientific Reports*, *12*(1), 1–8. https://doi.org/10.1038/s41598-021-03562-8.

Yang, X., Lv, K., Wang, M., & Zhou, G., (2017). Investigation of viruses infecting rice in Southern China using a multiplex RT-PCR assay. *Crop Protection*, *91*, 8–12. https://doi.org/https://doi.org/10.1016/j.cropro.2016.09.012.

Yoon, J. Y., Kim, S., Choi, G. S., & Choi, S. K., (2015). A simple and reliable molecular detection method for *Tomato yellow leaf curl virus* in *Solanum lycopersicum* without DNA extraction. *Research in Plant Disease*, *21*(3), 180–185. https://doi.org/10.5423/rpd.2015.21.3.180.

Yu, C., Wu, J., & Zhou, X., (2005). Detection and subgrouping of *Cucumber mosaic virus* isolates by TAS-ELISA and immunocapture RT-PCR. *Journal of Virological Methods*, *123*(2), 155–161. https://doi.org/https://doi.org/10.1016/j.jviromet.2004.09.014.

Zechmann, B., & Zellnig, G., (2009). Rapid diagnosis of plant virus diseases by transmission electron microscopy. *Journal of Virological Methods*, *162*(1, 2), 163–169. https://doi.org/10.1016/j.jviromet.2009.07.032.

Zhang, Y., Singh, K., Kaur, R., & Qiu, W., (2011). Association of a novel DNA virus with the grapevine vein-clearing and vine decline syndrome. *Phytopathology*, *101*(9), 1081–1090. https://doi.org/10.1094/PHYTO-02-11-0034.

Zhu, H., Cen, H., Zhang, C., & He, Y., (2016). Early detection and classification of tobacco leaves inoculated with *Tobacco mosaic virus* based on hyperspectral imaging technique. In: *2016 ASABE Annual International Meeting* (p. 1). ASABE. https://doi.org/https://doi.org/10.13031/aim.20162460422.

Zhu, H., Chu, B., Zhang, C., Liu, F., Jiang, L., & He, Y., (2017). Hyperspectral imaging for presymptomatic detection of tobacco disease with successive projections algorithm and machine-learning classifiers. *Scientific Reports*, *7*(1), 4125. https://doi.org/10.1038/s41598-017-04501-2.

Zsögön, A., Čermák, T., Naves, E. R., Notini, M. M., Edel, K. H., Weinl, S., Freschi, L., et al., (2018). De novo domestication of wild tomato using genome editing. *Nature Biotechnology*, *36*(12), 1211–1216. https://doi.org/10.1038/nbt.4272.

## CHAPTER 3

# Efficient, Rapid, and Sensitive Detection of Plant Viruses with Next-Generation CRISPR-Based Diagnostic Methods

SACHIN ASHOK BHOR,[1] SHAIKHUL ISLAM,[2]
ANANDRAO ASHOK PATIL,[3] and GOKUL UTTAMGIR GOSAVI[4]

[1]*Laboratory of Plant Molecular Biology and Virology, Faculty of Agriculture, Ehime University, Matsuyama, Ehime, Japan*

[2]*Plant Pathology Division, Bangladesh Wheat and Maize Research Institute, Nashipur, Dinajpur 5200, Bangladesh*

[3]*Department of Biology, University of Washington, Seattle, Washington, USA*

[4]*Institute of Plant Protection, The Graduate School of Chinese Academy of Agricultural Sciences, Beijing, China*

### ABSTRACT

More than 1,000 phytopathogenic viruses and viroids cause severe diseases in plants, leading to annual yield losses of 10–15%. Therefore, early detection of pathogens can provide growers with sufficient time to develop disease prevention strategies. Various molecular diagnostic methods are commonly used to detect plant viruses. However, these existing diagnostic methods have limitations and disadvantages. As a result, there is still a need for simple, precise, inexpensive, rapid, large-scale, and field-deployable diagnostic tools to detect viruses in field crops.

Currently, CRISPR and CRISPR/Cas systems have emerged as extraordinary tools for plant virus detection. Next-generation CRISPR platforms, such

Detection and Management of New and Emerging Mystery Plant Virus Sources.
Alireza Golnaraghi, Rajarshi Kumar Gaur, and Rakesh Kumar Verma (Eds.)

as CRISPR/Cas12a-based diagnostic methods, can simultaneously detect multiple RNA viruses and viroids. Additionally, the modification of the *in vitro* Specific CRISPR-based Assay for Nucleic acids detection (iSCAN) method has resulted in a one-pot assay called iSCAN one-pot (iSCAN-OP), which can rapidly and sensitively detect PVX, PVY, and TMV.

Recently, CRISPR/Cas12a coupled with LAMP has shown promise as a sensitive, in-field, inexpensive, quick, and easy-to-interpret diagnostic platform for the rapid detection of plant DNA viruses. In this chapter, we provide a brief overview of CRISPR/Cas technology, its principles, and components. We then focus on the applications of CRISPR/Cas technology and its variants (Cas12, Cas13, etc.) in the detection of plant viruses. Furthermore, we highlight the limitations and advantages of these diagnostic methods compared to conventional methods. Finally, we conclude this chapter by discussing the challenges in adopting these methods at the field level, as well as the advances and prospects of this novel technology.

## 3.1 INTRODUCTION

Since ancient times, plant diseases have had a negative impact on agricultural production. Worldwide, annual crop yield losses due to plant diseases and pests are estimated to be between 20% and 40%. Among these, viruses are the second largest contributor. Plant viruses affect growth, reduce the photosynthetic activity of leaves, and decrease the quality and quantity of yield, resulting in significant financial losses (Savary et al., 2012). To date, more than 25 families of identified plant viruses have been found to infect diverse crop species, leading to substantial economic losses (He & Krainer, 2020).

Plant viruses can be transmitted through seeds, infected planting material, insect vectors, and contaminated agricultural tools. Once infected, viruses cause phenotypic changes in plants such as yellowing or chlorosis, mosaic patterns, stunting, and necrosis, which hinder both vegetative and reproductive growth (Seo et al., 2018). Unlike other plant pathogens, plant viruses incorporate themselves into the plant's genome, making them resistant to chemical treatments. Currently, there are very few treatment options available to control viral diseases, especially under field conditions. Therefore, early detection of causal agents is crucial in preventing the further spread of disease (Adams et al., 2009).

Accurate detection of plant viruses requires sensitive, specific, and time-saving diagnostic methods. As the field of diagnostics continues to evolve, several innovative techniques and methods are being developed. However, most of the available virus detection methods are expensive, time-consuming, labor-intensive, and cannot be easily deployed at the field level. Additionally, the presence of emerging virus species and mixed virus infections pose significant challenges in plant-based viral diagnostics (Rubio et al., 2020). Nowadays, electron microscopy, serological (ELISA, western blot, etc.), PCR (RT-qPCR), biosensors, optical sensors, and molecular biology-based methods are widely used (Massart et al., 2014). However, the use of these methods has been limited due to the time required to complete the assay, the need for sophisticated laboratory equipment and expertise, as well as issues with the accuracy and specificity of test results (Fang & Ramasamy, 2015). To enhance the sensitivity and specificity of detection, the combination of several diagnostic techniques is widely accepted. Furthermore, advancements in the field of molecular biology and genetics have allowed us to develop more sensitive, real-time, cost-effective, field-deployable virus diagnostic tools.

Since the discovery of clustered regularly interspaced short palindromic repeats (CRISPRs), the potential of different Cas effectors has been harnessed to develop methods for detecting viruses and other pathogens as well (Aman et al., 2020a, b; Li et al., 2019b). The development of DNA endonuclease-targeted CRISPR *trans* reporter (DETECTR) (Chen et al., 2018) and specific high-sensitivity enzymatic reporter unlocking (SHERLOCK) systems (Gootenberg et al., 2017) has demonstrated a diverse range of novel diagnostic platforms for rapid, affordable, precise, and accurate diagnosis of the target pathogen. Most of these CRISPR-based methods involve the isothermal amplification of the target region and subsequent recognition of the target by CRISPR/Cas systems (Cas12 and Cas13). The presence of the target sequence is indicated by collateral cleavage of reporter DNA or RNA molecules (Kellner et al., 2019). The signal readouts are observed and quantified fluorescently, colorimetrically, or electrochemically.

We begin this chapter by providing a brief overview of the origin, discovery, and classification of CRISPR technology. We then elaborate on the limitations and challenges faced by existing virus diagnostic methods, emphasizing the need for innovative, rapid, efficient, and field-deployable diagnostic tools. The subsequent section describes the currently available CRISPR/Cas system-based tools for virus diagnosis and their properties, followed by a detailed explanation of nucleic acid detection mechanisms.

We present several case studies of CRISPR/Cas system-based detection of different DNA and RNA viruses using various isothermal amplification methods. Finally, we conclude this chapter with a discussion of the advantages and current challenges of these innovative tools, as well as their prospects in light of technological advancements for plant virus detection.

## 3.2 CRISPR TECHNOLOGY: DISCOVERY, COMPONENTS, PRINCIPLE, AND CLASSIFICATION

### *3.2.1 DISCOVERY OF CRISPR*

Clustered regularly interspaced short palindromic repeats, also known as CRISPRs, are repeating DNA sequences in the genomes of prokaryotes, such as bacteria and archaea. In 1987, a Japanese scientist, Yoshizumi Ishino, and his team accidentally identified CRISPRs in *E. coli*. Later, while analyzing gene(s) responsible for the conversion of alkaline phosphatase, they cloned those unusual series of repeated sequences (Ishino et al., 2018). However, due to insufficient DNA sequence data, they were unable to identify the function of those arrays. Similarly, Sola & coworkers (1993) discovered that different strains of *Mycobacterium tuberculosis* harbor different spacer sequences between the repetitive DNA sequences. Furthermore, they used the genotyping technique to characterize *M. tuberculosis* based on those spacer DNA sequences (Sola et al., 2015). Subsequently, similar arrays of unusual series of repeated sequences were identified and characterized in several other bacteria and archaea. However, the term CRISPR was first referred to by Francisco Mojica and Ruud Jansen (Morange, 2015).

Initially, CRISPR systems were thought to be innovative DNA repair mechanisms in approximately 90% of thermophilic archaea and 50% of bacteria (Makarova et al., 2002). In the early 2000s, it was suggested that these spacer sequences were similar to nucleic acid sequences found in viruses, bacteriophages, and plasmids. Moreover, it was revealed that viruses cannot infect bacteria that harbor these homologous spacer sequences, indicating that these sequences play a role in the adaptive immune system of prokaryotes (Ishino et al., 2018). In general, when a prokaryotic organism is attacked by a virus, the spacer sequences in CRISPR arrays are transcribed to synthesize short CRISPR RNA (crRNA), which further guides the CRISPR-associated sequence (Cas) protein to cleave complementary DNA (cDNA) or RNA viral sequences (specific to the CRISPR Cas system type). Similarly,

CRISPR/Cas systems function as an adaptive immune response to prevent repeated infections in prokaryotes by the same virus.

### 3.2.2 COMPONENTS AND PRINCIPLES OF CRISPR/CAS SYSTEMS

Cas proteins are endonucleases with high selectivity and specific programmable activities. The CRISPR array and Cas proteins may vary in size, structural and biochemical composition, and actions across different species (Mohanraju et al., 2016). The CRISPR array consists of repetitions of 25–35 DNA base pairs (bp) interspersed with a 30–40 bp spacer sequence, which represents the recorded portion of DNA from the invader. These spacers retain the memory of invasions to provide defense and can be passed on to offspring during microbial cell division. The CRISPR/Cas system functions similarly to RNA interference (RNAi), where a group of proteins silences genes by cleaving specific mRNA transcripts (Makarova et al., 2002).

Each CRISPR array encodes the Cas proteins, which have helicase and endonuclease activities. When the microorganism is invaded for the first time, the CRISPR array encodes a Cas enzyme that recognizes and integrates a portion of the foreign genetic material (known as protospacer) into a spacer sequence of the CRISPR array. Protospacer adjacent motifs (PAMs) are necessary components of CRISPR/Cas systems. PAMs are short sequences, typically 2–6 base pairs long, in the viral genome located adjacent to sequences targeted by the Cas nucleases. Cas nucleases are very specific in recognizing PAMs and cannot cleave foreign DNA sequences unless a PAM is present. The Cas9 effector protein from *S. pyogenes* recognizes NGG as a PAM (Anders et al., 2014; Lander, 2016; Jiang & Doudna, 2017). PAMs play a significant role in confirming that only foreign viral nucleic acids are cleaved, but not CRISPR arrays (Shah et al., 2013).

From the protospacer sequence stored at the time of the first infection, a guide crRNA is synthesized. It is then utilized by the Cas enzyme to form a ribonucleoprotein (RNP) interference complex that recognizes and specifically cleaves foreign nucleic acid (Marraffini et al., 2015; Leon et al., 2018; Hille et al., 2018). The Cas9 proteins have two nuclease domains, namely HNH and RuvC. The HNH-like nuclease domain has a target DNA strand that is complementary to the guide RNA (gRNA) sequence, while the RuvC-like nuclease domain cleaves the non-target strand (Jinek et al., 2012; Gao et al., 2017; Jiang & Doudna, 2017). The cleavage by both domains at the target site creates a DNA double-stranded break (DSB), which is further used to introduce modifications by non-homologous end joining (NHEJ) or homology dependent recombination (HDR) mechanisms (Symington & Gautier, 2011).

A single guide RNA (sgRNA) consists of crRNA, which has a specific sequence for the target DNA, and tracrRNA, which interacts with the Cas effector protein. It binds to the recombinant Cas protein to form double-strand breaks (DSBs) in the target region through endonuclease activity. Additionally, the DNA repair mechanism known as NHEJ can introduce errors by adding or deleting nucleotides, ultimately disrupting the function of the targeted gene. This is the fundamental principle behind CRISPR/Cas-mediated targeted gene disruption.

### *3.2.3 CLASSIFICATION OF CRISPR/CAS SYSTEMS*

Based on the use of specific Cas proteins and the mechanism of interference, CRISPR/Cas systems are classified into two distinct classes: 6 types and several subtypes (Mohanraju et al., 2016). The class 1 system comprises multi-subunit crRNA-Cas proteins, whereas the class 2 system includes single multidomain-crRNA-Cas proteins. The class 1 system is found in bacteria, archaea, etc., whereas the class 2 system is only present in bacteria. Cas enzymes belonging to type I, III, and IV in class 1 systems, and type II, V, and VI in class 2 systems, are well studied (Koonin & Makarova, 2019). The differences in CRISPR/Cas systems (Classes 1 and 2) and the specific activities of the Cas proteins have been studied, and their potential in the development of biosensing systems has been explored. In this chapter, we have discussed the specific activities of Cas effector proteins belonging to class 2 systems (Cas9, Cas12, Cas13, and Cas14), which have been extensively used for nucleic acid biosensing applications (Li et al., 2019b).

The Class 2 CRISPR/Cas system is composed of three different types of enzymes: II (e.g., Cas9), V (e.g., Cas12), and VI (e.g., Cas13), along with various subtypes such as V-A (Cas12a), V-B (Cas12b), V-C (Cas12c), VI-A (Cas13a), and VI-B (Cas13b). These enzymes have evolved through different evolutionary pathways. The evolutionary development of these systems has resulted in a wide range of specific enzymatic activities, all of which are programmable endonucleases. Additionally, all of these Cas endonucleases utilize the guide crRNA to recognize the target nucleic acid. Endonucleases belonging to types II and V have been programmed to recognize target DNA sequences, while Type VI recognizes target RNA sequences (Shmakov et al., 2017) (Figure 3.1).

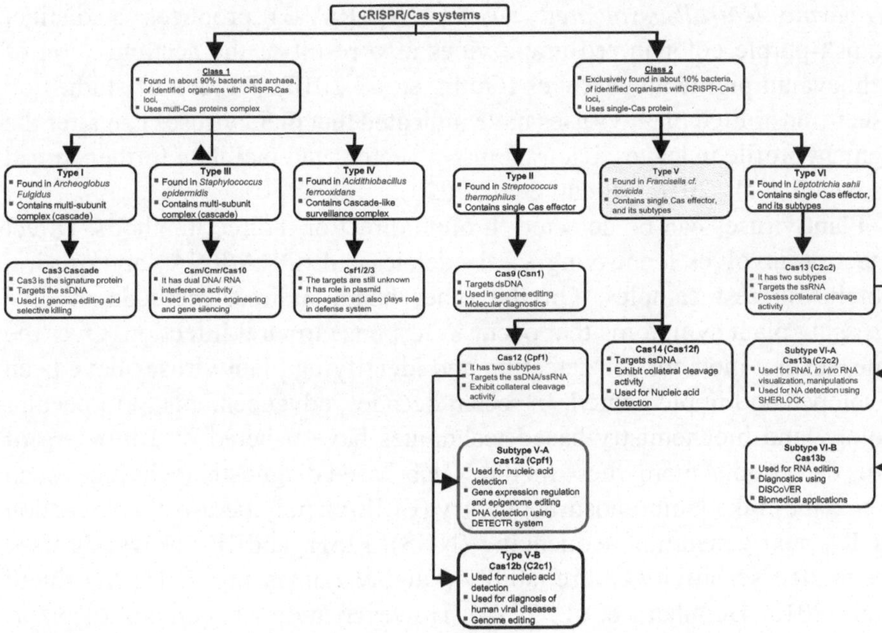


**FIGURE 3.1** General classification of CRISPR/Cas systems with their emphasized applications. The blue-bordered box indicates that Cas proteins are utilized for virus detection potential in plant virology.

*Source:* Adapted and modified from: Bhardwaj et al. (2022); and Stout et al. (2017).

## 3.3 CONVENTIONAL PLANT-VIRUS DIAGNOSIS METHODS: LIMITATIONS AND CHALLENGES

Plant viruses consist of one or more nucleic acids (either DNA or RNA) which are protected by a coat protein (CP) (Smith, 1977). They lack most of the cellular components like any other living organisms and therefore must parasitize a host cell and hijack the host's transcription and translation machinery to replicate and survive (Hull, 2013). Virus replication can occur in most plant cells, which serve as a "niche" (Hipper et al., 2013). Virus infection in healthy plants leads to a variety of physiological and biochemical alterations, collectively resulting in disease. At the physiological and biochemical level, unbalanced hormones and amino acids, chloroplast degradation leading to yellowing and decreased photosynthetic ability of leaves, abnormal cell structure, and reduced nutrient uptake by roots all contribute to the inability of virus-infected plants to grow and develop normally (Mauck et al., 2014; Jaime et al., 2019). The apparent physiological changes observed and/or detected during the diseased state of plants are called disease symptoms.

*Grapevine leafroll-associated virus 3* (GLRaV-3) produces a distinct reddish-purple color in red grape vines as a result of the accumulation of anthocyanin pigments in leaves (Gutha et al., 2010). Results of studies on insect-transmitted plant viruses have indicated that plant viruses can alter the pigment profile in leaves to attract insect vectors and facilitate further spread (Maxwell et al., 2016; Moeini et al., 2020).

Plant viruses can be detected through direct or indirect methods. Direct detection involves identifying viral nucleic acid (DNA/RNA) and/or viral proteins in test samples. On the other hand, indirect methods involve assessing plant symptoms that occur in response to viral infection. Over the time, several methods for detecting and identifying plant viruses have been developed and implemented. In recent decades, advancements in molecular biology and biochemistry-based techniques have ushered in a new era of virus detection. Among these methods, lab-based diagnostic techniques such as enzyme-linked immunosorbent assay (ELISA), polymerase chain reaction (PCR), next-generation sequencing (NGS), FISH, and IF are widely used due to their sensitivity and reliability (Fang & Ramasamy, 2015; Martinelli et al., 2015; Boonham et al., 2014). However, indirect methods of plant-virus detection include the traditional approach of visually assessing virus symptoms in the field or utilizing optical sensors to evaluate virus-induced phenotypic changes (Wang et al., 2022).

The direct methods of virus diagnosis are classified into two major groups: serological and nucleic acid-based (NA-based). Antibodies generated from an animal's natural immune response are used to detect viruses. These antibodies are further combined with an enzyme that specifically binds to viral proteins, producing distinct reaction patterns that can be easily noticed and measured spectrophotometrically (Torrance & Jones, 1981; Nakane & Pierce, 1966). Later, an improved version of ELISA called double-antibody sandwich (DAS) was developed, which effectively quantifies virus concentration in a given sample (Clark & Adams, 1977). The salient features of this method are its economy and suitability for a large number of samples. Further advancements in the development of serology-based virus diagnostic tools resulted in lateral flow devices (LFDs). Antibodies specific to viruses are immobilized on a nitrocellulose membrane (NCM) along with nanoparticles (NPs) (colored), leading to visible results (Danks & Barker, 2000; Maheshwari et al., 2018; Selvarajan et al., 2020). This method is rapid and field-deployable, hence widely accepted. In Immunofluorescence, antibodies specific to viruses are labeled with fluorescent dyes and used to bind with virus particles (antigens), and the fluorescence is observed under a microscope (Boine et al., 2012). This is a lab-based technique and expensive, but it provides information about the intracellular spread of virus particles.

Since 1979, NA-based virus detection methods have been used. Saiki et al. (1985) developed a PCR-based method for plant virus detection, which has made a significant contribution to the field of plant virus diagnosis. Further advances and developments in PCR-based techniques for plant virus detection have resulted in reverse transcription PCR (RT-PCR), quantitative PCR (qPCR), and loop-mediated isothermal amplification (LAMP). PCR-based methods are more sensitive and effective compared to others. Nowadays, RT-PCR is the most commonly used molecular technique for the detection of viruses because it is capable of detecting viruses even at very low concentrations in test samples (Scagliusi et al., 2009; Mekuria et al., 2003). Several studies support the notion that, for plant virus detection, RT-PCR is more sensitive than ELISA (McGavin et al., 2010; Lima et al., 2013). Notomi & coworkers (2000) developed the LAMP technique, which does not require a thermocycler. It is a quick, easy, low-cost, and simple method that has recently become popular for plant virus detection.

NGS or high-throughput sequencing is used to sequence the entire viral genome. This platform is useful for the detection and study of plant viruses, as well as the discovery of unknown viruses (Barba et al., 2014; Wu et al., 2020; Pecman et al., 2017). In fluorescence *in situ* hybridization (FISH), fluorescently labeled probes are utilized to detect viral nucleic acids (Rudkin & Stoller, 1977). FISH can detect and localize the spread of viruses in plant and vector tissues (Kliot et al., 2014; Shargil et al., 2015). Flow cytometry (FCM) has the potential to simultaneously detect multiple viruses, measure genome size, and provide gene expression information (D'Hondt et al., 2011). However, lab-based methods require special precautions during sampling and subsequent steps to avoid cross-contamination, making them laborious and time-consuming. Additionally, these methods demand expensive and sophisticated equipment setups, as well as costly materials for sample preparations. Considering the cost, it is not economically feasible for a large number of samples.

Visual inspections of infected plants are the most common and traditional method for plant virus detection. It is easy to identify viruses, especially when plant-produced symptoms are disease-specific (Naidu & Hughes, 2001). However, diagnosing viral diseases based on symptoms alone is challenging because several viruses may infect a host plant simultaneously, which can modify symptom expression patterns (Wu et al., 2020). Various methods, such as visible-light imaging, thermal imaging, chlorophyll fluorescence imaging, and hyperspectral imaging, are used for visual inspection of virus symptoms. However, these methods can be influenced by environmental and biological factors (Mutka & Bart, 2015). In some cases, viruses may not produce visible disease symptoms in host plants, making virus detection difficult. To address this issue, biological indexing or the use of an indicator plant has been employed. These indicator plants are susceptible to selected viruses and develop typical viral disease symptoms when challenged with virus inoculum (Smith, 1977). Biological indexing has been used as a complementary test method to lab-based molecular methods (Wolfenden, 2018). However, this method has limitations, such as the varying time required from virus inoculation to the development of visible symptoms, which can range from days to months. Additionally, the symptoms produced by indicator plants can be influenced by environmental and biological factors (Legrand, 2015). Therefore, questions have been raised regarding the suitability of variety or cultivar as an indicator (Mehetre et al., 2021).

For effective viral disease management and complete virus eradication from the field, there is an urgent need to develop methods that would identify viruses in real-time and with great accuracy. Various plant viruses and diagnostic methods have been developed to date (Rubio et al., 2020). However, the practical use of these methods has been hindered by several limitations, such as cost, time requirements, the need for equipment and facilities, unavailability of skilled staff, and lower accuracy rates (Fang & Ramasamy, 2015). To overcome these issues, CRISPR-based techniques are the best choice for developing diagnostic tools for virus and other pathogen detection (Aman et al., 2020a; Li et al., 2019b). Initially used for gene editing, CRISPR has been further utilized in various diagnostic platforms to detect several pathogens quickly and with the highest precision, as demonstrated by the DETECTOR and SHERLOCK systems (Chen et al., 2018; Gootenberg et al., 2017). These diagnostic systems employ isothermal amplification of target nucleic acid and CRISPR/Cas-mediated target recognition, followed by collateral cleavage of the target nucleic acid, indicating the presence of target DNA or RNA in a given test sample (Kellner et al., 2019).

## 3.4 CRISPR/CAS TOOLS FOR PLANT VIRUS DETECTION

### *3.4.1 CRISPR/CAS9*

Cas9 is an RNA-guided DNA endonuclease of bacterial origin that plays an active role in the CRISPR adaptive immune system. This enzyme was previously known by various names such as Csn1, Csx12, Cas5, etc. It serves a crucial function in the bacterial immune system by defending against attacking pathogens like viruses and phages. While Cas9 is a component of the bacterial immune system, it has now been utilized as a groundbreaking genome engineering tool for introducing site-directed double-strand breaks (DSB) in a sequence-specific manner. These DSBs are guided to the target region by a gRNA (Jinek et al., 2012; Maeder et al., 2013). Reprogramming Cas9 for genome editing purposes involves the assembly of Cas9, tracrRNA, and crRNA. The crRNA and tracrRNA are co-expressed as sgRNAs (~20 nucleotides) upstream of the PAM sequence (Jinek et al., 2012). By altering the sequence of crRNA, CRISPR/Cas9 can be used to target different sites within the same genome. The recognition of the crRNA:tracrRNA:target DNA complex is facilitated by the recognition lobe (REC) present in Cas9, while the PAM interaction (PI) domain of the Cas9 effector is responsible for PAM sequence recognition (Figure 3.2(A)). The cleavage of the target and non-target strands through blunt-ended double-stranded breaks (DSBs) upstream of the PAM sequence is carried out by the HNH and RuvC domains, respectively (Shahid et al., 2021). The properties of Cas9 effector proteins, such as target nucleic acid, DSB structure, nuclease domains, PAM sequence, etc., are listed in Table 3.1.

The CRISPR/Cas9 system has been extensively studied and utilized for nucleic acid detection in test samples from various sources. Several nucleic acid detection methods have been developed using the CRISPR/Cas9 system in combination with PCR techniques, such as the CRISPR/Cas9-triggered isothermal exponential amplification reaction (CAS-EXPAR), the CRISPR-typing PCR (ctPCR) (Wang et al., 2018), and Cas9-based reverse PCR (CARP) (Zhang et al., 2018; Huang et al., 2018; Chang et al., 2019). These techniques have demonstrated high sensitivity and target specificity, but their application has been limited to laboratory settings. The challenges associated with designing reverse PCR primers, time consumption, the need for expensive and sophisticated equipment, and the use of enzymes in the PCR reaction mixture that are sensitive to RNase have hindered their wider use. Furthermore, these methods have not been employed for the detection of plant pathogens, such as viruses, in field or laboratory settings. Therefore, an improved version of these techniques, known as CRISPR/Cas9-mediated lateral flow nucleic acid assay (CASLFA), has been developed as a rapid and highly sensitive technique. Additionally, the CRISPR/Cas9 system has been coupled with recombinase polymerase amplification (RPA) for target gene amplification and visual detection of the target using a lateral flow assay (LFA) at room temperature (Wang et al., 2020).

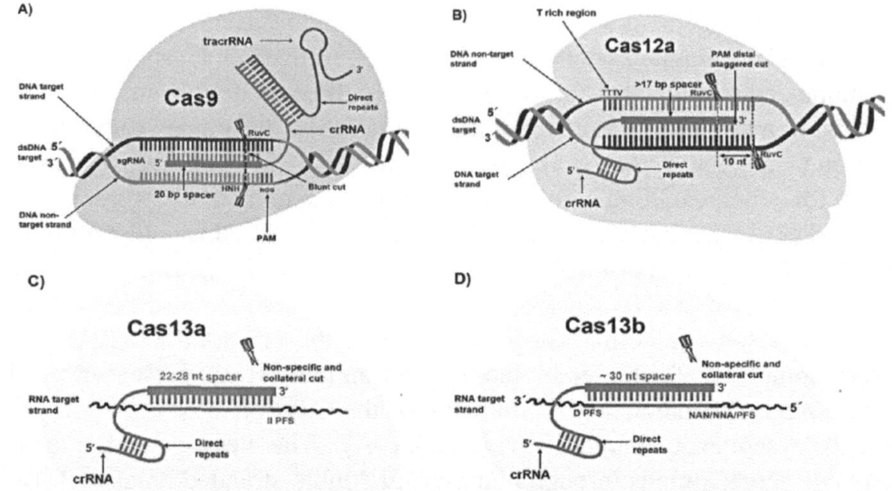


**FIGURE 3.2** Schematic representation of different Cas proteins widely used for genome editing, to engineer antiviral resistance in plants, and virus detection in plants.

*Note:* (A) The CRISPR-Cas9-based genome editing is facilitated through an effector Cas9, the crRNA, and tracrRNA. The crRNA and tracrRNA are jointly expressed as sgRNAs (~20 nt) upstream of the PAM sequence. The recognition of crRNA:tracrRNA:target DNA complex and PAM sequence recognition take place to introduce DSBs upstream of the PAM sequence. (B) The CRISPR/Cas12a is used for the detection of viral nucleic acids. CRISPR/Cas12a-based detection of nucleic acids is mediated through Cas12a and a single crRNA. The Cas12a:crRNA complex formation takes place, binding at the target region on the target strand, which is downstream of the PAM sequence. Further, ~25 nt long gRNA sequence is recognized by PAM, which leads to specific DNA-binding and DSBs are introduced at the target region, cleaving the target strand 18 nt downstream of PAM by the Nuc domain, while the non-target strand is cleaved by RuvC domain 23 nt downstream of the PAM, leaving staggered end cut. (C) and (D) CRISPR/Cas13a and b are used for sensing ssRNA using a single Cas effector protein (Cas13) and a crRNA. The Cas13a/Cas13b crRNA complex formation takes place, and it binds to the recognition site based on sequence complementarity ssRNA, directed by the gRNA. The non-specific, collateral cleavage of ssRNA is brought by 2X HEPN domains. Both Cas proteins have protospacer flanking sequence (PFS) which are different, and their spacer lengths are also different. [PAM: protospacer adjacent motif; and PFS: protospacer flanking sequence].

*Source:* Adapted and modified from: Shahid et al. (2021); and Tang & Fu (2018).

Dead or dCas9 was prepared by mutating nuclease domains (HNH and RuvC) of SpCas9 (Jiang & Daudna, 2017). Although it has lost nuclease activity, it proved to be useful for target DNA detection through sgRNA. Due to this specific activity, it has been successfully used for DNA detection

**TABLE 3.1** Properties of Class 2 CRISPR/Cas Nucleases

| Effector | Cas9 | Cas12a | Cas12b | Cas13a | Cas13b |
|---|---|---|---|---|---|
| Subtype | II-A | V-A | V-B | VI-A | VI-B |
| Former name | – | Cpf1 | C2c1 | C2c2 | – |
| Nuclease domain(s) | RuvC, HNH | RuvC, NuC | RuvC | 2xHEPN | 2xHEPN |
| Target | dsDNA | dsDNA (trans cleavage-dsDNA/ ssDNA) | dsDNA | ssRNA | ssRNA |
| tracrRNA requirement | Yes | No | Yes | No | No |
| Pre-crRNA processing by | RNase III | WED III | – | Helical 1 or HEPN 2 | – |
| Spacer length | 18–24 nt | 17–24 nt | 18–24 nt | 22–28 nt | ~ 30 nt |
| Cut nature (overhang) | Staggered (5′1-or 3-nt overhang) | Staggered (5-nt overhang), collateral activity for ssDNA | Staggered (staggered (5–7 nt overhang), collateral activity for ssDNA) | Collateral cleavage activity | Collateral cleavage activity |
| [a]PAM/PFS | 3′GC-rich PAM | 5′AT-rich PAM | 5′AT-rich PAM | 3′PFS: non-G | 3′PFS: [b]NAN/NNA; 5′PFS: non C |
| [c]Average Size (aa) | 1,228 | 1,268 | 1,129 | 1,228 | 1,128 |

[a]*PAM*: protospacer adjacent motif; and PFS: protospacer-flanking sequence; [b]N: A, C, G, T (or U); and [c]aa: amino acids.

*Source:* Adapted and modified from: Koonin & Makarova (2019); Tang & Fu (2018); and Chen et al. (2022).

in various CRISPR/Cas-based diagnostics. Most of these platforms were exploited for biomedical applications (Koo et al., 2018; Zhang et al., 2017; Guk et al., 2017; Hajian et al., 2019; Xu et al., 2020). The potential of CRISPR/dCas9-based biosensing platforms in plant virology is yet to be tested.

### *3.4.2 CRISPR/CAS12A*

The Cas12a endonucleases were formerly known as Cpf1. According to the general classification, they belong to class-II, type V CRISPR system and are composed of a Cas12a-Cas4-Cas1-Cas2-CRISPR array. Both Cas9 and Cas12a have similar shape and size, with Cas9 consisting of 1,228 amino acids and Cas12a consisting of 1,268 amino acids. The catalytic activity of

Cas9 endonuclease requires two nuclease domains, HNH and RuvC, while Cas12a has a single RuvC nuclease domain. The distinct nuclease domain of Cas12a is inserted at a similar but not identical position within the RuvC-like domain of Cas9. Cas12a is capable of cleaving both RNA and DNA. Initially, the DNA cleavage process occurs, and the crRNA produced in the first reaction is subsequently used for RNA cleavage (Zetsche et al., 2015). Another notable difference between Cas9 and Cas12a endonucleases is that Cas12a only requires crRNA in its active state, whereas Cas9 requires both crRNA and tracrRNA (Table 3.1). The PAMs for Cas9 and Cas12a are located at opposite ends of their respective protospacer sequences (Figures 3.2(A) and 3.2(B)). For efficient and specific DNA cleavage, shorter PAM sequences (mainly T-dependent), such as 5′-YTN-3′, 5′-TTN-3′, or 5′-TTTN-3′, are required (Yamano et al., 2016). In the case of DNA, the target and non-target strands are cleaved only by a single RuvC domain (Swarts et al., 2017) and not by both RuvC nuclease domains (Yamano et al., 2016). For example, the cleavage site of *Francisella novicida* Cas12a (FnCas12a) is located 23 bases downstream of the PAM on the target strand, or 18 bases downstream on the non-target strand. The Cas12a:crRNA complex is recognized by the REC lobe of Cas12a nuclease. This complex then binds to the target DNA strand at the designated target region, which is located downstream of the PAM sequence (Figure 3.2(B)). The PAM sequence, usually around 23–25 nt in length, is recognized by the PI domain, facilitating a specific DNA-binding mechanism. To introduce a double-strand break (DSB) at the target region, the NUC domain cleaves the target strand 18 nt downstream of the PAM sequence, while the RuvC domain cleaves the non-target strand 23 nt downstream of the PAM (Shahid et al., 2021). Compared to Cas9, Cas12a endonuclease exhibits higher efficiency due to the generation of a 5′ overhang of 4 or 5 nt during DSBs, which further stimulates NHEJ (Zetsche et al., 2015; Fonfara et al., 2016; Tang & Fu, 2018). Further studies have shown that the 5′ sticky ends are preferred for homology-dependent recombination (HDR) mechanisms (Vu et al., 2017; Schmidt et al., 2019).

The collateral cleavage activity is a unique and characteristic feature of Cas12a effector proteins, which has been exploited in the development of simple, rapid, reliable, ultrasensitive, and specific nucleic acid detection techniques. *Lachnospiraceae bacterium* Cas12a protein (LbCas12a) has been used to develop diagnostic platforms such as the one-hour low-cost multipurpose highly efficient system (HOLMES) and the DETECTR (Li et al., 2018a, b; Chen et al., 2018). Another diagnostic approach based on CRISPR/Cas12a:RPA was developed to detect *Mycobacterium tuberculosis* with the highest sensitivity (Ai et al., 2019). CRISPR/Cas12a-based diagnostic platforms have provided a visual interpretation of the results through turbidity, blue light, or LFAs (Bai et al., 2019; Wang et al., 2019). Moreover, this diagnostic approach was combined with filter paper-based DNA extraction and LFAs with RPA-LbCas12a for the detection of the rice blast pathogen (*Magnaporthe oryzae*) (Zhang et al., 2020). Additionally, several diagnostic tools using isothermal amplification-coupled CRISPR/Cas12a-based platforms for the detection of plant viruses and viroids have been developed (Table 3.2).

### 3.4.3 CRISPR/CAS12B

Li et al. (2019a) reported the trans-cleavage activity of Cas12b against collateral single-stranded DNA (ssDNA). Cas12b effector proteins, formerly known as C2c1, belong to Class 2, type V-B CRISPR system. It also has a single RuvC nuclease domain and produces a 5′ overhang after cleavage. The distinct feature of this effector protein is that it requires tracrRNA (Table 3.1). This specific activity of the thermophilic *Alicyclobacillus acidoterrestris* Cas12b (AacCas12b) endonucleases was utilized to develop a nucleic acid detection platform, namely HOLMESv2 (Li et al., 2019a). A thermophilic Cas12b effector is necessary because the LAMP reaction takes place isothermally but at a higher temperature (60–68°C). In HOLMESv2, LAMP is combined for the amplification of the target and the AacCas12b effector protein, for visible interpretation of nucleic acid detection. Compared to HOLMES, this is a single-step method performed in a single tube. Also, Cas12b has a unique ability of target preference than Cas12a and robust ssDNA trans-cleavage ability (Li et al., 2019a). Yet, there are no reports about the development of CRISPR/Cas12b nuclease-based tools for plant virus diagnosis.

### 3.4.4 CRISPR/CAS13A

Most of the CRISPR/Cas systems cleave dsDNA, while type III and VI-A specifically target RNA (Shmakov et al., 2017). This group of Cas effector proteins was formerly known as C2c2. Cas13 and its subtypes, namely Cas13a, are effector proteins belonging to class 2, type VI-A CRISPR systems that do not have DNase domains like Cas9, Cas12a, and Cas12b. As an alternative to this, Cas13a endonucleases have two HEPN domains located on the exterior of Cas13a (Table 3.1) (Anantharaman et al., 2013). Due to the two ribonuclease domains, Cas13a targets pre-crRNA, leading to crRNA maturation (East-Seletsky et al., 2016), and another ribonuclease domain is involved in pre-crRNA processing (Liu et al., 2017a, b). Most of the Cas13a loci consist of the Cas13a gene and CRISPR array (Shmakov et al., 2015). Later, Cas13a (LshCas13a) emerged as a novel programmable RNA cleaving endoribonuclease that cleaves entire ssRNA by recognizing the 22–28 nt long target sequence complementary to stem-loop (at least 24 nt long) containing crRNA spacer sequence (Abudayyeh et al., 2016). For Cas13a-based cleavage, the target sequence needs to be flanked by a single nucleotide (A, U, or C) protospacer-flanking site (PFS) at the 3′-end (Figure 3.2(C)). More specifically, the binding of LshCas13a and crRNA produces a duplex structure that further binds with the target sequence. This binding preferentially cleaves the loop region of ssRNA either at nucleotide-uridine or -adenosine in the absence of tracrRNA (East-Seletsky et al., 2017). Once the LshCas13a:crRNA duplex is activated by the target ssRNA, it can further cleave another ssRNA as well but in a non-specific mode, which is termed as "collateral effect."

Using collateral effect and recombinase polymerase-based isothermal amplification has opened ample opportunities for precise, ultra-highly sensitive CRISPR-based diagnostics (CRISPRdx), also known as specific high sensitivity enzymatic reporter unlocking (SHERLOCK) (Gootenberg et al., 2017). Additionally, SHERLOCK has proven its potential for ultra-highly sensitive detection, discrimination, identification, genotyping, and many other purposes. Within a short period, SHERLOCK has been improved and named SHERLOCKv2 (Gootenberg et al., 2018). The latest version of SHERLOCK provides advantages in terms of sensitivity up to zeptomolar level, suitability for using lateral flow strips, detection of multiple targets by simultaneously using Cas12a, Cas13, and Csm6, and a platform for quantitative detection. Several diagnostic tools based on Cas13a are being used for biomedical applications; however, their potential from a plant virology perspective is yet to be fully explored.

### *3.4.5 CRISPR/CAS13B*

Cas13b, formerly known as c2c6, belongs to Class 2, Type VI-B CRISPR system, composed of the Cas13b-Cas4-CRISPR array. Cas13b is an example of an RNA-guided RNA-cleavage CRISPR system with a collateral effect.

Similar to Cas13a, this effector protein has dual HEPN domains located on the outer side (Figure 3.2(D)). CRISPR/Cas13b, alone or in combination with Csx27, exhibits greater specificity than Cas13a. This improved specificity may be attributed to RNA cleavage being dependent on both sides of PFS, with a nucleotide D (A, U, or G) towards the 5′-end and NAN/NNA at the 3′-end (Table 3.1). Compared to the Cas13a subtype, Cas13b nucleases are CRISPR-associated RNA-guided RNases, containing two variants of crRNA (Smargon et al., 2017). Accessory proteins Csx27 and Csx28 differentially regulate the activity of Cas13b. The potential of Cas13b in the development of nucleic acid diagnostics and RNA editing has already been tested (Cox et al., 2017). Both CRISPR/Cas13a and Cas13b-based systems have not yet been used for plant virus detection. However, it is feasible to develop field-deployable diagnostic tools for RNA viruses and possibly for DNA viruses as well (Ali et al., 2021).

## 3.5 MECHANISMS OF CRISPR/CAS SYSTEM-BASED DETECTION OF NUCLEIC ACIDS

### *3.5.1 CRISPR/CAS9-BASED NUCLEIC ACID DETECTION*

In bacteria, the CRISPR/Cas9 system is naturally utilized to protect itself from foreign DNA that enters the bacterial cell via horizontal gene transfer. Initially, a unique, non-coding, trans-activating CRISPR RNA (tracrRNA) hybridizes with the precursor CRISPR transcript (pre-crRNA). This hybridization reaction is carried out through repeat sequences, which play a role in post-transcriptional RNA processing and endonucleolytic cleavage by RNaseIII, resulting in the production of crRNA. Furthermore, this crRNA undergoes structural modifications to form a dual RNA hybrid structure (Jiang & Doudna, 2017). The dual RNA guide leads Cas9 for dsDNA cleavage, which is complementary to the target sequence found next to PAM and cleaves each strand with the action of two different endonucleases, namely HNH or RuvC (Figure 3.3). To simplify the process, sgRNAs were engineered by combining crRNA and tracrRNA sequences. The PAM sequence NGG is required for the CRISPR/Cas9 system, which occurs randomly in DNA, allowing the design of desired sgRNA (Wang et al., 2019). Target DNA strand recognition by sgRNA activates Cas9 proteins, which generate DSBs in the target DNA strand. This characteristic feature of Cas9 has been used for the development of various biosensing platforms (Wang et al., 2019; Zhang et al., 2018).

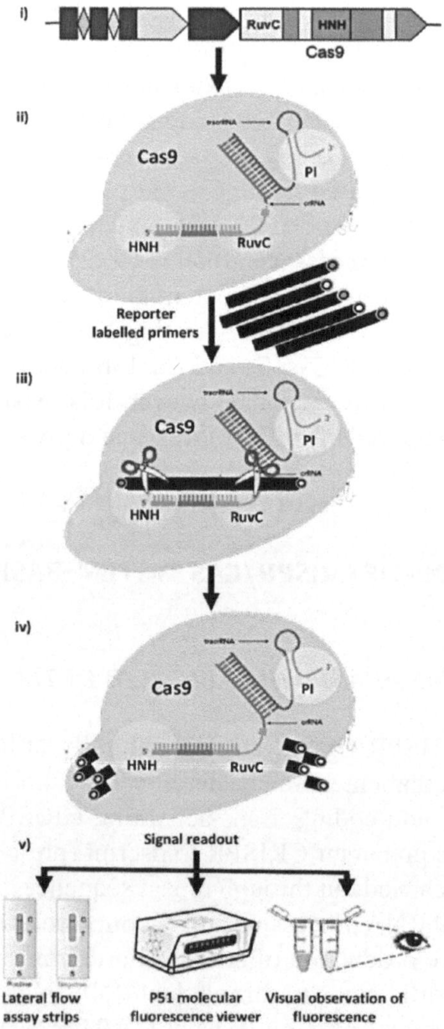


**FIGURE 3.3** Schematic representation of the mechanism of CRISPR/Cas9-based DNA detection. (i) Genomic architecture of CRISPR/Cas9 system and its main components. (ii) artificially designed gRNAs enable Cas9 protein to recognize Cas9 protein to target and activate endonuclease activity. (iii) the target nucleic acid (DNA) is amplified using isothermal amplification methods (LAMP/RPA/RCA) by using primers with a fluorescent quencher attached to them. (iv) the amplified DNA is recognized by the Cas9 protein, which cleaves the target DNA. By the cleavage activity of Cas9, it produces a blunt end by releasing the fluorophores for signal readout. (v) The fluorescent signals were quantified/observed using Lateral Flow Assay (LFA) strips, using a P51 molecular fluorescence viewer, or visually observed by the naked eye.

*Source*: Adapted and modified from: Bhardwaj et al., (2022).

### 3.5.2 CRISPR/CAS12-BASED NUCLEIC ACID DETECTION

Cas12 is the type V effector, which contains a single RNA-guided RuvC nuclease domain. The Cas12 effector proteins are structurally similar to Cas9, encompassing a lone RuvC domain for endonuclease activity, which allows Cas12 to cleave target dsDNA by producing sticky ends (Zetsche et al., 2015). Both Cas12a and Cas12b have been extensively studied and exploited for their biosensing potential. A notable difference is that Cas12b requires tracrRNA for catalyzing the reaction, whereas this is not the case with Cas12a. However, the endonuclease activity in both Cas12a and Cas12b is initiated by a PAM sequence located upstream of its target. The crRNA binds to the target sequence of the protospacer, which is adjacent to the PAM sequence. Using two $Mg^{2+}$ ions, the RuvC endonuclease cleaves the non-target DNA strand within the protospacer region, particularly 18 nt away from the PAM sequence (Figure 3.4) (Nalefski et al., 2021). The cleavage of dsDNA produces cohesive ends with a 5–7 nt overhang (Yan et al., 2019; Rusk, 2019; Strich & Chertow, 2019).

### 3.5.3 CRISPR/CAS13-BASED NUCLEIC ACID DETECTION

Cas13 effector proteins belong to Class 2, Type VI CRISPR/Cas system, which targets ssRNA. Compared to other CRISPR/Cas systems, Cas13 does not require tracrRNA, which is a distinctive feature of Cas13's HEPN-binding domain (Figure 3.5). Studies have shown that Cas13a possesses dual RNase activities: (i) for crRNA maturation; and (ii) for RNA-directed single-stranded RNA cleavage (Abudayyeh et al., 2016; East-Seletsky et al., 2016). The dual RNase activity is contributed by the REC and NUC lobes, which cleave pre-crRNA and target RNA after recognition, respectively (East-Seletsky et al., 2016). The target RNA cleavage mechanism for Cas13a involves recognition and attachment of the Cas13a effector to a PFS, which is present just behind the complementary spacer sequence on the 3′-end of ssRNA. In contrast, the cleavage mechanism differs for Cas13b, where the PFS is located adjacent to the complementary spacer sequence present on the 5′- and 3′-ends of the protospacer. The cleavage pattern of Cas13 endonucleases is distinct, inducing blunt-end breakages on the target strand and simultaneously cleaving non-target ssRNA in the vicinity (Hu et al., 2022; Kostyusheva et al., 2021). Recent studies have reported that the endonuclease activity depends on – (a) conserved catalytic residues in two HEPN domains; and (b) preferential bias toward uracil (Abudayyeh et al., 2016).

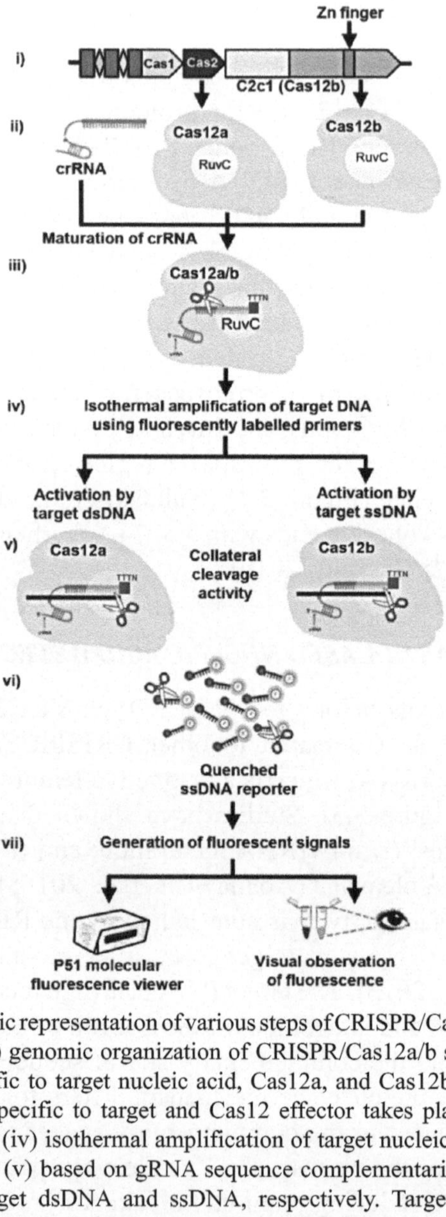


**FIGURE 3.4** Schematic representation of various steps of CRISPR/Cas12a/b-based detection of viral nucleic acid. (i) genomic organization of CRISPR/Cas12a/b systems, (ii) artificially designed gRNAs specific to target nucleic acid, Cas12a, and Cas12b effector proteins. (iii) Assembly of gRNAs specific to target and Cas12 effector takes place and further crRNA maturation takes place. (iv) isothermal amplification of target nucleic acid using fluorescent reporter labeled oligos. (v) based on gRNA sequence complementarity, Cas12a and Cas12b effectors recognize target dsDNA and ssDNA, respectively. Target recognition activates their endonuclease activity and cleaves the reporter-labeled target DNA by non-specifically collateral cleavage activity. (vi) due to collateral cleavage activity, it cuts designed quenched molecular reporters which produce fluorescence signals. (vii) the fluorescent signals can be visualized by the naked eye or by using a hand-held, P51 molecular fluorescence viewer.

*Source:* Adapted and modified from: Bhardwaj et al. (2022).

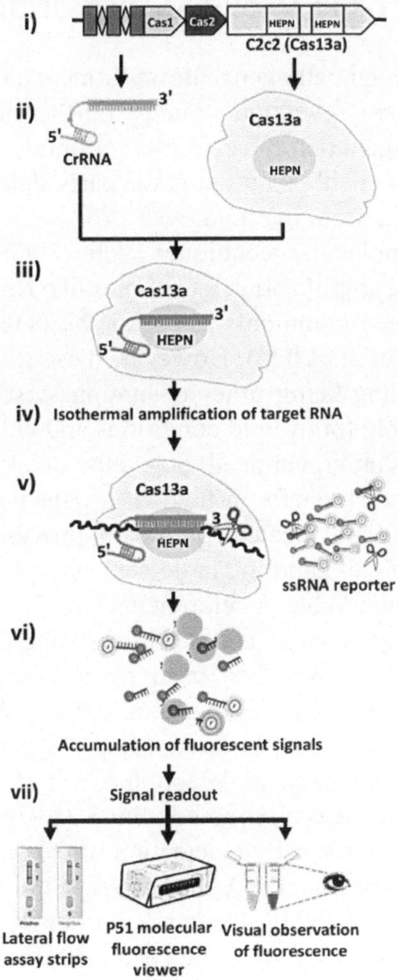


**FIGURE 3.5** The schematic representation illustrates the various steps involved in the CRISPR/Cas13a-based detection of viral nucleic acid.

*Note:* These steps include – (i) genomic organization of CRISPR/Cas13a; (ii) artificially designed gRNAs specific to the target nucleic acid and Cas13 effector; (iii) assembly of gRNAs specific to the target and Cas13a effector, leading to the maturation of crRNA; (iv) isothermal amplification of the target nucleic acid using fluorescent reporter-labeled oligos; (v) recognition of the target ssRNA by Cas13a and the effector, based on gRNA sequence complementarity. Target recognition activates their endonuclease activity, resulting in non-specific collateral cleavage of the reporter-labeled target RNA; (vi) collateral cleavage activity leads to the cleavage of designed quenched molecular reporters, generating fluorescence signals; and (vii) the fluorescent signals can be visualized either by the naked eye or by using a hand-held P51 molecular fluorescence viewer.

*Source:* Adapted and modified from: the works of Bhardwaj et al. (2022); and Jiao et al. (2021).

## 3.6 CRISPR/CAS FOR PLANT VIRAL DISEASE DIAGNOSIS

Timely detection of plant pathogens allows farmers ample time to decide on a suitable plant disease prevention strategy. Furthermore, for plant viruses, there is no chemical agent that can effectively limit the spread of the disease. Therefore, the only available solution is the early detection and destruction of virus-infected plants from the field.

To date, various molecular techniques such as PCR, RT-PCR, and field-deployable isothermal amplification techniques like LAMP, RT-LAMP, RPA, and NASBA have been commonly used for the detection of plant pathogenic viruses (Wang et al., 2020). However, these techniques have several disadvantages, including being time-consuming, costly, less sensitive, and accurate, and unsuitable for in-field conditions and bulk plant samples.

Therefore, there is an urgent need to develop quick, easy-to-operate, and cost-effective diagnostic tools for in-field virus detection. These tools should be equipped with less sophisticated equipment, provide accurate and precise outputs, and be capable of handling large-scale samples in the field.

In recent years, remarkable developments have been made in virus diagnostic methods through nucleic acid detection using CRISPR/Cas systems. The CRISPR/Cas system was first used for the diagnosis of pathogenic strains with a dsDNA genome (Zhang et al., 2017). Further developments in CRISPR/Cas-based diagnostics have tested the potential of Cas12a and Cas13a nucleases for nucleic acid detection (Chen et al., 2018; Gootenberg et al., 2017). Cas13a nucleases, coupled with SHERLOCK, have been used to develop rapid and ultra-sensitive detection (up to attomolar (aM) concentration) of pathogen-specific DNA and RNA. Similarly, a Cas12a-based detection method called DETECTR has been developed to detect dsDNA targets through fluorescent signals (Chen et al., 2018). Furthermore, Cas12a, utilizing the HOLMES detection method coupled with LAMP, has resulted in a rapid and highly sensitive detection platform for target DNA and RNA (Li et al., 2018). While the target is DNA, the target nucleic acid is amplified using recombinase polymerase amplification (RPA) reaction. For the target RNA, the sample is amplified using reverse-transcription-RPA (RT-RPA) reactions. These RPA products are then subjected to transcription that synthesizes RNA using a T7 RNA polymerase enzyme. The collateral cleavage of those RNA transcripts is performed using Cas12/13 endonucleases in the presence of a quenchable ssRNA reporter, and the fluorescence is quantified. However, all these techniques were specifically developed for the detection of human viruses, microorganisms, and transgenes (Chaijarasphong et al.,

2019; Myhrvold et al., 2018; Zhang et al., 2020). Although there have been tremendous developments in CRISPRdx, the potential of these diagnostic tools has not been fully utilized in the context of plant virology. However, several recent studies report the development and application of inexpensive, field-deployable CRISPR-based diagnostic systems for robust detection of plant viruses (Table 3.2).

### 3.6.1 LAMP-COUPLED CRISPR/CAS12A-BASED ASSAY FOR THE DETECTION OF PLANT DNA-VIRUSES

Mahas & colleagues (2021) developed a diagnostic method for geminiviruses, namely *tomato yellow leaf curl virus* (TYLCV; ssDNA) and *tomato leaf curl New Delhi virus* (ToLCNDV; ssDNA). Initially, they designed LAMP primers in the most conserved region of the CP gene in both virus genomes. LAMP primers were used to PCR amplify a conserved region of their genome, and Cas12a crRNAs efficiently targeted the amplified target region along with the PAM sequence (5′-TTTN-3′) that specifically cleaves the TOLCNDV-A ssDNA component and TYLCV ssDNA genomic component. LAMP is usually performed at 65°C, whereas Cas12a endonucleases have shown optimal activity at 35°C. Therefore, they divided this assay into two steps (LAMP followed by Cas12a-based viral detection) (Figure 3.6). They used HEX fluorophore in the reporter molecule, which is released via collateral cleavage activity of Cas12a, leading to the production of a bright fluorescence signal upon excitation by a light-emitting diode (LED) (Chen et al., 2018). This real-time, in-tube fluorescence can be easily viewed using an affordable, easy-to-use, handheld P51 Molecular Fluorescence Viewer. Using this system, they estimated the sensitivity of virus detection to be as little as 100 aM, along with the greatest specificity (no cross-reactivity) within ~1 h (Mahas et al., 2021).

To test the newly developed diagnostic system, DNA extracted from the upper leaves of *N. benthamiana* plants previously infected with ToLCNDV and TYLCV was used through Agrobacterium-mediated transformation. This diagnostic method was able to detect viral sequences quickly, and the results were compared with conventional PCR. Additionally, the potential of this system to detect both viral sequences (ToLCNDV and TYLCV) in tomato plants was examined. The assay successfully and precisely detected both viruses. With further improvements, this LAMP-coupled CRISPR/Cas12a virus detection method could be developed into a field-deployable diagnostic test.

**TABLE 3.2** CRISPR/Cas-based Detection of Different Plant Viruses/Viroids

| Plant Virus | Family | Viral Genome | Host Plant(s) | Targeted Genomic Region | Pre-Amplification Method | CRISPR/Cas-based Diagnostic System | Readout | Assay Temperature (°C) | Viral Nucleic Acid Detection Sensitivity | Time Required (min.) | References |
|---|---|---|---|---|---|---|---|---|---|---|---|
| Tomato yellow leaf curl virus (TYLCV) | Geminiviridae | ssDNA | Tomato, *N. benthamiana* | CP | LAMP | CRISPR/Cas12a | Fluorescent | 60 | 100 aM | ~ 60 | Mahas et al. (2021) |
| Tomato leaf curl New Delhi virus (ToLCNDV) | Geminiviridae | ssDNA | Tomato, *N. benthamiana* | CP | LAMP | CRISPR/Cas12a | Fluorescent | 60 | 100 aM | ~ 60 | Mahas et al. (2021) |
| Beet necrotic yellow vein virus (BNYVV) | Bunyaviridae | ssRNA | Sugar beet | RNA-1 | RT-RPA | CRISPR/Cas12a | Fluorescent | 42 | 0.1 pM | 60 | Ramachandran et al. (2021) |
| Potato virus X (PVX) | Alphaflexiviridae | ssRNA | *N. benthamiana* | CP | RT-RPA | CRISPR/Cas12a | Fluorescent | 37 | *Up-to picomolar range | 30 | Aman et al. (2020b) |
| Potato virus Y (PVY) | Potyviridae | ssRNA | *N. benthamiana* | CP | RT-RPA | CRISPR/Cas12a | Fluorescent | 37 | *Up-to picomolar range | 30 | Aman et al. (2020b) |
| Tobacco mosaic virus (TMV) | Virgaviridae | ssRNA | *N. benthamiana* | CP | RT-RPA | CRISPR/Cas12a | Fluorescent | 37 | *Up-to picomolar range | 30 | Aman et al. (2020b) |
| Tomato mosaic virus (ToMV) | Virgaviridae | ssRNA | Tomato | ORF1 | RT-PCR | CRISPR/Cas12a | Fluorescent | 37 | 15–30 ng of PCR products | ~120 | Alon et al. (2021) |
| Tomato brown rugose fruit virus (ToBRFV) | Virgaviridae | ssRNA | Tomato | ORF1 | RT-PCR | CRISPR/Cas12a | Fluorescent | 37 | 15–30 ng of PCR products | ~120 | Alon et al. (2021) |
| Tomato brown rugose fruit virus (ToBRFV) | Virgaviridae | ssRNA | Tomato, *N. benthamiana* | MP-ORF | RT-LAMP | CRISPR/Cas12a | Fluorescent and colorimetric | 37 | $10^{-4}$ ng viral RNA/reaction dilution | ~60 | Bernabé-Orts et al. (2022) |
| Apple necrotic mosaic virus (ApNMV) | Bromoviridae | ssRNA | Apple | – | RT-RPA | CRISPR/Cas12a with oligonucleotide conjugated AuNPs. | Colorimetric | 37 | #0.01–1 fM (2,500 viral RNA copies/reaction) | 30 | Jiao et al. (2021) |
| Apple chlorotic leaf spot virus (ACLSV) | Betaflexiviridae | ssRNA | Apple | – | RT-RPA | CRISPR/Cas12a with oligonucleotide conjugated AuNPs. | Colorimetric | 37 | #0.01–1 fM (2,500 viral RNA copies/reaction) | 30 | Jiao et al. (2021) |
| Apple stem grooving virus (ASGV) | Betaflexiviridae | ssRNA | Apple | – | RT-RPA | CRISPR/Cas12a with oligonucleotide conjugated AuNPs. | Colorimetric | 37 | #0.01–1 fM (250 viral RNA copies/reaction). | 30 | Jiao et al. (2021) |

**TABLE 3.2** *(Continued)*

| Plant Virus | Family | Viral Genome | Host Plant(s) | Targeted Genomic Region | Pre-Amplification Method | CRISPR/Cas-based Diagnostic System | Readout | Assay Temperature (°C) | Viral Nucleic Acid Detection Sensitivity | Time Required (min.) | References |
|---|---|---|---|---|---|---|---|---|---|---|---|
| Apple stem pitting virus (ASPV) | Betaflexiviridae | ssRNA | Apple | – | RT-RPA | CRISPR/Cas12a with oligonucleotide conjugated AuNPs. | Colorimetric | 37 | #0.01–1 fM (250 viral RNA copies/reaction). | 30 | Jiao et al. (2021) |
| Apple scar skin viroid (ASSVd) | Pospiviroidae | ssRNA | Apple | – | RT-RPA | CRISPR/Cas12a with oligonucleotide conjugated AuNPs. | Colorimetric | 37 | #0.01–1 fM (2,500 viral RNA copies/reaction). | 30 | Jiao et al. (2021) |

*The detection limit can be increased up to the femtomolar range if reaction time increased.
#Virus RNA transcript detection sensitivity was similar to RT-qPCR.

*Abbreviations:* MP: Movement protein; ORF: open reading frame; ssRNA: single-stranded RNA; ssDNA: single-stranded DNA; CP: coat protein; aM: attomolar; pM: picomolar; fM: femtomolar; LAMP: loop-mediated isothermal amplification; RT-LAMP: reverse transcription loop-mediated isothermal amplification; RT-RPA: reverse transcription-recombinase polymerase amplification; RT-PCR: reverse transcription-polymerase chain reaction; and AuNPs: gold nanoparticles.

*Source:* Adapted and modified from: Sharma et al. (2021); Bhat et al. (2022); and Chen et al. (2022).

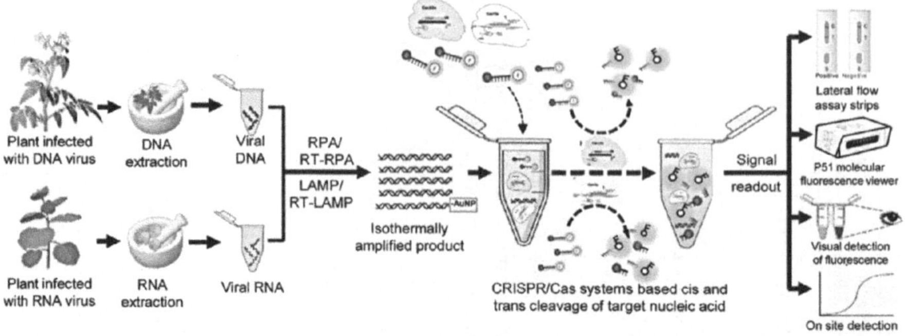


**FIGURE 3.6** A schematic flow diagram of viral nucleic acid detection using CRISPR/Cas-based tools is presented.

*Note:* The systemic leaves from plants infected with DNA or RNA viruses are utilized for total DNA/RNA extraction. Subsequently, virus-specific nucleic acids are amplified through different isothermal amplification methods (RPA/LAMP for DNA viruses; RT-RPA/RT-LAMP for RNA viruses) or using RT-PCR with target-specific oligonucleotides. The resulting amplicon is then subjected to CRISPR/Cas system-based detection in the presence of quenched ssDNA molecular reporters. The collateral cleavage activity of Cas proteins cleaves the reporter molecules, generating fluorescent signals that can be detected and/or quantified using the P51 molecular fluorescence viewer, colorimetrically using lateral flow assay strips, or visualized under UV or LED lights.

*Source:* This approach has been modified and adapted from studies by Bhat et al. (2022); Mahas et al. (2021); Jiao et al. (2021); Ramachandran et al. (2021); Aman et al. (2022b); Alon et al. (2021); and Bernabé-Orts et al. (2022).

### 3.6.2 RT-RPA-COUPLED CRISPR/CAS12A-BASED ASSAY FOR DIAGNOSIS OF PLANT RNA-VIRUSES

The collateral cleavage activities of CRISPR/Cas systems, namely Cas12a, Cas13 (targets ssRNA), and Cas14 (targets ssDNA), have recently been utilized for the development of SHERLOCK, HOLMES, HOLMESv2, DETECTR, etc., platforms, which can precisely, rapidly, and sensitively detect target nucleic acids. Aman et al. (2020a) reported the RT-RPA-coupled CRISPR/Cas12a-based single-step assay named iSCAN (*in vitro* Specific CRISPR-based Assay for Nucleic acid detection One-Pot (iSCAN-OP), which proved to be easy to use, rapid, and efficient for *potato virus X* (PVX; +ve sense ssRNA), *potato virus Y* (PVY; +ve sense ssRNA), and *tobacco mosaic virus* (TMV; +ve sense ssRNA) RNA detection. It includes total RNA extraction from systemic leaves of *N. benthamiana* previously infected with PVX, PVY, and TMV. RPA primers specific to viral-CPs were designed and used for amplification. RT-RPA pre-amplified products were then subjected to collateral cleavage by a *trans* cleavage activity of LbCas12a (Figure 3.6). The reaction mixture included Cas12a, crRNAs, and Hex-labeled ssDNA reporter (for TMV, FAM-labeled reporter) (Aman et al., 2020b). As a result, the collateral cleavage of the ssDNA reporter molecule takes place, leading to the release of fluorescent signals (Chen et al., 2018), which can be quantified by combining this assay with a commercially available, handheld, affordable P51 molecular fluorescence viewer (Figure 3.6). The detection sensitivity of this assay for viral RNA was recorded up to a picomolar concentration. Additionally, when multiple viruses infected plant material was subjected to this assay, this platform could detect the presence of specific viral RNA, proving specificity (no cross-reactivity) and multiplexed detection of diverse viruses in a case of mixed infection. The key features of this diagnostic method are: (i) the entire assay can be performed at a single temperature; (ii) the assay can be performed in one pot; (iii) the assay completes within 30 minutes; and (iv) the results can be visualized with an affordable, easy-to-operate, commercially available handheld P51 Molecular Fluorescence Viewer that enables real-time, rapid, in-field RNA virus diagnosis.

Similarly, Jiao & colleagues (2021) have also developed an RT-RPA-coupled CRISPR/Cas12a method using gold nanoparticles (AuNPs) for the detection of RNA viruses in apple, including economically important viruses and/or viroids. These viruses include *apple necrotic mosaic virus* (ApNMV), *apple stem pitting virus* (ASPV), *apple stem grooving virus* (ASGV), *apple chlorotic leaf spot virus* (ACLSV), and *apple scar skin viroid* (ASSVd). Total RNA was extracted from systemic leaves and specific primers flanking the unique genomic region of each virus were used for amplification through RT-RPA, followed by conversion to dsDNA. The resulting amplicon was then combined with the LbCas12a system along with a pre-designed, virus-specific crRNA. This activates the collateral *trans* cleavage activity of Cas12a, leading to the degradation of non-specific ssDNA-linker. A complementary ssDNA-linker, designed to hybridize with DNA:1/2AuNP, is added to the reaction mixture. Further cleavage of the ssDNA-linker inhibits the accumulation of DNA:1/2AuNP, preventing a change in the original color (red) of the reaction mixture, indicating a positive result. If the virus nucleic acid is not present, the ssDNA-linker remains intact and crosslinks with DNA:AuNPs, resulting in a colorless reaction mixture, indicating a negative test result (Jiao et al., 2021). Both the colors of different reaction mixtures can be observed without the use of any instruments. The results of this diagnostic method were equally as sensitive as RT-qPCR, which could detect up to 250–2,500 viral RNA copies. The entire protocol can be completed in less than 1 hour, making it significantly simpler and less expensive. Similarly, rapid, efficient, and on-site detection of *beet necrotic yellow vein virus* (BNYVV) was reported by Ramachandran et al. (2021), with initial amplification by RT-RPA followed by LbCas12a-based detection using released fluorescence signals.

### 3.6.3 RT-LAMP AND RT-PCR COUPLED CRISPR/CAS12A-BASED ASSAY FOR DIAGNOSIS OF PLANT RNA-VIRUSES

*Tomato brown rugose fruit virus* (ToBRFV) is an emergent virus with a significant impact on solanaceous crops such as tomato and pepper. The total RNA was extracted from symptomatic tomato leaves, and reverse transcription (RT)-LAMP was utilized to amplify the desired nucleic acid. Subsequently, the RNA-guided LbCas12a endonuclease recognized the target region of the RNA through specific base pairing (~20 bp) with the target sequence and guide-RNA. The recognition of the target nucleic acid sequence triggers collateral cleavage activity of the LbCas12a nuclease, resulting in non-specific, *trans* cleavage of the ssDNA reporter (Figure 3.6). This activity allows for quantification of readouts using a molecule attached to the 3′-end. Furthermore, when using a fluorophore, the endonuclease activity causes the fluorochrome to dissociate from the quencher at the 5′-end of the oligonucleotide, enabling easy detection of fluorescence. On the other hand, the use of a biotin ligand for detecting LbCas12a activity has facilitated colorimetric results that can be observed with the naked eye (Table 3.2) (Bernabé-Orts et al., 2021).

Furthermore, for the detection of ToBRFV and *tomato mosaic virus* (ToMV) in tomato plants, Alon et al. (2021) used RT-PCR instead of isothermal amplification using RT-LAMP (Table 3.2) to convert viral RNA from crude leaf extract into dsDNA. The dsDNA then serves as a substrate for LbCas12a-mediated collateral cleavage. The fluorescent signal can be quantified using the P51 Molecular Fluorescence Viewer.

### 3.6.4 ADVANTAGES OF DIFFERENT CRISPR/CAS-BASED VIRUS DIAGNOSTIC PLATFORMS

Within a short period, CRISPR/Cas-based diagnostic methods have advanced from laboratory-based virus detection methods to field-deployable diagnostic tools. The reasons behind their rapid evolution are reliable, cost-effective, real-time, simple, rapid, and highly sensitive detection at the field level. Additionally, CRISPR-based methods could provide single-base resolution, the least detection sensitivity up to femtomolar (fM), and in some cases up to aM (Yin et al., 2021). These diagnostic platforms can detect a few viral nucleic acid copies per reaction and differentiate between virus isolates or mutations occurring therein (Bai et al., 2019; Gootenberg et al., 2017). Furthermore, it is possible to combine these diagnostic methods with other techniques to obtain different signal readouts (fluorescent, colorimetric, electrochemical) (Ding et al., 2020; Fozouni et al., 2021). The LFA coupled with CRISPR/Cas systems can increase their use specifically in point-of-care testing (POCT). In field conditions, plants can be attacked by various pathogens simultaneously. Therefore, it is difficult to use an appropriate, pathogen/virus-specific diagnostic tool. Multiplex virus detection is an additional advantage of the CRISPR-based diagnostic method, where multiple concurrent RNA viruses can be detected using RT-RPA coupled with CRISPR/Cas12a/AuNP, as reported by Jiao et al. (2021). Furthermore, these diagnostic methods are suitable for large-scale screening and bulk testing of samples due to their low cost and time-saving nature. They can also tolerate various crude sample extracts without compromising their sensitivity, as mentioned by Schermer et al. (2020). While most CRISPR-based diagnostic methods require pre-amplification of the target, Fozouni et al. (2021) have developed an amplification-free CRISPR/Cas13a-based diagnostic method that can directly and quantitatively detect SARS-CoV-2 RNA in clinical test samples.

## 3.7 LIMITATIONS AND CHALLENGES OF DIFFERENT CRISPR/CAS-BASED VIRUS-DIAGNOSIS METHODS

Besides the several advantages of CRISPR/Cas-based diagnostics, there are still some bottlenecks in the adoption of these tools. One of the major limiting factors is the requirement for the initial amplification of targets. Additionally, the primers designed for pre-amplification of the target nucleic acid region and guide-RNA, as well as crRNAs designed for the virus nucleic acid detection, need to be lab tested and their reaction conditions (reaction buffer, reaction time, reaction composition, temperature, the abundance of reaction inhibitors, DNase/RNase contamination, nucleic acid extraction method, reaction volume, etc.) must be standardized (Kaminski et al., 2021). To use CRISPRdx in diverse domains (food safety, biomedical, plant protection, etc.), standard operation protocols established by the authorized agencies would boost its actual use and market expansion in the coming years. Recently, a single pot-one step detection assay (iSCAN-OP) coupled with CRISPR/Cas12a has significantly reduced the complexities of using CRISPRdx (Aman et al., 2020b). Conversely, the requirement of separate steps for sample preparation and the use of incubating devices at higher temperatures remains an obstacle and restricts their use for in-field detection of plant viruses. Therefore, to increase the acceptability and diverse applications of these on-site diagnostic platforms, further technical developments are needed to overcome existing limitations.

For the endonuclease activity of Cas12a to target dsDNA, it requires PAM sequences that specifically recognize the target sequence. Conversely, it limits the choice of the target nucleic acid region. In the case of short nucleic acid sequence detection, it is not possible to use Cas12a-based diagnostic tools (Yin et al., 2021). Using a CRISPR/Cas-based assay for quantitative testing is challenging to quantify the targets with a high titer because of the higher sensitivity of the system and the limited reporter molecules available; it would easily reach the signal plateau (Li et al., 2019). Additionally, due to the pre-amplification step, the original concentration of nucleic acids in a sample might sometimes mislead the quantitation result (Kebschull et al., 2015). Therefore, there is a need for an internal control to interpret results in an absolute or relative manner. CRISPRdx have shown that pretreatment of the complex, raw sample is necessary for following rounds of virus detection. As a result, the need for new Cas-effector proteins of a more robust and tolerant type, improved sample preparation procedures compatible with CRISPRdx, or more appropriate, easy-to-use, tailored devices adaptive to CRISPR-based diagnostic tools has been identified (Yin et al., 2021). The detection of nucleic acid without contamination is crucial for viral diagnostics. However, the pre-amplification step significantly contributes to contamination due to the generation of nucleic acids, which reduces sensitivity. Therefore, there is an immediate need to develop contamination-free detection or highly specific gene amplification methods (Hu et al., 2021). To enable the application of field-deployable virus diagnostics, the development of a simplified procedure and the combination of CRISPR/Cas with other platforms such as mobile phone technology, LFA, nanotechnology, microfluidics, biochips, cloud computing, and artificial intelligence (AI) will be an effective approach (Ibrahim et al., 2020).

## 3.8 CONCLUSION AND FUTURE PROSPECTUS

In conclusion, emerging plant viruses cause severe losses and have a negative impact on global food security. Therefore, there is an urgent need to develop rapid, sensitive, specific, and field-deployable diagnostic tools that can contribute to effectively managing and preventing the further spread of viral diseases. CRISPR/Cas systems, besides their use as excellent genome editing tools, have recently been developed as a novel and highly efficient diagnostic technique (Mahas et al., 2021; Aman et al., 2020; Ramachandran et al., 2021; Alon et al., 2021; Jiao et al., 2021; Bernabé-Orts et al., 2022). CRISPR/Cas12 and CRISPR/Cas13-based diagnostic systems utilize their *trans* cleavage activity, which forms the foundation for molecular diagnostics. Currently, several direct or indirect viral diagnostic methods have been developed. However, most of them are expensive, time-consuming, laborious, and require a laboratory setting. To overcome the limitations of conventional virus diagnostic methods, CRISPR/Cas-based techniques such as SHERLOCK, DETECTR, HOLMES, HOLMESv2, etc., have been developed. For the detection of plant viral diseases, isothermal amplification techniques such as LAMP, RT-LAMP, RPA, and RT-RPA are performed to obtain dsDNA or ssDNA. These amplified products were further subjected to Cas12/Cas13-based cleavage to produce signal readouts. These methods have virus nucleic acid detection sensitivity ranging from femtomolar (fM) to attomolar (aM) concentration with single-base resolution. The results can be obtained in the form of fluorescent, colorimetric, or electrochemical readouts. Moreover, multiple virus detection using CRISPR/Cas12a/AuNPs has been achieved with the greatest specificity and at a single temperature within a short time (Jiao et al., 2021). CRISPR/Cas12a and Cas13a-based methods are affordable and do not require any sophisticated equipment. Despite several benefits, there are some obstacles to using CRISPR/Cas-based diagnostics for plant viruses. In summary, CRISPR/Cas-based virus detection is a novel, rapidly developing, and expanding platform that has revolutionized the approach to diagnosing plant-infecting viruses by combining already available isothermal amplification methods. Without a doubt, CRISPR/Cas-based diagnostics will become an essential toolkit for plant pathologists. CRISPR/Cas systems have immense, yet-to-be-exploited potential in the field of diagnostics. Moving forward, the combinatorial and multifaceted use of biomaterials, bioengineering, cloud computing, and AI will make these systems extraordinary.

## KEYWORDS

- **biosensors**
- **Cas12**
- **Cas13**
- **Cas9**
- **CRISPR/Cas**
- **DNA viruses**
- **food security**
- **gene/genome editing**
- **genetic engineering**
- **molecular diagnostics**
- **nucleic acid detection**
- **plant viruses**
- **RNA viruses**
- **viroids**

## REFERENCES

Abudayyeh, O. O., Gootenberg, J. S., Konermann, S., Joung, J., Slaymaker, I. M., Cox, D. B. T., Shmakov, S., et al., (2016). C2c2 is a single-component programmable RNA-guided RNA-targeting CRISPR effector. *Science, 353*, aaf5573. https://doi.org/10.1126/science.aaf5573.

Adams, I. P., Glover, R. H., Monger, W. A., Mumford, R., Jackeviciene, E., Navalinskiene, M., Samuitiene, M., & Boonham, N., (2009). Next-generation sequencing and metagenomic analysis: A universal diagnostic tool in plant virology. *Mol. Plant Pathol., 10*, 537–545. https://doi.org/10.1111/j.1364-3703.2009.00545.x.

Ai, J. W., Zhou, X., Xu, T., Yang, M., Chen, Y., He, G. Q., Pan, N., et al., (2019). CRISPR-based rapid and ultra-sensitive diagnostic test for *Mycobacterium tuberculosis*. *Emerg. Microbes Infect., 8*, 1361–1369. https://doi.org/10.1080/22221751.2019.1664939.

Ali, Z., & Magdy, M. M., (2021). CRISPR/Cas systems versus plant viruses: Engineering plant immunity and beyond, *Plant Physiol., 186*(4), 1770–1785, https://doi.org/10.1093/plphys/kiab220.

Alon, D. M., Hak, H., Bornstein, M., Pines, G., & Spiegelman, Z., (2021). Differential detection of the *Tobamo viruses Tomato mosaic virus* (ToMV) and *Tomato brown rugose fruit virus* (ToBRFV) using CRISPR/Cas12a. *Plants, 10*(6), 1256. https://doi.org/10.3390/plants10061256.

Aman, R., Mahas, A., & Mahfouz, M., (2020a). Nucleic acid detection using CRISPR/Cas biosensing technologies. *ACS Synth. Biol., 9*, 1226–1233. https://doi.org/10.1021/acssynbio.9b00507.

Aman, R., Mahas, A., Marsic, T., Hassan, N., & Mahfouz, M. M., (2020b). Efficient, rapid, and sensitive detection of plant RNA viruses with one-pot RT-RPA–CRISPR/Cas12a assay. *Front Microbiol., 11*, 3277. https://doi.org/10.3389/fmicb.2020.610872.

Anantharaman, V., Makarova, K. S., Burroughs, A. M., Koonin, E. V., & Aravind, L., (2013). Comprehensive analysis of the HEPN superfamily: Identification of novel roles in intra-genomic conflicts, defense, pathogenesis, and RNA processing. *Biol Direct., 8*, 15. https://doi.org/10.1186/1745-6150-8-15.

Anders, C., Niewoehner, O., Duerst, A., & Jinek, M., (2014). Structural basis of PAM-dependent target DNA recognition by the Cas9 endonuclease. *Nature, 513*, 569–573. https://doi.org/10.1038/nature13579.

Bai, J., Lin, H., Li, H., Zhou, Y., Liu, J., Zhong, G., Wu, L., et al., (2019). Cas12a-based on-site and rapid nucleic acid detection of African swine fever. *Front. Microbiol., 10*, 2830. https://doi.org/10.3389/fmicb.2019.02830.

Barba, M., Czosnek, H., & Hadidi, A., (2014). Historical perspective, development, and applications of next-generation sequencing in plant virology. *Viruses, 6*, 106–136. https://doi.org/10.3390/v6010106.

Bernabé-Orts, J. M., Hernando, Y., & Aranda, M. A. (2022). Toward a CRISPR-based point-of-care test for tomato brown rugose fruit virus detection. *PhytoFrontiers™. 2*(2), 92–100. https://doi.org/10.1094/PHYTOFR-08-21-0053-TA.

Bhardwaj, P., Kant, R., Behera, S. P., Dwivedi, G. R., & Singh, R., (2022). Next-generation diagnostic with CRISPR/Cas: Beyond nucleic acid detection. *Int. J. Mol. Sci., 23*, 6052. https://doi.org/10.3390/ijms23116052.

Bhat, A., Aman, R., & Mahfouz, M., (2022). Onsite detection of plant viruses using isothermal amplification assays. *Plant Biotechnol. J.* Accepted Author Manuscript. https://doi.org/10.1111/pbi.13871.

Boine, B., Kingston, R. L., & Pearson, M. N., (2012). Recombinant expression of the coat protein of *Botrytis virus X* and development of an immunofluorescence detection method to study its intracellular distribution in *Botrytis cinerea. J. Gen. Virol., 93*, 2502–2511. https://doi.org/10.1099/vir.0.043869-0.

Boonham, N., Kreuze, J., Winter, S., Van, D. V. R., Bergervoet, J., Tomlinson, J., & Mumford, R., (2014). Methods in virus diagnostics: From ELISA to next generation sequencing. *Virus Res., 186*, 20–31. https://doi.org/10.1016/j.virusres.2013.12.007.

Chaijarasphong, T., Thammachai, T., Itsathitphaisarn, O., Sritunyalucksana, K., & Suebsing, R., (2019). Potential application of CRISPR/Cas12a fluorescence assay coupled with rapid nucleic acid amplification for detection of white spot syndrome virus in shrimp. *Aquaculture, 512*, 734340. https://doi.org/10.1016/j.aquaculture.2019.734340.

Chang, W., Liu, W., Liu, Y., Zhan, F., Chen, H., Lei, H., & Liu, Y., (2019). Colorimetric detection of nucleic acid sequences in plant pathogens based on CRISPR/Cas9 triggered signal amplification. *Microchim. Acta., 186*, 1–8. https://doi.org/10.1007/s00604-019-3348-2.

Chen, B., Li, Y., Xu, F., & Yang, X., (2022). Powerful CRISPR-based biosensing techniques and their integration with microfluidic platforms. *Frontiers in Bioengineering and Biotechnology, 10*, 851712. https://doi.org/10.3389/fbioe.2022.851712.

Chen, J. S., Ma, E., Harrington, L. B., Da Costa, M., Tian, X., Palefsky, J. M., & Doudna, J. A., (2018). CRISPR/Cas12a target binding unleashes indiscriminate single-stranded DNase activity. *Science, 360*, 436–439. https://doi.org/10.1126/science.aar6245.

Clark, M. F., Adams, A. N., Graham, F. L., Smiley, J., Russell, W. C., & Nairn, R., (1977). Characteristics of the microplate method of enzyme-linked immunosorbent assay for the detection of plant viruses. *J. Gen. Virol., 34*, 475–483. https://doi.org/10.1099/0022-1317-34-3-475.

Cox, D. B., Gootenberg, J. S., Abudayyeh, O. O., Franklin, B., Kellner, M. J., Joung, J., & Zhang, F., (2017). RNA editing with CRISPR/Cas13. *Science, 358*(6366), 1019–1027. https://doi.org/10.1126/science.aaq0180.

D'Hondt, L., Höfte, M., Van, B. E., & Leus, L., (2011). Applications of flow cytometry in plant pathology for genome size determination, detection, and physiological status. *Mol. Plant Pathol., 12*, 815–828. https://doi.org/10.1111/j.1364-3703.2011.00711.x.

Danks, C., & Barker, I., (2000). On-site detection of plant pathogens using lateral-flow devices. *EPPO Bull., 30*, 421–426. https://doi.org/10.1111/j.1365-2338.2000.tb00922.x.

Ding, X., Yin, K., Li, Z., Lalla, R. V., Ballesteros, E., Sfeir, M. M., & Lui, C., (2020). Ultrasensitive and visual detection of SARS-CoV-2 using all-in-one dual CRISPR/Cas12a assay. *Nat. Commun., 11*, 4711. https://doi.org/10.1038/s41467-020-18575-6.

East-Seletsky, A., O'Connell, M. R., Burstein, D., Knott, G. J., & Doudna, J. A., (2017). RNA targeting by functionally orthogonal type VI-A CRISPR/Cas enzymes. *Mol. Cell, 66*(3), 373–383.e3. https://doi.org/10.1016/j.molcel.2017.04.008.

East-Seletsky, A., O'Connell, M. R., Knight, S. C., Burstein, D., Cate, J. H. D., Tjian, R., & Doudna, J. A., (2016). Two distinct RNase activities of CRISPR-C2c2 enable guide-RNA processing and RNA detection. *Nature, 538*, 270–273. https://doi.org/10.1038/nature19802.

Fang, Y., & Ramasamy, R. P., (2015). Current and prospective methods for plant disease detection. *Biosensors, 5*, 537–561. https://doi.org/10.3390/bios5030537.

Fonfara, I., Richter, H., Bratovič, M., Le Rhun, A., & Charpentier, E., (2016). The CRISPR-associated DNA-cleaving enzyme Cpf1 also processes precursor CRISPR RNA. *Nature, 532*, 517–521. https://doi.org/10.1038/nature17945.

Fozouni, P., Son, S., Díaz, De. L. D. M., Knott, G. J., Gray, C. N., D'Ambrosio, M. V., Zhao, C., et al., (2021). Amplification-free detection of SARS-CoV-2 with CRISPR/Cas13a and mobile phone microscopy. *Cell, 184*(2), 323–333.e9. https://doi.org/10.1016/j.cell.2020.12.001.

Gao, L., Cox, D. B. T., Yan, W. X., Manteiga, J. C., Schneider, M. W., Yamano, T., Nishimatsu, H., Crosetto, N., & Zhang, F., (2017). Engineered Cpf1 variants with altered PAM specificities. *Nat. Biotechnol., 35*, 789–792. https://doi.org/10.1038/nbt.3900.

Gootenberg, J. S., Abudayyeh, O. O., Kellner, M. J., Joung, J., Collins, J. J., & Zhang, F., (2018). Multiplexed and portable nucleic acid detection platform with Cas13, Cas12a and Csm6. *Science, 360*, 439–444. https://doi.org/10.1126/science.aaq0179.

Gootenberg, J. S., Abudayyeh, O. O., Lee, J. W., Essletzbichler, P., Dy, A. J., Verdine, J. J., Donghianichole, N., et al., (2017). Nucleic acid detection with CRISPR/Cas13a/C2c2. *Science, 356*, 438–442. https://doi.org/10.1126/science.aam9321.

Guk, K., Keem, J. O., Hwang, S. G., Kim, H., Kang, T., Lim, E. K., & Jung, J., (2017). A facile, rapid and sensitive detection of MRSA using a CRISPR-mediated DNA FISH method, antibody-like dCas9/sgRNA complex. *Biosens. Bioelectron., 95*, 67–71. https://doi.org/10.1016/j.bios.2017.04.016.

Gutha, L. R., Casassa, L. F., Harbertson, J. F., & Naidu, R. A., (2010). Modulation of flavonoid biosynthetic pathway genes and anthocyanins due to virus infection in grapevine (*Vitis vinifera* L.) leaves. *BMC Plant Biol., 10*, 187. https://doi.org/10.1186/1471-2229-10-187.

Hajian, R., Balderston, S., Tran, T., De Boer, T., Etienne, J., Sandhu, M., Wauford, N. A., et al., (2019). Detection of unamplified target genes via CRISPR–Cas9 immobilized on a graphene field-effect transistor. *Nat. Biomed. Eng., 3*, 427–437. https://doi.org/10.1038/s41551-019-0371-x.

He, S., & Krainer, K. M., (2020). Pandemics of people and plants: Which is the greater threat to food security? *Mol. Plant., 13*, 933–934. https://doi.org/10.1016/j.molp.2020.06.007.

Hille, F., Richter, H., Wong, S. P., Bratovič, M., Ressel, S., & Charpentier, E., (2018). The biology of CRISPR/Cas: Backward and forward, *Cell, 172*, 1239–1259. https://doi.org/10.1016/j.cell.2017.11.032.

Hipper, C., Brault, V., Ziegler-Graff, V., & Revers, F., (2013). Viral and cellular factors involved in phloem transport of plant viruses. *Front. Plant Sci., 4*, 154. https://doi.org/10.3389/fpls.2013.00154.

Hu, B., Wang, Y., Li, N., Zhang, S., Luo, G., & Huang, Z., (2021). Highly convenient and highly specific-and-sensitive PCR using Se-atom modified dNTPs. *Chem. Commun., 57*, 57–60. https://doi.org/10.1039/D0CC06172G.

Hu, F., Liu, Y., Zhao, S., Zhang, Z., Li, X., Peng, N., & Jiang, Z., (2022). A one-pot CRISPR/Cas13a-based contamination-free biosensor for low-cost and rapid nucleic acid diagnostics. *Biosensors & Bioelectronics, 202*, 113994. https://doi.org/10.1016/j.bios.2022.113994.

Huang, M., Zhou, X., Wang, H., & Xing, D., (2018). Clustered regularly interspaced short palindromic repeats/Cas9 triggered isothermal amplification for site-specific nucleic acid detection. *Anal. Chem., 90*, 2193–2200. https://doi.org/10.1021/acs.analchem.7b04542.

Hull, R., (2013). *Plant Virology* (5th ed.). Elsevier: London, UK; Academic Press: London, UK.

Ibrahim, A. U., Al-Turjman, F., Sa'id, Z., & Ozsoz, M., (2020). Futuristic CRISPR-based biosensing in the cloud and internet of things era: An overview. *Multimedia Tools and Applications*, 1–29. https://doi.org/10.1007/s11042-020-09010-5.

Ishino, Y., Krupovic, M., & Forterre, P., (*2018*). History of CRISPR/Cas from encounter with a mysterious repeated sequence to genome editing technology. *J. Bacteriol., 200*(7), e00580-17. https://doi.org/10.1128/JB.00580-17.

Jaime, C., Muchut, S. E., Reutemann, A. G., Gieco, J. O., & Dunger, G., (2019). Morphological changes, alteration of photosynthetic parameters and chlorophyll production induced by infection with alfalfa dwarf virus in *Medicago sativa* plants. *Plant Pathol., 69*, 393–402. https://doi.org/10.1111/ppa.13109.

Jiang, F., & Doudna, J. A., (2017). CRISPR/Cas9 structures and mechanisms. *Annu. Rev. Biophys., 46*, 505–529. https://doi.org/10.1146/annurev-biophys-062215-010822.

Jiao, J., Kong, K., Han, J., Song, S., Bai, T., Song, C., Wang, M., et al., (2021). Field detection of multiple RNA viruses/viroids in apple using a CRISPR/Cas12a-based visual assay. *Plant Biotechnol. J., 19*, 394–405. https://doi.org/10.1111/pbi.13474.

Jinek, M., Chylinski, K., Fonfara, I., Hauer, M., Doudna, J. A., & Charpentier, E., (2012). A programmable dual-RNA–guided DNA endonuclease in adaptive bacterial immunity. *Science, 337*, 816–821. https://doi.org/10.1126/science.1225829.

Kaminski, M. M., Abudayyeh, O. O., Gootenberg, J. S., Zhang, F., & Collins, J. J., (2021). CRISPR-based diagnostics. *Nat. Biomed Eng., 5*, 643–656. https://doi.org/10.1038/s41551-021-00760-7.

Kebschull, J. M., & Zador, A. M., (2015). Sources of PCR-induced distortions in high-throughput sequencing data sets. *Nucleic Acids Res., 43*(21), e143. https://doi.org/10.1093/nar/gkv717.

Kellner, M. J., Koob, J. G., Gootenberg, J. S., Abudayyeh, O. O., & Zhang, F., (2019). Sherlock: Nucleic acid detection with CRISPR nucleases. *Nat. Protoc., 14*, 2986–3012. https://doi.org/10.1038/s41596-019-0210-2.

Kliot, A., Kontsedalov, S., Lebedev, G., Brumin, M., Cathrin, P. B., Marubayashi, J. M., Škaljac, M., et al., (2014). Fluorescence in situ hybridizations (FISH) for the localization of viruses and endosymbiotic bacteria in plant and insect tissues. *J. Vis. Exp., 84*, e51030. https://doi.org/10.3791/51030.

Koo, B., Kim, D., Kweon, J., Jin, C. E., Kim, S. H., Kim, Y., & Shin, Y., (2018). CRISPR/dCas9-mediated biosensor for detection of tick-borne diseases. *Sens. Actuators B Chem., 273*, 316–321. https://doi.org/10.1016/j.snb.2018.06.069.

Koonin, E. V., & Makarova, K. S., (2019). Origins and evolution of CRISPR/Cas systems. *Phil. Trans. R. Soc.*, B3742018008720180087. https://doi.org/10.1098/rstb.2018.0087.

Kostyusheva, A., Brezgin, S., Babin, Y., Vasilyeva, I., Glebe, D., Kostyushev, D., & Chulanov, V., (2021). CRISPR/Cas systems for diagnosing infectious diseases. *Methods, 203*, 431–446. https://doi.org/10.1016/j.ymeth.2021.04.007.

Lander, E. S., (2016). The heroes of CRISPR. *Cell, 164*, 18–28. https://doi.org/10.1016/j.cell.2015.12.041.

Legrand, P., (2015). Biological assays for plant viruses and other graft-transmissible pathogens diagnoses: A review. *EPPO Bull., 45*, 240–251. https://doi.org/10.1111/epp.12222.

Leon, L. M., Mendoza, S. D., & Bondy-Denomy, J., (2018). How bacteria control the CRISPR/Cas arsenal. *Curr. Opin. Microbiol., 42*, 87–95. https://doi.org/10.1016/j.mib.2017.11.005.

Li, L., Li, S., Wu, N., Wu, J., Wang, G., Zhao, G., & Wang, J., (2019a). HOLMESv2: A CRISPR/Cas12b-assisted platform for nucleic acid detection and DNA methylation quantitation. *ACS Synth. Biol., 8*(10), 2228–2237. https://doi.org/10.1021/acssynbio.9b00209.

Li, S. Y., Cheng, Q. X., Liu, J. K., Nie, X. Q., Zhao, G. P., & Wang, J., (2018a). CRISPR/Cas12a has both cis- and trans-cleavage activities on single-stranded DNA. *Cell Res., 28*, 491–493. https://doi.org/10.1038/s41422-018-0022-x.

Li, S. Y., Cheng, Q. X., Wang, J. M., Li, X. Y., Zhang, Z. L., Gao, S., Cao, R. B., et al., (2018b). CRISPR/Cas12a-assisted nucleic acid detection. *Cell Discov., 4*, 1–4. https://doi.org/10.1038/s41421-018-0028-z.

Li, Y., Li, S., Wang, J., & Liu, G., (2019b). CRISPR/Cas systems towards next-generation biosensing, *Trends Biotechnol., 37*, 730–743. https://doi.org/10.1016/j.tibtech.2018.12.005.

Lima, J. A. A., Nascimento, A. K. Q., Radaelli, P., Silva, A. K. F., & Silva, F. R., (2013). A technique combining immunoprecipitation and RT-PCR for RNA plant virus detection. *J. Phytopathol., 162*, 426–433. https://doi.org/10.1111/jph.12208.

Liu, L., Li, X. Y., Ma, J., Li, Z. Q., You, L. L., Wang, J. Y., Wang, M., et al., (2017a). The molecular architecture for RNA-guided RNA cleavage by Cas13a. *Cell, 170*(4), 714–726. e10. https://doi.org/10.1016/j.cell.2017.06.050.

Liu, L., Li, X., Wang, J., Wang, M., Chen, P., Yin, M., Li, J., Sheng, G., & Wang, Y., (2017b). Two distant catalytic sites are responsible for C2c2 RNase activities. *Cell, 168*, 121–134. https://doi.org/10.1016/j.cell.2016.12.031.

Maeder, M. L., Linder, S. J., Cascio, V. M., Fu, Y., Ho, Q. H., & Joung, J. K., (2013). CRISPR RNA–guided activation of endogenous human genes. *Nat. Meth., 10*, 977–979. https://doi.org/10.1038/nmeth.2598.

Mahas, A., Hassan, N., Aman, R., Marsic, T., Wang, Q., Ali, Z., & Mahfouz, M. M., (2021). LAMP-coupled CRISPR/Cas12a module for rapid and sensitive detection of plant DNA viruses. *Viruses, 13*(3), 466. https://doi.org/10.3390/v13030466.

Maheshwari, Y., Vijayanandraj, S., Jain, R. K., & Mandal, B., (2018). Field-usable lateral flow immunoassay for the rapid detection of a macluravirus, large cardamom chirke virus. *J. Virol. Methods, 253*, 43–48. https://doi.org/10.1016/j.jviromet.2017.12.009.

Makarova, K. S., Aravind, L., Grishin, N. V., Rogozin, I. B., & Koonin, E. V., (2002). A DNA repair system specific for thermophilic *Archaea* and bacteria predicted by genomic context analysis. *Nucleic Acids Research, 30*(2), 482–496. https://doi.org/10.1093/nar/30.2.482.

Marraffini, L. A., (2015). CRISPR/Cas immunity in prokaryotes. *Nature, 526*, 55–61. https://doi.org/10.1038/nature15386.

Martinelli, F., Scalenghe, R., Davino, S., Panno, S., Scuderi, G., Ruisi, P., Villa, P., et al., (2015). Advanced methods of plant disease detection: A review. *Agron. Sustain. Dev., 35*, 1–25. https://doi.org/10.1007/s13593-014-0246-1.

Massart, S., Olmos, A., Jijakli, H., & Candresse, T., (2014). Current impact and future directions of high throughput sequencing in plant virus diagnostics. *Virus Res., 188*, 90–96. https://doi.org/10.1016/j.virusres.2014.03.029.

Mauck, K. E., De Moraes, C. M., & Mescher, M. C., (2014). Biochemical and physiological mechanisms underlying effects of *Cucumber mosaic virus* on host-plant traits that mediate transmission by aphid vectors. *Plant Cell Environ., 37*, 1427–1439. https://doi.org/10.1111/pce.12249.

Maxwell, D. J., Partridge, J. C., Roberts, N. W., Boonham, N., & Foster, G. D., (2016). The effects of plant virus infection on polarization reflection from leaves. *PLoS One, 11*, e0152836. https://doi.org/10.1371/journal.pone.0152836.

McGavin, W. J., Cock, P. J. A., & Macfarlane, S. A., (2010). Partial sequence and RT-PCR diagnostic test for the plant rhabdovirus *Raspberry vein chlorosis virus*. *Plant Pathol., 60*, 462–467. https://doi.org/10.1111/j.1365-3059.2010.02387.x.

Mehetre, G. T., Leo, V. V., Singh, G., Sorokan, A., Maksimov, I., Yadav, M. K., Upadhyaya, K., et al., (2021). Current developments and challenges in plant viral diagnostics: A systematic review. *Viruses, 13*(3), 412. https://doi.org/10.3390/v13030412.

Mekuria, G., Ramesh, S., Alberts, E., Bertozzi, T., Wirthensohn, M., Collins, G., & Sedgley, M., (2003). Comparison of ELISA and RT-PCR for the detection of *Prunus necrotic ring spot virus* and *Prune dwarf virus* in almond (*Prunus dulcis*). *J. Virol. Methods, 114*, 65–69. https://doi.org/10.1016/j.jviromet.2003.08.014.

Moeini, P., Afsharifar, A., Homayoonzadeh, M., & Hopkins, R. J., (2020). Plant virus infection modifies plant pigment and manipulates the host preference behavior of an insect vector. *Èntomol. Exp. Appl., 168*, 599–609. https://doi.org/10.1111/eea.12944.

Mohanraju, P., Makarova, K. S., Zetsche, B., Zhang, F., Koonin, E. V., & Van Der, O. J., (2016). Diverse evolutionary roots and mechanistic variations of the CRISPR/Cas systems. *Science, 80*, 353. https://doi.org/10.1126/science.aad5147.

Morange, M., (2015). What history tells us XXXVII. CRISPR/Cas: The discovery of an immune system in prokaryotes. *J. Biosciences., 40*, 221–223. https://doi.org/10.1007/s12038-015-9532-6.

Mutka, A. M., & Bart, R. S., (2015). Image-based phenotyping of plant disease symptoms. *Front. Plant Sci., 5*, 734. https://doi.org/10.3389/fpls.2014.00734.

Myhrvold, C., Freije, C. A., Gootenberg, J. S., Abudayyeh, O. O., Metsky, H. C., Durbin, A. F., Kellner, M. J., et al., (2018). Field-deployable viral diagnostics using CRISPR/Cas13. *Science, 360*, 444–448. https://doi.org/10.1126/science.aas8836.

Naidu, R. A., & Hughes, J. D. A., (2001). Methods for the detection of plant virus diseases. In: *Proceedings of the Conference on Plant Virology in Sub Saharan Africa Organized by IITA* (pp. 233–253). Ibadan, Nigeria.

Nakane, P. K., & Pierce, G. B., (1966). Enzyme-labeled antibodies: Preparation and application for the localization of antigens. *J. Histochem. Cytochem., 14*, 929–931. https://doi.org/10.1177/14.12.929.

Nalefski, E. A., Patel, N., Leung, P. J. Y., Islam, Z., Kooistra, R. M., Parikh, I., Marion, E., et al., (2021). Kinetic analysis of Cas12a and Cas13a RNA-Guided nucleases for development of improved CRISPR-Based diagnostics. *iScience*, *24*, 102996. https://doi.org/10.1016/j.isci.2021.102996.

Notomi, T., Okayama, H., Masubuchi, H., Yonekawa, T., Watanabe, K., Amino, N., & Hase, T., (2000). Loop-mediated isothermal amplification of DNA. *Nucleic Acids Res., 28*, E63. https://doi.org/10.1093/nar/28.12.e63.

Pecman, A., Kutnjak, D., Gutiérrez-Aguirre, I., Adams, I., Fox, A., Boonham, N., & Ravnikar, M., (2017). Next generation sequencing for detection and discovery of plant viruses and viroids: Comparison of two approaches. *Front. Microbiol., 8*, 1998. https://doi.org/10.3389/fmicb.2017.01998.

Ramachandran, V., Weiland, J. J., & Bolton, M. D., (2021). CRISPR-based isothermal next-generation diagnostic method for virus detection in sugarbeet. *Front Microbiol., 12*, 1760. https://doi.org/10.3389/fmicb.2021.679994.

Rubio, L., Galipienso, L., & Ferriol, I., (2020). Detection of plant viruses and disease management: Relevance of genetic diversity and evolution. *Front. Plant. Sci., 11*, 1092. https://doi.org/10.3389/fpls.2020.01092.

Rudkin, G. T., & Stollar, B. D., (1977). High resolution detection of DNA–RNA hybrids in situ by indirect immunofluorescence. *Nature, 265*, 472–473. https://doi.org/10.1038/265472a0.

Rusk, N., (2019). Spotlight on Cas12. *Nat. Methods., 16*, 215. https://doi.org/10.1038/s41592-019-0347-5.

Saiki, R. K., Scharf, S., Faloona, F., Mullis, K. B., Horn, G. T., Erlich, H. A., & Arnheim, N., (1985). Enzymatic amplification of beta-globin genomic sequences and restriction site analysis for diagnosis of sickle cell anemia. *Science, 230*, 1350–1354. https://doi.org/10.1126/science.2999980.

Savary, S., Ficke, A., Aubertot, J. N., & Hollier, C., (2012). Crop losses due to diseases and their implications for global food production losses and food security. *Food Secur., 4*, 519–537. https://doi.org/10.1007/s12571-012-0200-5.

Scagliusi, S. M., Basu, S., Gouvêa, J. A. D., & Vega, J., (2009). Comparison of two diagnostic methods for evaluation of *Sugarcane yellow leaf virus* concentration in Brazilian sugarcane cultivars. *Funct. Plant Sci. Biotechnol., 3*, 26–30.

Schermer, B., Fabretti, F., Damagnez, M., Di Cristanziano, V., & Heger, E., (2020). Rapid SARS-CoV-2 testing in primary material based on a novel multiplex RT-LAMP assay. *PloS One, 15*. https://doi.org/10.1371/journal.pone.0238612.

Schmidt, C., Pacher, M., & Puchta, H., (2019). DNA break repair in plants and its application for genome engineering. In: Kumar, S., Barone, P., & Smith, M., (eds.), *Transgenic Plants* (pp. 237–266). Humana Press, New York, USA. https://doi.org/10.1007/978-1-4939-8778-8_17.

Selvarajan, R., Kanichelvam, P. S., Balasubramanian, V., & Subramanian, S. S., (2020). A rapid and sensitive lateral flow immunoassay (LFIA) test for the on-site detection of *Banana bract mosaic virus* in banana plants. *J. Virol. Methods, 284*, 113929. https://doi.org/10.1016/j.jviromet.2020.113929.

Seo, J. K., Kim, M. K., Kwak, H. R., Choi, H. S., Nam, M., Choe, J., Choi, B., et al., (2018). Molecular dissection of distinct symptoms induced by *Tomato chlorosis virus* and Tomato yellow leaf curl virus based on comparative transcriptome analysis. *Virol., 516*, 1–20. https://doi.org/10.1016/j.virol.2018.01.001.

Shah, S. A., Erdmann, S., Mojica, F. A. M., & Garrett, R. A., (2013). Protospacer recognition motifs: Mixed identities and functional diversity. *RNA Biology*., *10*(5), 891–899. https://doi.org/10.4161/rna.23764.

Shahid, M. S., Sattar, M. N., Iqubal, Z., Raza, A., & Al-Sadi, A. M., (2021). Next-generation sequencing and the CRISPR/Cas nexus: A molecular plant virology perspective. *Front. Microbiol., 11*. https://doi.org/10.3389/fmicb.2020.609376.

Shargil, D., Zemach, H., Belausov, E., Lachman, O., Kamenetsky, R., & Dombrovsky, A., (2015). Development of a fluorescent in situ hybridization (FISH) technique for visualizing CGMMV in plant tissues. *J. Virol. Methods., 223*, 55–60. https://doi.org/10.1016/j.jviromet.2015.07.014.

Sharma, K. S., Gupta, O. P., Pathaw, N., Sharma, D., Maibam, A., Sharma, P., Sanasam, J., et al., (2021). CRISPR/Cas-led revolution in diagnosis and management of emerging plant viruses: New avenues toward food and nutritional security. *Frontiers in Nutrition, 8*. https://doi.org/10.3389/fnut.2021.751512.

Shmakov, S., Abudayyeh, O. O., Makarova, K. S., Wolf, Y. I., Gootenberg, J. S., Semenova, E., Minakhin, L., et al., (2015). Discovery and functional characterization of diverse class 2 CRISPR/Cas systems. *Mol Cell., 60*(3), 385–397. https://doi.org/10.1016/j.molcel.2015.10.008.

Shmakov, S., Smargon, A., Scott, D., Cox, D., Pyzocha, N., Yan, W., Abudayyeh, O. O., et al., (2017). Diversity and evolution of class 2 CRISPR/Cas systems. *Nat. Rev. Microbiol., 15*(3), 169–182. https://doi.org/10.1038/nrmicro.2016.184.

Smargon, A. A., Cox, D. B., Pyzocha, N. K., Zheng, K., Slaymaker, I. M., Gootenberg, J. S., Abudayyeh, O. A., et al., (2017). Cas13b is a type VI-B CRISPR-associated RNA-guided RNase differentially regulated by accessory proteins Csx27 and Csx28. *Mol. Cell, 65*(4), 618–630.e7. https://doi.org/10.1016/j.molcel.2016.12.023.

Smith, K. M., (1977). Introduction. In: *Plant Viruses* (pp. 1–5). Springer: Dordrecht, The Netherlands.

Sola, C., Abadia, E., Le Hello, S., & Weill, F. X., (*2015*). High-throughput CRISPR typing of *Mycobacterium tuberculosis* complex and *Salmonella enterica* serotype *Typhimurium*. *Methods Mol. Biol. 2015,* 1311. https://doi.org/10.1007/978-1-4939-2687-9_6.

Stout, E., Klaenhammer, T. R., & Barrangou, R., (2017). CRISPR-Cas technologies and applications in food bacteria. *Annual Rev. of Food Sci. Tech., 8*, 413–437. https://doi.org/10.1146/annurev-food-072816-024723.

Strich, J. R., & Chertow, D. S., (2019). CRISPR/Cas biology and its application to infectious diseases. *J. Clin. Microbiol., 57*, e01307–18. https://doi.org/0.1128/JCM.01307-18.

Symington, L. S., & Gautier, J., (2011). Double-strand break end resection and repair pathway choice. *Annu. Rev. Genet., 45*, 247–271. https://doi.org/10.1146/annurevgenet-110410-132435.

Tang, Y., & Fu, Y., (2018). Class 2 CRISPR/Cas: An expanding biotechnology toolbox for and beyond genome editing. *Cell Biosci., 8*, 59. https://doi.org/10.1186/s13578-018-0255-x.

Torrance, L., & Jones, R. A. C., (1981). Recent developments in serological methods suited for use in routine testing for plant viruses. *Plant Pathol., 30*, 1–24. https://doi.org/10.1111/j.1365-3059.1981.tb01218.x.

Vu, G. T., Cao, H. X., Fauser, F., Reiss, B., Puchta, H., & Schubert, I., (2017). Endogenous sequence patterns predispose the repair modes of CRISPR/Cas9–induced DNA double–stranded breaks in *Arabidopsis thaliana*. *Plant J., 92*, 57–67. https://doi.org/10.1111/tpj.13634.

Wang, B., Wang, R., Wang, D., Wu, J., Li, J., Wang, J., Liu, H., & Wang, Y., (2019). Cas12aVDet: A CRISPR/Cas12a-based platform for rapid and visual nucleic acid detection. *Anal. Chem., 91*, 12156–12161. https://doi.org/10.1021/acs.analchem.9b01526.

Wang, Q., Zhang, B., Xu, X., Long, F., & Wang, J., (2018). CRISPR-typing PCR (ctPCR), a new Cas9-based DNA detection method. *Sci. Rep., 8*, 1–13. https://doi.org/10.1038/s41598-018-32329-x.

Wang, X., Xiong, E., Tian, T., Cheng, M., Lin, W., Sun, J., & Zhou, X., (2020). Clustered regularly interspaced short palindromic repeats/Cas9-mediated lateral flow nucleic acid assay. *ACS Nano, 14*(2), 2497–2508. https://doi.org/10.1021/acsnano.0c00022.

Wang, Y. M., Ostendorf, B., Gautam, D., Habili, N., & Pagay, V., (2022). Plant viral disease detection: From molecular diagnosis to optical sensing technology—A multidisciplinary review. *Remote Sens., 14*, 1542. https://doi.org/10.3390/rs14071542.

Wolfenden, R., Henderson, C., & Dennien, S., (2018). Innovating new virus diagnostics and planting bed management in the Australian sweet potato industry. *Hort Innovation*. North Sydney, Australia.

Wu, Q., Habili, N., Constable, F., Al Rwahnih, M., Goszczynski, D. E., Wang, Y., & Pagay, V., (2020). Virus pathogens in Australian vineyards with an emphasis on shiraz disease. *Viruses, 12*, 818. https://doi.org/https://doi.org/10.3390/v12080818.

Xu, X., Luo, T., Gao, J., Lin, N., Li, W., Xia, X., & Wang, J., (2020). CRISPR-assisted DNA detection, a novel dCas9-based DNA detection technique. *The CRISPR Journal., 3*(6), 487–502. https://doi.org/10.1089/crispr.2020.0041.

Yamano, T., Nishimasu, H., Zetsche, B., Hirano, H., Ian, M., Li, Y., Fedorova, I., et al., (2016). Crystal structure of Cpf1 in complex with guide RNA and target DNA. *Cell, 165*(4), 949–962. https://doi.org/10.1016/j.cell.2016.04.003.

Yan, F., Wang, W., & Zhang, J., (2019). CRISPR/Cas12 and Cas13: The lesser known siblings of CRISPR/Cas9. *Cell Biol. Toxicol., 35*, 489–492. https://doi.org/10.1007/s10565-019-09489-1.

Yin, L., Yin, M. S., Ye, S., Liu, G., & Ma, L., (2021). CRISPR/Cas based virus detection: Recent advances and perspectives. *Biosensors Bioelectr., 193*, 113541. https://doi.org/10.1016/j.bios.2021.113541.

Zetsche, B., Gootenberg, J. S., Abudayyeh, O. O., Slaymaker, I. M., Makarova, K. S., Essletzbichler, P., Volz, S. E., et al., (2015). Cpf1 is a single RNA-guided endonuclease of a class 2 CRISPR/Cas system. *Cell, 163*, 759–771. https://doi.org/10.1016/j.cell.2015.09.038.

Zhang, B., Wang, Q., Xu, X., Xia, Q., Long, F., Li, W., Shui, Y., et al., (2018). Detection of target DNA with a novel Cas9/sgRNAs-associated reverse PCR (CARP) technique. *Anal. Bioanal. Chem., 410*, 2889–2900. https://doi.org/10.1007/s00216-018-0873-5.

Zhang, Y. M., Zhang, Y., & Xie, K., (2020). Evaluation of CRISPR/Cas12a-based DNA detection for fast pathogen diagnosis and GMO test in rice. *Mol. Breeding, 40*, 11. https://doi.org/10.1007/s11032-019-1092-2.

## CHAPTER 4

# Identification and Diagnosis of Plant Viruses Using Different Strategies

A. A. KHAN,[1] S. AHMAD,[2] A. SAJJAD,[3] and A. K. TIWARI[4]

[1]*Department of Botany, Gandhi Faiz-E-Aam College, Shahjahanpur (Affiliated with M.J.P.R.U. Bareilly), Uttar Pradesh, India*

[2]*Central Lab Sugarcane Research Institute, Shahjahanpur, Uttar Pradesh, India*

[3]*Department of Botany, Gandhi Faiz-E-Aam College, Shahjahanpur (Affiliated with M.J.P.R.U. Bareilly), Uttar Pradesh, India*

[4]*Central Laboratory, Sugarcane Research Institute, Shahjahanpur, Uttar Pradesh, India*

## ABSTRACT

Plant viruses are critical for plants, causing significant economic losses in Indian agriculture. New viruses are frequently discovered in plants. This research explores the different types of viruses found in plants and their characteristics. The identification of plant viruses has been done by focusing on various tests, including precipitation/agglutination tests, enzyme-linked immunosorbent assays (ELISA), and immunoblotting. Additionally, this research reviews the use of molecular tools and techniques such as PCR, RT-PCR, and Q-PCR for diagnosing plant diseases. Virus diagnostic methods are categorized into two aspects: biological aspects related to the virus's interaction with its host and/or vector (e.g., symptomatology and transmission assays), and intrinsic virus properties (coat protein (CP) and nucleic acid). This research demonstrates how these diagnostic approaches offer improved flexibility, sensitivity, and

Detection and Management of New and Emerging Mystery Plant Virus Sources.
Alireza Golnaraghi, Rajarshi Kumar Gaur, and Rakesh Kumar Verma (Eds.)

specificity for the quick identification of plant viral infections. Therefore, this research highlights the different types of plant viruses and their diagnostic techniques, which can enhance the sustainability of agriculture and crops.

## 4.1 INTRODUCTION

Viruses that infect plants are known as plant viruses. Viruses are parasitic parasites that can only develop and multiply in the presence of a live host. Viruses enter plant cells through plasmodesmata and spread through the plant via phloem. Plant viruses consist of two components: a protein coat and a nucleic acid core. The nucleic acid is the most infectious part of the virus; once inside the plant cell, the virus sheds its protein coat and replicates independently. While viruses cannot be transmitted between humans and plants, physical contact between humans and plants can spread plant viral infections. Infected seeds, grafting, wind, splashing, pollination, and sap leakage can all transmit viruses. Unlike human cells, plant cells are unable to recover from viral infections while still alive. Plant viruses significantly reduce agricultural output, resulting in a significant economic impact on farmers. The tobacco mosaic virus (TMV) was the first virus to be identified. Viruses that infect plants are classified into 73 genera and 49 families.

Virus particles are so small that only an electron microscope (EM) can see them. The majority of plant viruses have a rod-like or isometric (polyhedral) shape. Examples of long flexible rod-shaped viruses include potato virus (PVY), cucumber mosaic virus (CMV), and tomato mosaic virus (TMV). A protein sheath or coat, either RNA or DNA, surrounds the nucleic acid core of a virus, known as the capsid. All viruses harmful to both humans and animals have a membrane surrounding their capsid, which allows for easy mobilization of virus cells. The internal parts of cells are protected by a strong cell wall, so plant viruses must enter through a wound. For example, when plants grow lateral roots, they may create wounds themselves (Pallás et al., 2018). Agronomic or horticultural practices, as well as other mechanical methods, fungal, nematode, or parasitic plant diseases, or insects, can all cause these wounds. In rare situations, the organism that created the wound may be able to host and spread the virus. Infection spreads through the use of microorganisms as vectors. The two most frequent mechanisms of plant virus transmission are mechanical transmission and insect vector transmission (Rubio et al., 2020). Artificial plant reproduction, such as budding, grafting, or cuttings, is a major route for viral disease transmission. Plant virologists use grafting and budding to research viral propagation and identification. The seedling offspring of a virus-infected plant is generally, but not always, virus-free, depending on the plant species and viral type (James et al., 2006). Insects are the primary vectors of viral transmission in the wild. The most prevalent and commercially important plant virus vectors include aphids, planthoppers, leafhoppers, whiteflies, and mealybugs. All of them have piercing sucking mouthparts. Pollen grains and seeds can also be used to transfer plant viruses (Naidu, 2003) (Figure 4.1).

**FIGURE 4.1** Tomato plant infected with TMV.

## 4.2 STRUCTURE OF PLANT VIRUSES

Viruses are such minute creatures that can only be observed with high magnification. A virus's shape is dictated by the proteins that encase its DNA. When virus particles come into contact with one another, they self-assemble. Rod-shaped viruses account for more than half of all plant viruses. Although the size of the particle is largely governed by the genome, it is normally between 300 and 500 nanometers in height and has a radius of 7–10 nm. Protein monomers can be stacked around the periphery in a circular pattern to form a disc. The discs are stacked near the viral genome, forming a tube with a central area for the nucleic genome. Isometric particles are the second most common type of plant virus structure, measuring 25–50 nm in diameter. Some viruses have dual-coated proteins that are structurally linked to create an inverted cone structure (Hema & Konakalla, 2021). The Geminiviridae family consists of three genera of fragments that resemble two connected geometric entities. In addition to the envelope protein, a small number of plant viruses have a lipid envelope. This is formed when uncontrolled proliferation buds off from the plant's cell membrane (Massart et al., 2014) (Figure 4.2).

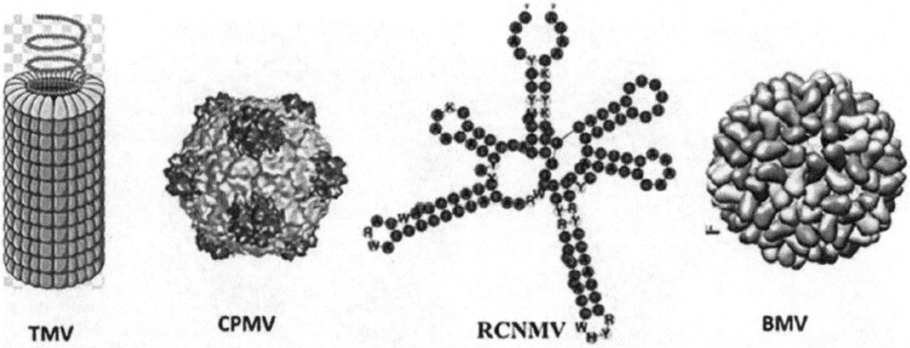


**FIGURE 4.2** Structure of plant virus.

## 4.3 ATTACKING MECHANISM OF PLANT VIRUSES

Viruses are necessary parasites, meaning they cannot survive or proliferate without the help of a living host. When a viral particle enters a damaged cell, it sheds its protein sheath, and the nucleic acid in the virus particle promotes replication of the virus particle and related proteins, resulting in the formation of additional virus particles (García-Arenal & McDonald, 2003) (Figure 4.3).

Plant viruses move from cell to cell via plasmodesmata, which are cytoplasmic "bridges" that connect cells, and then spread to infected plants through the phloem. Plant viruses cause sickness by disrupting normal cellular processes and redirecting photosynthates as they grow, although the intricacies of plant virus reproduction are extensive and beyond the scope of this information. It is important to note that many plant species can be infected with the virus without showing any signs or symptoms

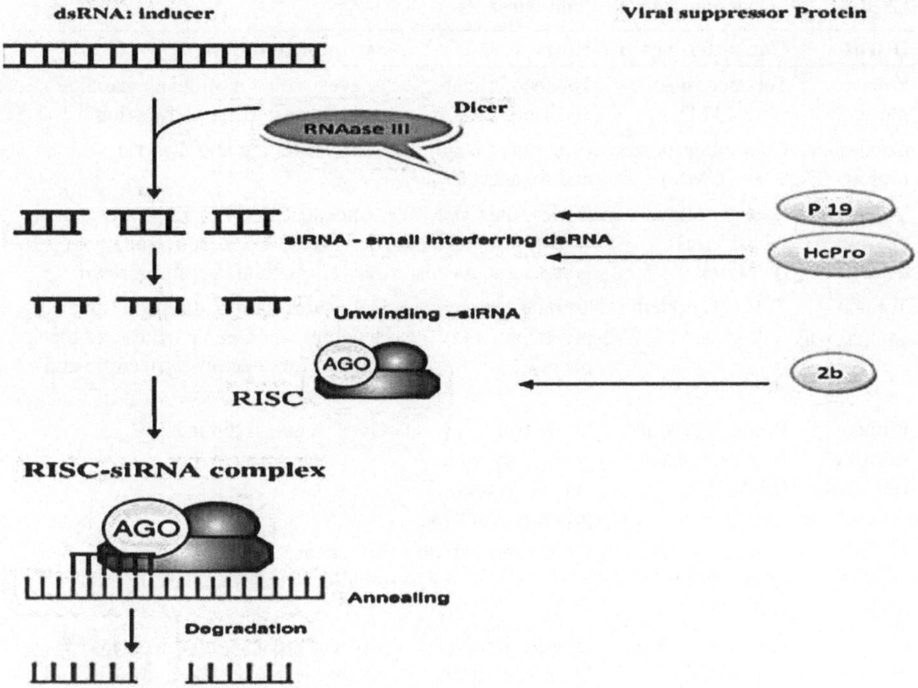


**FIGURE 4.3** Attacking mechanism of virus on the plant cell.

(Rettcher et al., 2015). Infections that have not yet manifested are referred to as latent infections. In their natural state, many viruses, such as CMV and CPMV, exist as a complex of multiple component particles, each with its own nucleic acid core. Multi-component viruses require the presence of all components in a plant to infect and replicate. For instance, tobacco ring spot virus affects the buds in soybeans (Table 4.1).

## 4.4 IDENTIFICATION OF PLANT VIRUS

To determine the type of plant disease, it is important to identify the pathogen and the corresponding effects caused by it, followed by the sources of transmission. Therefore, various researchers have proposed a series of tests to identify the class of plant virus and other characteristics of the disease.

**TABLE 4.1** Characteristics of Plant Diseases

| Disease | Causative Agent | Hosts | Symptoms and Signs |
|---|---|---|---|
| Tobacco mosaic | Tobacco mosaic virus (TMV) | Tobacco, tomato, and weeds | Leaves with a mottled appearance (mosaic pattern) and dwarfing. |
| Cucumber mosaic | Cucumber mosaic virus (CMV) | Cucumber, bean, and tobacco | Leaf yellowing and dwarfing. |
| Barley yellow dwarf | Barley yellow dwarf virus (BYDV) | Barley, oats, rye, wheat; also pasture grasses and weeds. | Concentric necrotic rings are seen on the leaves; the necrotic zone begins to get yellow then becomes brown. |
| Tomato spotted wilt | Tomato spotted wilt virus (TSWV) | Tomato, pepper, pineapple, and peanut | Delayed foliation, diseased leaves develop pale green patches and black rings before becoming necrotic and falling off. |
| Prunus necrotic ring spot | Prunus necrotic ring spot virus (PNRV) | Stone fruits, e.g., cherry, almond, peach, apricot, plum, and others. | Growth gets inhibited |
| Potato spindle tuber | Potato spindle tuber viroid (PSTV) | Potato and tomato | Tubers are spindle-shaped and smaller in size than healthy tubers. |
| Citrus exocortis | Citrus exocortis viroid (CEV) | Orange, lemon, lime, and other citrus plants. | Vertical breaks with thin strips of partly loosened bark, and a cracked, scaly look are all signs of infection. |

### *4.4.1 CHLOROPLAST AGGLUTINATION AND TUBE PRECIPITIN TESTS*

This test was first conducted and gained recognition in the 1960s to aid in the detection of filamentous viruses, such as Potato virus, during the growth of potato seeds. A drop of the plant's extract in its unfiltered form, along with an antiserum of the respective virus, is observed under a waxed microscope slide. Positive reactions are observed when aggregates are generated by chloroplast agglutination in response to the interaction between the virus and the antibody. However, this test has a major limitation in detecting isometric viruses. To overcome this limitation, the tube-precipitin test was chosen as it is equally effective in identifying viruses of various morphologies (Gaafar & Ziebell, 2020). The standard operating procedure involves using 7 mm thin-walled glass tubes with 0.5 mL of antiserum diluted to form a saline compound, with a maximum of three dilutions. To obtain accurate results, the volume of the original virus or plant extract should be equivalent and diluted accordingly, depending on the virus genus (Naidu & Hughes, 2001) (Figure 4.4).

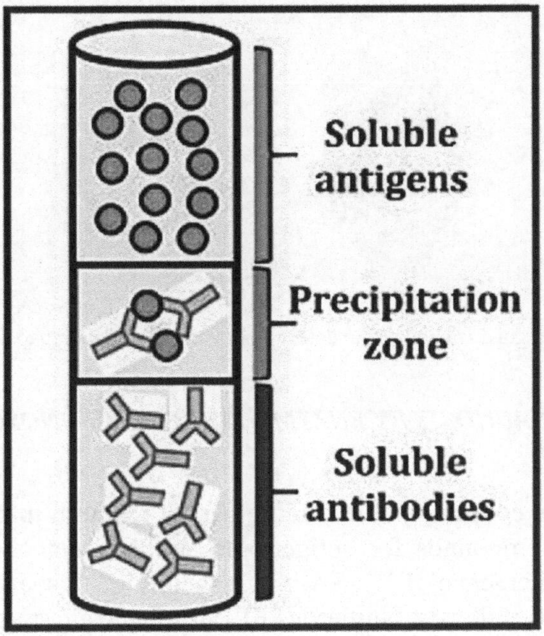


**FIGURE 4.4** Tube precipitin tests.

### 4.4.2 AGAR-GEL DOUBLE DIFFUSION TESTS

Due to the ease of conducting such tests in the semisolid phase with a minimum of 0.7% to 1.0% agar gel, this test is commonly known as the Ouchterlony test. It is notably used for the identification and detection of isometric viral infections (Kutnjak et al., 2021). In this test, the antibody is placed in the central well of a Petri dish gel, while the antigen, in its original form, such as an untouched virus sample or infected plant symptomatic parts, is placed in the four to eight outside wells as required (Tatineni et al., 2020).

The advantage of this test is that it can also be used to titrate virus formation and determine the virus's serological family. Additionally, it can recognize the effect as a unique precipitin area located halfway between the antibody and the antigen wells. However, a disadvantage is that filamentous viruses do not respond to agar gel due to their chemical nature and other factors (Figure 4.5).

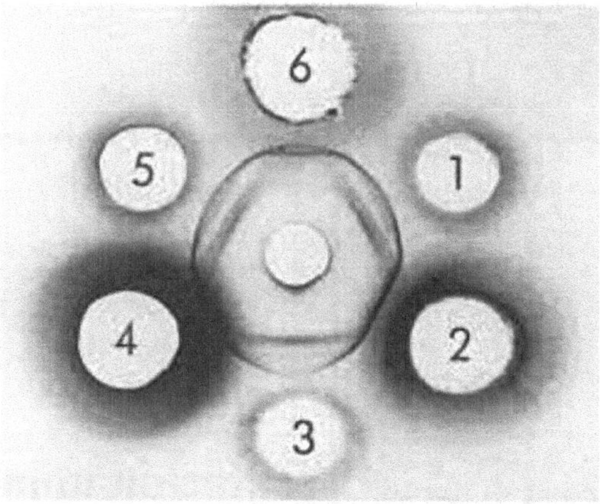


**FIGURE 4.5** Agar gel diffusion test to detect plant virus effect leading to cell wall disruption.

### *4.4.3 SERODIAGNOSIS BY ENZYME-LINKED IMMUNOSORBENT ASSAY (ELISA)*

Enzyme-protein coupling is the leading part of research that represents the utilization of compounds for antigen and antibody detection, paving the basis for the emergence of ELISA. Since then, virus disease serodiagnosis has evolved tremendously. Numerous ELISA diagnostic starting procedures are not applicable now since ELISA is highly sensitive, consumes minimal quantities of antiserum, delivers accurate results, and is an important test for screening large amounts of data that can be semi-automated (Dodds et al., 1984). ELISA is sometimes used to distinguish other analogous viruses and strains by utilizing virus or strain-specific monoclonal antibodies, respectively. The ongoing availability of high-quality antisera, as well as their validation and cost, has been a major issue in serodiagnosis, especially for viruses that emerge in low quantities, such as begomoviruses, closteroviruses, and luteoviruses (Sankaran et al., 2010). The use of monoclonal antibodies (MABs) and the functioning of antisera to target cloned viral coat proteins (CPs) can address the demand for high-quality antisera against such viruses, but expense and need remain a concern.

#### *4.4.3.1 DOUBLE ANTIBODY SANDWICH (DAS) ENZYME-LINKED IMMUNOSORBENT ASSAY (ELISA)*

The double antibody sandwich ELISA (DAS-ELISA) is a regulatory element used to bind proteins with specific viruses and antibodies. It is commonly performed on 96-well polystyrene microtiter plates. To conduct a DAS-ELISA, follow the procedures below:

1. **Trapping Antibodies:** Isolate IgG from a monoclonal antibody against the virus to be identified and place it in the wells of the microtiter plate, where it binds to the well walls. Treat the surfaces of the walls not coated with IgG with a non-reacting protein solution, such as bovine serum albumin.
2. **Antigen Binding:** Introduce the antigen of the virus under investigation, which will bind to the previously coated IgGs. This can be done using a plant extract or a purified formulation of the symptomatic portion (Biswas et al., 2016).
3. **Immunoprobe:** Introduce IgGs conjugated with the commonly used enzyme alkaline phosphatase, which will react with the virus of interest.
4. **Substrate Reaction:** Introduce the substrate para-nitrophenyl phosphate, which will cause color to develop during the enzyme's action.
5. **Color Evaluation:** Terminate the enzyme-substrate reaction using sodium hydroxide. The color intensity can be visually evaluated on a preset scale or using an ELISA reader (NaOH).

DAS-ELISA is often used not only for viral detection but also for virus quantification. Numerous ELISA variations have been developed for a range of applications.

#### *4.4.3.2 AFFIRMER PROTEIN-BASED ENZYME-LINKED IMMUNOSORBENT ASSAY (ELISA)*

AFPs are unique antibody mimics that adhere to antigens like antibodies. Chemo-recognition sites are positioned within two variable loops, each containing up to nine amino acids. Unlike antibodies, they are easy to manufacture and stable, making them suitable diagnostic agents. AFP-based DIBA has been proven to identify viral infections in plants (Dodds et al., 1984). Using a biotin-streptavidin linkage ELISA, it detected the virus in infected leaf samples without interfering with the four other viruses. Unlike DAS-ELISA, which involves incubation, conjugated homologous antibody, and substrate, Direct ELISA detects viral particles directly on microliter plates (Gergerich & Dolja, 2006). ELISA for viral detection is called indirect

ELISA since it does not need conjugated homologous antibodies. To detect the virus, the unconjugated homologous antibody generated in the first animal (rabbit or other) must be followed by the conjugated antibody produced in the second animal (goat or other). It is made up of three antibodies: a trapping antibody, a detecting antibody (from the same or a different antiserum), and an immunoprobe (specific to the Fc part of the detecting IgG produced in another animal). Less sensitive than DAS-ELISA. Virus-specific conjugates are not required.

## 4.5 DIAGNOSIS OF DISEASES DUE TO PLANT VIRUS

With the introduction of nucleic acid-based (NA-based) technologies, the demand for precise and sensitive detection of plant viruses and viroids has been a major concern. Since viruses were discovered to be nucleoproteins, nucleic acid analysis has been used to differentiate viruses based on their number, type, and nucleotide ratio. Plant viruses are classified into five classes based on their nucleotide ratio. The advent of dot blot hybridization and gel electrophoresis has made nucleo-diagnosis (NA-based diagnostics) prominent, especially for begomoviruses. In 1986, Kary Mullis et al. developed the PCR process, which earned Kary Mullis the Nobel Prize in Chemistry in 1993.

### *4.5.1 POLYMERASE CHAIN REACTION (PCR)*

Due to its high precision, ease of use, and broad application, PCR has established itself as a critical diagnostic tool. It is used for the large-scale diagnosis and classification of viruses, as well as for promoting the discovery of new viruses. PCR is routinely employed to detect and identify plant-infecting DNA and RNA viruses. The detection of RNA viruses begins with the transcription of the virus into complementary DNA (cDNA), which is then amplified. The reverse transcription polymerase chain reaction (RT-PCR) has been extensively used for diagnosing plant diseases (Jeong et al., 2014). Immune capture polymerase chain reaction (IC-PCR) and immune precipitation polymerase chain reaction (IP-PCR) have been shown to be successful in detecting viruses in trace amounts. In IC-PCR, virus particles trapped by antibodies are utilized for PCR, while in IP-PCR, viral antibody precipitate is used for PCR. Both IC-PCR and IP-PCR improve detection efficiency. To avoid the need for nucleic acid reactions and a heat cycler in the PCR test, a portable fluorescence-based real-time isothermal reverse transcription recombine polymerase amplification assay system has been developed. This system can detect viruses in leaf extracts. This technology could be used for on-the-job diagnostics and certification. PCR-based diagnostics could potentially play an important role in smartphone-based agricultural aid for monitoring and managing serious viral infections, such as Cassava mosaic virus. While PCR can also be used for quantification, the introduction of real-time quantitative polymerase chain reaction (Q-PCR) has added a new dimension to pathogen detection and quantification. However, the high cost and reliance on complex equipment and reagents limit the diagnostic use of real-time Q-PCR (Zaitlin & Palukaitis, 2000). Nested PCR (nPCR) was developed to improve the detection of viruses present in trace amounts. In these cases, the amplicons from the initial PCR cycle are used in a second PCR cycle to enhance detectability. Viruses such as Prunus necrotic ring spot virus, prune dwarf virus (PDV), plum pox virus (PPV), and citrus tristeza virus (CTV) have been identified using nested PCR (Figure 4.6).

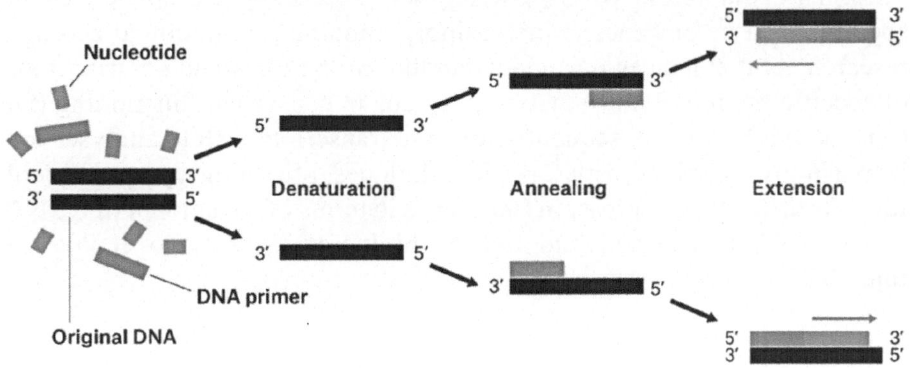


**FIGURE 4.6** Steps involved in PCR test.

In a thermal cycle, PCR aids in amplifying or generating identical DNA sequences. The synthesis of two primers with 10-to-20-nt sequences that complement the two ends of the detected segment of DNA is easy to conduct. PCR plays a major role in the performance of gene-related modification tests and the strain classification process. Multiplex PCR (M-PCR) reliably detects numerous infections. This method has been used to discover an RNA virus in apples, DNA viruses in rice, banana, tomato, citrus, and a viral phytoplasma infection in tomato. However, multiplex RT-PCR is more successful when detecting two to five pathogens. This is the most common multiplex diagnostic approach.

### 4.5.2 *QUANTITATIVE-PCR (Q-PCR)*

In microbiology laboratories, this procedure is used to determine the number of genes and transcripts in a desired sample. The specificity of any Q-PCR test is determined by the structure of the primers used, which helps identify taxonomic and functional gene markers present in the sample (Joo-Jin, 2014). Q-PCR has been proven to be highly precise and accurate in tracking phylogenetic and functional gene changes over time and space, across various environmental or experimental conditions (Maliogka et al., 2018). The quantitative data collected can also be used to compare variations in gene abundances and/or levels of gene expression (measured in terms of transcript quantities) with variations in abiotic or biotic variables, as well as variations in biological activities and process rates. The availability of Q-PCR datasets describing the abundance of specific bacteria or genes is increasingly important in microbial ecology research, as it enhances our understanding of the roles and contributions of specific microbial and functional groups in ecosystem functioning (Di Carli et al., 2012). Consequently, reverse transcription (RT) analyses are increasingly combined with Q-PCR techniques, providing a powerful tool for measuring gene expression (measured in terms of the amount of r-RNA and m-RNA transcripts) and linking biological activity to ecological function.

## 4.6 IMMUNIZATION MEASURES

The most widely used immunization approach is backcrossing, also known as plant breeding, to introduce resistance mechanisms from domesticated or wild species into susceptible crops. Resistant dominant genes that encode local components necessary for viral infection generate passive resistance (Massart et al., 2019). Active resistance is created by dominant alleles that encode resistance proteins capable of recognizing specific patterns or structures in the viral genome (virulence determinants) and eliminating intracellular pathogens through a hypersensitive response (HR), thereby limiting disease spread to adjacent cells and preventing bacterial overgrowth.

Plant breeders often aim for absolute resistance, where the virus cannot propagate throughout the plant. To evaluate resistance in germ lines and cultivars, ELISA and molecular hybridization techniques have been employed due to the large number of plants that need to be investigated. If achieving comprehensive tolerance is not possible, breeding for partial or complete resistance (reducing viral accumulation) or sensitivity (minimizing virus damage without affecting virus reproduction) may be considered. The most effective method for assessing relative resistance levels is Q-PCR. Time-course testing can identify both relative resistance and tolerance, which are measured by observing symptoms as an indicator of damage (Yadav & Khurana, 2016). Survival studies can also be conducted to determine levels of resistance or tolerance based on the likelihood of remaining healthy or experiencing no symptoms.

Many crops and viruses avoid producing resistant cultivars due to a lack of resistance mechanisms detected in genetically comparable kin. Another approach would be to utilize artificial evolution to increase the selectivity or spectrum of existing resistant strains, allowing them to provide barriers to change other viruses or viral strains (Cunniffe et al., 2015). This strategy uses a high-error-rate PCR to create a huge number of arbitrary mutants out of resistance determinants, followed by the selection of those variants that display the necessary resistance qualities.

## 4.7 CONCLUSION

Plant viruses create significant economic losses and imperil the long-term viability of agriculture. International commerce, climate change, and the rapid evolution of viruses all contribute to the emergence of new viral diseases. Disease management is achieved through two methods: immunization (genetic immunity achieved through plant breeding, plant transformation, cross-protection, or other mechanisms) and prophylaxis (viral prevention) (using isolation or control of vectors). It is crucial to promptly and accurately identify the causative agent for disease treatment. Virus diagnosis involves classifying a virus that has infected a plant sample into a species, which is a group of viruses with similar characteristics. Diagnostic specificity can range from higher taxonomic levels such as genera or species to lower taxonomic values. Diagnostic techniques should be optimized for reliability by accurately identifying the specific class of virus and distinguishing it from other classes of viruses.

## KEYWORDS


- **agricultural sustainability**
- **diagnosis tests**
- **diagnostic methods**
- **enzyme-linked immunosorbent assay**
- **plant diagnosis**
- **plant viruses**
- **quantitative-PCR**
- **reverse transcriptase PCR**
- **virus identification**


## REFERENCES


Biswas, K., Tarafdar, A., & Kumar, R., (2016). Detection of plant viruses for their management: Recent trends. *Plant Pathogen Interaction: Recent Trends* (pp. 77–92). Sharma Publications & Distributors, India.

Cunniffe, N. J., Koskella, B., Metcalf, C. J., Parnell, S., Gottwald, T. R., & Gilligan, C. A., (2015). Thirteen challenges in modeling plant diseases. *Epidemics, 10*, 6–10.

Di Carli, M., Benvenuto, E., & Donini, M., (2012). Recent insights into plant–virus interactions through proteomic analysis. *Journal of Proteome Research, 11*(10), 4765–4780.

Dodds, J. A., Morris, T. J., & Jordan, R. L., (1984). Plant viral double-stranded RNA. *Annual Review of Phytopathology, 22*(1), 151–168.

Gaafar, Y. Z., & Ziebell, H., (2020). Comparative study on three viral enrichment approaches based on RNA extraction for plant virus/viroid detection using high-throughput sequencing. *PLoS One, 15*(8), e0237951.

García-Arenal, F., & McDonald, B. A., (2003). An analysis of the durability of resistance to plant viruses. *Phytopathology, 93*(8), 941–952.

Gergerich, R. C., & Dolja, V. V., (2006). Introduction to plant viruses, the invisible foe. *The Plant Health Instructor*, 478.

Hema, M., & Konakalla, N. C., (2021). Recent developments in detection and diagnosis of plant viruses. In: *Recent Developments in Applied Microbiology and Biochemistry* (pp. 163–180). Academic Press.

James, D., Varga, A., Pallas, V., & Candresse, T., (2006). Strategies for simultaneous detection of multiple plant viruses. *Canadian Journal of Plant Pathology, 28*(1), 16–29.

James, W. C., (1974). Assessment of plant diseases and losses. *Annual Review of Phytopathology, 12*(1), 27–48.

Jeong, J. J., Ju, H. J., & Noh, J., (2014). A review of detection methods for the plant viruses. *Research in Plant Disease, 20*(3), 173–181.

Kutnjak, D., Tamisier, L., Adams, I., Boonham, N., Candresse, T., Chiumenti, M., De Jonghe, K., et al., (2021). A primer on the analysis of high-throughput sequencing data for detection of plant viruses. *Microorganisms, 9*(4), 841.

Maliogka, V. I., Minafra, A., Saldarelli, P., Ruiz-García, A. B., Glasa, M., Katis, N., & Olmos, A., (2018). Recent advances on detection and characterization of fruit tree viruses using high-throughput sequencing technologies. *Viruses, 10*(8), 436.

Massart, S., Chiumenti, M., De Jonghe, K., Glover, R., Haegeman, A., Koloniuk, I., Komínek, P., et al., (2019). Virus detection by high-throughput sequencing of small RNAs: Large-scale performance testing of sequence analysis strategies. *Phytopathology, 109*(3), 488–497.

Massart, S., Olmos, A., Jijakli, H., & Candresse, T., (2014). Current impact and future directions of high throughput sequencing in plant virus diagnostics. *Virus Research, 188*, 90–96.

Matthews, R. E., (2019). *Diagnosis of Plant Virus Diseases*. CRC Press.

Naidu°, R. A., & Hughes, J. D., (2003). Methods for the detection of plant virus diseases. In plant virology in Sub-Saharan Africa: *Proceedings of a Conference Organized by IITA: International Institute of Tropical Agriculture, Ibadan, Nigeria* (p. 233). IITA.

Nault, L. R., (1997). Arthropod transmission of plant viruses: A new synthesis. *Annals of the Entomological Society of America, 90*(5), 521–541.

Pallás, V., Sánchez-Navarro, J. A., & James, D., (2018). Recent advances on the multiplex molecular detection of plant viruses and viroids. *Frontiers in Microbiology*, 2087.

Rettcher, S., Jungk, F., Kühn, C., Krause, H. J., Nölke, G., Commandeur, U., Fischer, R., et al., (2015). Simple and portable magnetic immunoassay for rapid detection and sensitive quantification of plant viruses. *Applied and Environmental Microbiology, 81*(9), 3039–3048.

Roossinck, M. J., (2010). Lifestyles of plant viruses. *Philosophical Transactions of the Royal Society B: Biological Sciences, 365*(1548), 1899–1905.

Rubio, L., Galipienso, L., & Ferriol, I., (2020). Detection of plant viruses and disease management: Relevance of genetic diversity and evolution. *Frontiers in Plant Science*, 1092.

Sankaran, S., Mishra, A., Ehsani, R., & Davis, C., (2010). A review of advanced techniques for detecting plant diseases. *Computers and Electronics in Agriculture, 72*(1), 1–3.

Tatineni, S., Stewart, L. R., Sanfaçon, H., Wang, X., Navas-Castillo, J., & Hajimorad, M. R., (2020). Fundamental aspects of plant viruses– An overview on focus issue articles. *Phytopathology, 110*(1), 6–9.

Wheeler, B. E. J. (1969). *An Introduction to Plant Diseases*. Wiley and Sons, London, 374 p.

Yadav, N., & Khurana, S. M., (2016). Plant virus detection and diagnosis: Progress and challenges. In: *Frontier Discoveries and Innovations in Interdisciplinary Microbiology* (pp. 97–132). Springer, New Delhi.

Zaitlin, M., & Palukaitis, P., (2000). Advances in understanding plant viruses and virus diseases. *Annual Review of Phytopathology, 38*(1), 117–143.

CHAPTER 5

# Current Approaches and Prospective Strategies for Plant Virus Management

SOHAIL AKHTAR[1] and MUHAMMAD SHAFIQ SHAHID[2]

[1]*University of Agriculture Faisalabad, Sub-Campus Burewala, Pakistan*

[2]*Department of Plant Sciences, College of Agricultural and Marine Sciences, Sultan Qaboos University, Al-Khod, Muscat, Oman*

## ABSTRACT

Plant viruses are one of the most troublesome and challenging pathogens of food and fiber crops worldwide, posing a serious threat to sustainable agriculture. They cause more than US $30 billion in crop losses annually. Several factors, including international trade of biomaterials, insect vectors, climate change, and the ability of viruses to evolve quickly, are responsible for the frequent emergence of viral diseases. The challenge of dealing with these noxious entities has long been open, with very few success stories in the field. Since there is no viricide available for virus control, reliance is placed on the use of resistant plant varieties, vector control, and prophylactic measures to restrain virus spread. Plant disease resistance against viruses has been achieved through plant breeding technologies or advanced biotechnological tools such as RNA interference (RNAi) and clustered regularly interspaced short palindromic repeats (CRISPRs), which are based on a bacterial defense system and form the basis for CRISPR-Cas9 genome editing tools. Marker-assisted selection (MAS) has largely replaced trait-oriented conventional breeding as a more efficient and indirect method of selection. Biotechnological methods include genetic engineering and plant transformation protocols, which involve the development of resistance using RNA-based (e.g., RNAi) and protein-based approaches. In addition to these,

Detection and Management of New and Emerging Mystery Plant Virus Sources.
Alireza Golnaraghi, Rajarshi Kumar Gaur, and Rakesh Kumar Verma (Eds.)

genome editing technologies using CRISPR-Cas9 and nanotechnology are also being employed with greater potential for success. Prophylactic measures include quarantine (border control), the use of certified virus-free germplasm to prevent the entry of new viruses in a geographical area, control of insect vectors by chemical pesticides, and other cultural methods either to prevent the vectors from reaching the crop or to interfere in the transmission of viruses. Here, we have discussed all the strategies and protocols in practice for the better management of plant viruses. The current trends, future prospects, and a comparative analysis of all the approaches are the main focus of this chapter.

## 5.1 INTRODUCTION

Virus is a word of Latin origin, meaning poison. It is a submicroscopic, acellular, mesobiotic infectious agent that encapsidates its genomic contents (DNA or RNA) in a protein coat. Viruses are unable to capture or store energy, therefore, they are inactive outside the host and can only replicate inside the host cell. As a result, they are considered obligate parasites. According to Dimmock et al. (2007), "Viruses occur universally, but they can only be detected indirectly. Although they are well known for causing disease, most viruses coexist peacefully with their hosts." Martinus Beijerinck discovered the *tobacco mosaic virus* (TMV) for the first time in a tobacco plant showing mosaic symptoms.

Morphologically, viruses' range in size from as small as 12 nm to as large as 2,000 nm. They exist in a variety of shapes, including flexuous thread-like, cylindrical, helical, icosahedral, spherical, isometric polyhedral, and more complex structures. Similarly, genome size also varies from a few 100 bases (e.g., 2,700 bps of geminiviruses) to millions of bases (e.g., 1.2 Mb of mimiviruses). Viruses are one of the most abundantly found biological entities on Earth. It is estimated that there are $10^{31}$ viruses in total present on this planet (Breitbart & Rohwer, 2005). These noxious agents cause various diseases in plants, animals, and human beings.

In animals and humans, the spread and transmission of viruses mainly occur through direct contact. Other ways of virus entry into the body include inhalation, ingestion, sexual contact, and vector transmission, such as blood-sucking insects. In plants, one of the most common and efficient ways of transmission and spread is through insect vectors. Insects belonging to the order Homoptera constitute the largest group of virus-transmitting vectors, including aphids, leafhoppers, planthoppers, and whiteflies. Other ways of transmission include mechanical transmission through seed and pollen, fungus transmission, dodder transmission, and transmission through vegetative propagation.

Plant viruses are one of the most troublesome phytopathogens worldwide, causing more than US $30 billion in crop losses each year (Sastry & Zitter, 2014; Mumford et al., 2016). There is hardly a plant species (either monocotyledonous or dicotyledonous) that is out of the reach of viruses (Waterhouse et al., 2001). Viruses have caused several disastrous pandemics and epidemics, threatening food, fiber, and ornamental crop plants. Examples of plant viral pandemics on food crops include maize lethal necrosis virus (MLNV), sweet potato virus (SPV), citrus tristeza virus (CTV), rice tungro spherical virus (RTShV), banana bunchy top virus (BBTV), and PPV. Similarly, epidemics include wheat yellow dwarf virus (WYDV), potato tuber necrotic ringspot ($PVY^{NTN}$), pepino mosaic virus (PepMV), cucumber green mottle mosaic virus (CGMMV), wheat streak mosaic virus (WSMV), faba bean necrotic yellows virus (FBNYV), and tomato brown rugose fruit virus (ToBRFV) (Jones, 2021). The rate of emerging and re-emerging plant viral infections, in comparison to other phytopathogens including fungi and bacteria, has escalated in many cases of plant diseases (Anderson et al., 2004). High genetic variability of viruses is attributable to their rapid replication, large populations, multipartite genome particles, component capture, high recombination rate, and the role of insect vectors. Monoculturing and international trade of biomaterials are also contributing factors to the escalated global distribution of plant viruses. Mixed infections may result in additive synergistic interactions in a single plant, involving multiple viruses and viroids, which can lead to increased virulence and the emergence of new diseases (Syller, 2012; Moreno & Lopez-Moya, 2020). Additionally, compared to other plant pathogens, viruses have a high mutation rate due to their small genome size, multipartite genomic nature, component capture, and recombination (Nawaz-Ul-Rehman & Fauquet, 2009).

## 5.2 VIRUS DISEASE MANAGEMENT

Virus disease management is not easy due to the frequent and dynamic nature of virus epidemics, as well as the rapid evolution capacity of plant viruses. Two types of control measures are commonly practiced preventing these epidemics: immunization to develop plants resistant to virus infections, and prophylactic measures such as quarantine, certification, control of insect vectors, and removal of infected plants. Plant resistance is achieved through various strategies or techniques, including conventional breeding, marker-assisted selection (MAS) using sources of resistance with dominant and recessive resistance genes, RNA interference (RNAi), and genome editing technologies like CRISPR-Cas9.

## 5.3 IMMUNIZATION

### *5.3.1 PLANT BREEDING*

Plant breeding for virus resistance is one of the simplest and most efficient ways to prevent yield losses caused by virus diseases. Farmers have been practicing plant breeding for a long time, possibly since the beginning of agriculture. The process involves crossing a virus-resistant plant with another plant that lacks resistance but possesses desirable characteristics. For instance, in Pakistan, cotton varieties resistant to cotton leaf curl disease (CLCuD) were developed by crossing resistant sources (LRA 5166, Cedix, and CP15/2) with locally cultivated susceptible varieties (Rahman et al., 2002). Similarly, potato varieties FD3-9 and FD3-10, resistant to PVX, and FD1-8 and FD9619, resistant to potato virus Y (PVY), were developed using a similar crossing strategy (Ahmad et al., 2011). There are several methods of breeding for virus resistance, including backcrossing, inbreeding, hybrid breeding, and mutation breeding. Isogenic lines of rice resistant to rice stripe virus (RSV) were developed using backcrossing (Kang et al., 2019). Inbreeding refers to the fertilization of plants by themselves, done to produce an inbred variety that preserves the original traits. The discovery of novel sources of resistance in inbred lines facilitates the identification of virus resistance genes in different plants and their incorporation into breeding programs. In hybrid breeding, the two inbred varieties are crossed to produce an offspring with more stable features and hybrid vigor. In Kenya, resistant maize was developed against maize lethal necrosis (MLN) disease using a hybrid breeding approach. Around 50 elite maize inbred lines were selected, having a variable response to MLN. Crosses were made, and the hybrids produced were superior to the mean disease severity of the parents (Nyaga et al., 2020). Sometimes, naturally occurring genetic mutations appear to be the improvements that may be used to create a new variety. Mutations in eIF4Es confer loss-of-susceptibility to potyviruses and several other viruses (Hashimoto et al., 2016).

### *5.3.2 MARKER-ASSISTED SELECTION (MAS)*

Marker-assisted selection (MAS) is an indirect method of selection that involves the selection of certain markers/indicators (morphological, biochemical, or molecular) instead of selecting the trait of interest directly. For instance, when using MAS to choose individuals with disease resistance, the focus is on identifying a marker allele that is linked to disease resistance rather than evaluating the level of disease resistance itself. Trait-oriented conventional breeding has largely been replaced by MAS, which is a comparatively faster and indirect method of selection. MAS allows for accurate and cost-effective selection even at the seedling stage. It eliminates the need to test a crop against a specific pathogen through pathogenicity or infectivity assays.

Genetic markers (mostly DNA markers) are flags or signs that represent the presence or absence of a specific gene of interest in the genome. These markers are actually DNA sequences closely linked to a gene and are used for indirect selection. Instead of conducting infectivity bioassays at a later stage of the crop, Molecular Assisted Selection (MAS) allows for the early detection of resistance sources to plant viruses, such as resistance genes, susceptibility genes (S Genes), or other transgenes used for resistance. In the wheat breeding program of the International Maize and Wheat Improvement Center (CIMMYT) and some other countries, approximately 50 genes, including R genes for barley yellow dwarf virus (BYDV) and WSMV, have been selected for MAS. Markers for these genes and laboratory protocols have been optimized and are currently being used in breeding programs (Dubcovsky, 2011; Gupta et al., 2010).

## 5.4 PROPHYLACTIC APPROACHES

Prophylactic approaches include quarantine, seed certification, voluntary eradication of host plants, and other evasion or avoidance measures. These measures aim to prevent the contact of viruses with their hosts and to restrain the spread of the virus.

### *5.4.1 QUARANTINE AND SEED CERTIFICATION*

In history, there have been several instances where a virus of foreign origin has been introduced into an area where the host plants have never encountered that virus before. These exotic viruses have caused epidemics, resulting in significant losses to the crop hosts. For example, CLCuD in the Indian sub-continent, caused by the Cotton Leaf Curl Virus complex, has led to a massive decline in cotton production in Pakistan and India. Leaf curl begomoviruses appear to have been introduced for the first time in the sub-continent from Africa (Akhtar, 2014). Additionally, several geminiviruses of diverse origins have been reported from ornamental and vegetable crops in Oman, causing severe diseases and substantial damage to these crops. To prevent future losses, it is recommended to implement stricter phytosanitary measures (Khan et al., 2014).

Quarantine and seed certification are the primary regulatory controls imposed by state laws. Quarantine departments are established at entry points such as airports, seaports, and dry ports, as well as isolated regions within a country. These departments screen biological plant material for the presence of any new or suspected pathogens (viruses) before importation. To achieve effective quarantine screening, a highly sensitive and broad-spectrum method is required. High throughput sequencing, also known as next-generation sequencing (NGS), is the best available technique for detecting both known and unknown viruses in plant material (Villamore et al., 2019). Based on such screening, phytosanitary certification should be implemented for the importation of virus-free germplasm.

### *5.4.2 VIRUS-FREE PROPAGATING MATERIAL*

In the event that healthy plants are not available, infected plants are treated differently to eliminate the virus. Chemotherapy, thermotherapy, and tissue culture are some of the viable options that can be utilized to make a plant virus-free. These plants are then used to obtain virus-free seeds and vegetative propagating material.

Thermotherapy involves incubating the plants at high temperatures for a certain period of time to inactivate the virus. This can be achieved by disrupting the viral particles, breaking down the viral nucleic acids, or degrading the viral proteins (Hull, 2002). Tissue culture, on the other hand, utilizes apical meristems as explants, which are cultured in specially formulated nutrient media under sterile conditions. Apical meristems consist of rapidly dividing cells with a high metabolic rate, and viruses are unable to replicate in such cells. This is why the meristem region of most virus-infected plants remains virus-free and can be used for further virus-free culturing.

In chemotherapy, antiviral drugs such as salicylic acid (SA), ribavirin, and nucleoside analogs are used. These drugs may interrupt specific steps in the life cycle of the virus, thus inhibiting its replication and the infection process. Chemotherapeutic treatment of plant virus infections has a high cost that is unaffordable for field crops. Therefore, the treatment is mostly carried out in combination with other methods. For example, virus-free explants in tissue culture may also be prepared using this method (Kidulile et al., 2018). In addition to tissue culture, virus-free seed potatoes are being produced using aeroponics technology.

Recently, the technology has been beneficial in producing the first lot of virus-free seed potatoes at the Pakistan Agriculture Research Council (PARC) in collaboration with the Korean Government. The plants produced using such sanitation methods can finally be evaluated for the presence of the virus using real-time qPCR or any other highly sensitive technique.

### 5.4.3 CULTURAL METHODS TO REDUCE VIRUS INOCULUM

Roguing is an agronomic practice in which plants with undesirable characteristics are removed from the fields. Virus-infected plants of the main crop and weeds serve as a source of inoculum for secondary infection cycles. However, this practice is effective only in cases of low disease incidence or recent disease introduction into the field.

Viruses of annual crop plants, such as cucumber mosaic virus (CMV), can overwinter in perennial plants in or nearby the fields. These perennial plants serve as reservoirs for the viruses, which then induce primary infection in the following season. Similarly, volunteer plants that grow between planting seasons of tomatoes in sub-tropical areas can harbor viruses such as tomato mottle virus (TYLCV). These viruses cause disease in tomatoes when they are planted. Therefore, the eradication of such perennial and other volunteer hosts is sometimes sufficient for complete control of viral infections in the crop.

The incidence or spread of mechanically transmitted viruses, such as Tobamoviruses, can be controlled through sanitation methods. For example, wearing gloves and disinfecting hands, knives, and other agricultural tools (Agrios, 2005; Broadbent, 1976). In the fields, yellow-colored polyethylene traps are set up to capture incoming insects like aphids and other potential virus vectors. This practice can help reduce the virus inoculum to some extent. During the early stages of crop growth, yellow-colored polyethylene reflectant mulches are placed in crop rows to repel aphids, whiteflies, and thrips, thereby reducing the virus inoculum on the crop. However, the efficiency of these mulches decreases when the crop canopy covers them. Another important method is to plant taller trap crops (e.g., corn, rye) around the boundary line of shorter crops such as beans, peppers, and squash. This prevents aphids from directly reaching the cultivated crops, instead, they feed on the trap crops first. Since aphid-borne viruses are non-persistent, there is a probability that aphids, after feeding on corn or rye, will not carry any virus to infect the beans or other crops. This can also help reduce the virus inoculum on cultivated crops.

### *5.4.4 CHEMICAL CONTROL OF INSECT VECTORS*

Potential insect vectors involved in virus transmission, such as aphids, whiteflies, leafhoppers, and treehoppers, belong to the order Homoptera. The increased virus inoculum on cultivated crops depends on the infestation of insect vectors on these crops, either directly or indirectly through the feeding of these insects on the plants where they overwinter before entering the cultivated crop. Therefore, the chemical control of insect vectors before they enter the crop yields better results. Once they have arrived on the crop and started virus transmission, it becomes difficult to adequately control the spread of the virus. However, it may help in reducing the losses to some extent, thereby emphasizing the importance of insect control at any stage.

It is not always necessary to kill the insect vectors to prevent virus infections. Adequate reduction in virus inoculum can be achieved by simply disrupting the aphids' ability to transmit the virus. Spraying with fine-grade mineral oil several times a season interferes with the virus transmissibility of aphids and is not toxic to them. Successful control of certain aphid-borne viruses, such as PVY on pepper and CMV on cucumber and pepper, has been achieved through the spraying of mineral oils (Agrios, 2005).

## 5.5 RNA SILENCING

RNA silencing is a homology-dependent mechanism of gene silencing in which plant immunity is triggered by the introduction of a viral double-stranded RNA (dsRNA) molecule. In other words, we can say that the dsRNA molecule serves as a trigger for such an immune response. Contrary to pathogens like fungi and bacteria, for which the pattern recognition receptors (PRRs) are located on the plasma membrane (apoplast), PRRs for plant viruses are found intracellularly (Korner et al., 2013). After the virus enters the plant cell, its RNA starts replicating. The dsRNA formed as a replication intermediate is detected by the host RNase type-III-like enzyme called dicer-like proteins (DCLs) and degraded into 21–24 nucleotide fragments called short/small interfering RNAs (siRNAs) (Sharma et al., 2013). These siRNAs are then unwound, and the single-stranded siRNA molecules, known as guide RNAs (gRNAs), are loaded into a protein complex called the RNA-induced silencing complex (RISC). The other unbound strand is readily degraded. RISC is responsible for degrading any viral messenger RNA (mRNA) molecule that is complementary to the gRNA (Figure 5.1). The slicer protein, a core component of the RISC that is responsible for degrading the viral RNA, belongs to the Argonaute (AGO) family of proteins (Vaucheret, 2008). AGO proteins have a PAZ (Piwi/Argonaute/Zwille) domain that provides endonuclease activity to RISC. Once primary siRNAs are generated, there is a significant increase in the number of subsequent siRNAs, which ensures resistance signal locally as well as systemically (Kaldis et al., 2017). Similar to viruses, siRNAs move from cell to cell through plasmodesmata and systemically through phloem. The amplification in the number of siRNAs corresponds to the activity of host RNA-dependent RNA polymerases (RDRs) that convert viral mRNA into dsRNA in a siRNA-dependent and -independent manner. Their subsequent degradation by DCL proteins produces another set of secondary siRNAs different from the primary siRNAs, a process called transitive silencing (Sijen et al., 2001). As mentioned earlier, the viral transcript is converted into a double-stranded form because of host RDR activity and is then treated as an RNAi target. This process is also referred to as post-transcriptional gene silencing (PTGS). PTGS occurs against all DNA and RNA viruses. However, the process is slow and does not guarantee complete immunity against viral infections.

siRNA-directed DNA methylation, histone modification, and transcriptional gene silencing (TGS) are outcomes of the dsRNA trigger that have been observed in several cases. This process occurs in the nucleus and primarily targets the promoter regions. There is increasing evidence of the role of DNA methylation in response to biotic stresses (Xiao et al., 2021; Hewezi et al., 2017). The DNA methylation status of the host is reprogrammed upon viral invasion (Wang et al., 2019). In a study by Seemanpillai et al. (2003), transgenic plants containing a tomato leaf curl virus (ToLCV)-derived transgene promoter were prepared. Upon infection with ToLCV, siRNA-directed DNA methylation of the promoter region resulted in the silencing of the transgene.

Resistance is achieved by introducing DNA constructs into plants, with the ultimate goal of producing dsRNA that can trigger the RNAi mechanism. These DNA constructs for plant transformation can be designed using sense genes, antisense genes, or by creating hairpin RNAi constructs with inverted repeat (IR) sequences. Since RNAi is a mechanism that relies on homology, it is crucial to carefully select the target virus sequence to produce dsRNA. This selection should be based on conserved virus sequences to minimize the chances of frequent mutations. A recent example of such a construct is found in the work of Akhtar et al. (2021), where the beta-C1 gene of the betasatellite associated with cotton leaf curl begomovirus was targeted using a hairpin RNAi construct. The construct was then introduced into *Nicotiana benthamiana* plants, and the transgenic plants were subsequently exposed to the corresponding virus to observe the resistance response.

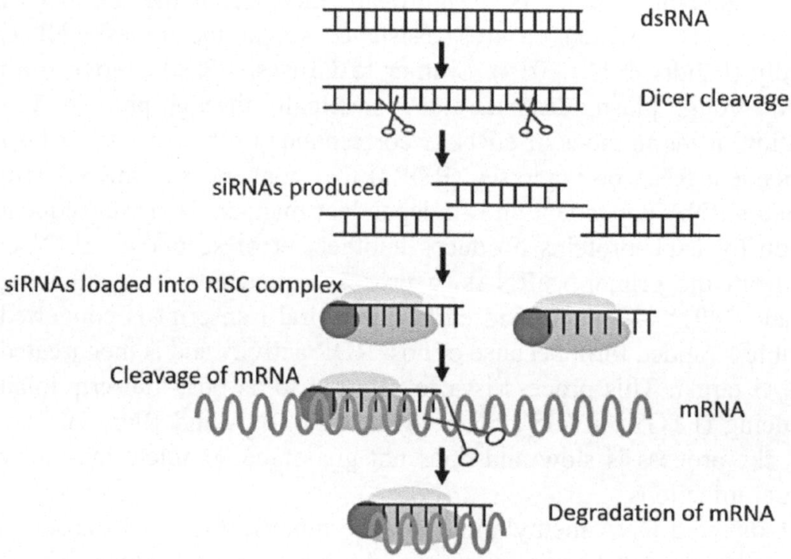


**FIGURE 5.1** The cellular mechanism of RNA interference involves the entry of dsRNA into the host cell, which serves as a trigger for dicer proteins. This leads to the production of short RNAs that are loaded into the RISC complex. The RISC complex then targets the invading mRNA by sequence complementarity and degrades it.

In plants, there are three natural RNA silencing pathways: (i) siRNAs; (ii) microRNAs (miRNAs); and (iii) DNA methylation and suppression of transcription. The pathway involving siRNAs has been discussed. miRNAs, on the other hand, are small RNAs that are approximately 20–24 nucleotides in length. They play a role in developmental regulation by silencing endogenous mRNAs. miRNAs are derived from a precursor transcript called pri-miRNA, which contains an inverted repeat structure with partially double-stranded regions transcribed by RNA polymerase II (pol II). The downstream processing of pri-miRNAs involves the use of a nuclear RNase-III like enzyme called DROSHA, which yields pre-miRNAs. These pre-miRNAs are further processed using several proteins, including Dicer and export proteins (such as HST, an exportin homolog), to produce mature miRNAs (Katiyar-Agarwal & Jin, 2010). One of the two strands of the mature miRNA duplexes is loaded onto RISC, while the other is destroyed. The rest of the function is similar to that of siRNAs. In addition to translational repression, miRNAs also induce DNA methylation (Bao et al., 2004).

For resistance against viruses in plants, the use of miRNAs has been a promising mechanism in practice for the last two decades (Jiang et al., 2022; Shahriari & Tahmasebi, 2018).

## 5.6 GENETIC RESISTANCE

Besides RNAi, there is a second layer of defense against plant viruses conferred by resistance genes. The use of genetic resistance is one of the most influential approaches used against plant viruses due to its target specificity, environmentally friendly outlook, reliability, and sustainability without additional labor and material costs (Maule et al., 2007). There are two types of genetic resistance based on the contribution of either dominant or recessive genes.

## 5.7 DOMINANT RESISTANCE

If the genes conferring resistance are dominant, the resistance is referred to as dominant resistance. It mainly involves the contribution of a single dominant resistance gene (R gene). This has become increasingly interesting to breeders and biotechnologists. Resistance genes from monocot and dicot crops, as well as their wild relatives, are introduced into susceptible plants to make them resistant. R gene-mediated resistance is primarily achieved through the interaction of the R gene-encoded protein of the host plant with the corresponding avirulence factor encoded by the Avr gene of the invading pathogen. This is characterized by a hypersensitive response (HR) and an extreme resistance (ER) response (Soosaar et al., 2005). The largest group of such genes consists of the NB-LRR type (nucleotide binding-leucine rich repeats). HR and the production of SA are commonly observed responses related to NB-LRR type-mediated resistance. NB-LRR proteins consist of three domains: the C-terminal domain characterized by leucine rich repeats (LRR) determines the specificity of the target protein, the nucleotide binding site (NBS) in the center for binding with and hydrolysis of ATP, and the coiled coil (CC) or toll and interleukin-1 receptor domain (TIR) at the N-terminal, which is only found in dicots from angiosperms (Moffett, 2009; Collier et al., 2011). These proteins recognize effectors and initiate effector/pathogen triggered immunity (ETI/PTI) (Jones & Dangl, 2006). The majority of the R genes belong to the CC-NB-LRR type, and only a few are of the TIR-NB-LRR type (Ronde et al., 2014). Defense induced through R genes is of a local nature, thus limiting the virus at the point of entry. However, the systemic effect of R genes mediated by SA signaling is also known for the N gene in tobacco and the Rx1 gene in potato, giving rise to systemic acquired resistance (SAR) (Liu et al., 2010; Delaney et al., 1994).

A few R genes have also been identified in *Arabidopsis thaliana* that do not belong to the NB-LRR type, such as *RTM1*, *RTM2*, and *RTM3*. RTM genes act by inhibiting the systemic spread of the virus, making them unable to be loaded into the phloem. In addition, no HR and production of SA responses were noted as observed in the case of NB-LRR type genes (Cosson et al., 2012).

Another R Gene, *Tm-1,* has been reported in wild tomato species that encodes a protein which interferes with the RNA replication process of tomato mosaic virus (ToMV). Again, no HR and SA production were reported for this gene. *Tm-1* seems to represent a housekeeping gene, as several homologs have been identified in other organisms including fungi and bacteria (Ishibashi et al., 2007, 2012).

Another typical example is the *Ty-1* gene in tomato that encodes RdRP and confers resistance to tomato yellow leaf curl virus (TYLCV). It converts viral RNA into a double-stranded form, which, as discussed earlier, serves as a trigger for RNAi (Verlaan et al., 2013). Tomato plants containing the *Ty-1* gene remain symptomless upon infection with TYLCV; however, a low virus titer is still found in the plants.

Examples of some of the R genes and their corresponding Avr genes of the pathogen are given in Table 5.1.

**TABLE 5.1** Dominant Resistance Genes Against Plant Viruses

| Host Plant | R Gene | Avr Gene | Virus* | References |
|---|---|---|---|---|
| *Capsicum annum* | L-locus | CP | TMV | Tomita et al. (2011) |
| *Phaseolus vulgaris* | PvVTT1 | BV1 | BDMV | Gururani et al. (2012) |
| Tomato | Ty-3 | – | TYLCV | Verlaan et al. (2013) |
| Tomato | Tm-1 | Replicase | ToMV | Ishibashi et al. (2012) |
| *Solanum peruvianum* | Sw5b | NSm | TSWV | Hallwass et al. (2014) |
| *Solanum tuberosum* | Rx1 | CP | PVX | Solomon-Blackburn & Barker (2001) |
| *Solanum tuberosum* | Rx2 | CP | PVX | Bendahmane et al. (2000) |
| *Solanum tuberosum* | Y-1 | – | PVY | Zvereva & Pooggin (2012) |
| *Glycine max* | Rsv1 | P3 + HC-Pro | SMV | Wen et al. (2013); Hayes et al. (2004) |
| *Brassica campestris* | BcTuR3 | Unknown | TuMV | Ma et al. (2010) |
| *Brassica rapa* | TuRB07 | Unknown | TuMV | Jin et al. (2014) |
| *Vigna mungo* | CYR1 | CP | MYMV | Maiti et al. (2012) |
| *Cucumis melo* | Pvr1-2 | Unknown | PRSV | Brotman et al. (2013) |
| *Nicotiana glutinosa* | N | p50 | TMV | Whitham et al. (1994) |

* *Bean dwarf mosaic virus* (BDMV); *mungbean yellow mosaic virus* (MYMV); *papaya ringspot virus* (PRSV); *potato virus X* (PVX); *potato virus* Y (PVY); *soybean mosaic virus* (SMV); *tobacco mosaic virus* (TMV); *tomato mosaic virus* (ToMV); *tomato spotted wilt virus* (TSWV); tomato yellow leaf curl virus (TYLCV); and turnip mosaic virus (TuMV).

## 5.8 RECESSIVE RESISTANCE

All plant viruses require interaction with host susceptibility factors, also known as S-genes, to cause infection. A compatible interaction between the S-gene and the virus leads to a successful infection, and vice versa. Since the susceptibility factors are in a dominant form and responsible for a compatible interaction with the virus, the resistance conferred is contributed by the recessive S Genes. This means that the loss or mutation of the necessary host factor required by the virus may induce resistance to that virus. Hence, this type of resistance is referred to as recessive resistance (Diaz-Pendon et al., 2004). Recessive resistance is more commonly found against plant viruses compared to other plant pathogens.

Plant viruses have a very small genome that is unable to cause infection in the host if not supported by host factors in providing necessary proteins. Most of the recessive S Genes have been observed to interact with potyviruses. Potyviruses, which harbor a single polyprotein, are known to interact with the eukaryotic translation initiation factors (eIFs) of the plants, particularly eIF4E, eIF4G, and their isoforms (Hashimoto et al., 2016). In plants, these proteins bind to the cap structure of the transcripts to make them translatable. However, transcripts of potyviruses do not contain the cap structure but have their own VPg (virus-protein genome linked). When potyviruses infect plants, they shut off the host cap-dependent translation and induce cap-independent translation. It has been observed in several potyviral infections that VPg and host-encoded eIF4E or eIF(iso)4E have a high affinity for each other. For example, the same interaction has been noted using the yeast two-hybrid (Y2H) system and enzyme-linked immunosorbent assay (ELISA) between *Arabidopsis* eIF(iso)4E and turnip mosaic virus (TuMV)/tobacco etch virus (TEV) VPg (Kang et al., 2005; Wittman et al., 1997). In this way, a selection pressure develops in the host to escape the interaction of VPg and eIF4E, thus inducing recessive resistance. The first known natural recessive resistance gene used against PVY in pepper was *pvr2*, which encodes a variant of eIF4E (Ruffel et al., 2002). The same variant pvr2 was shown to confer resistance to other potyviruses, including Pepper veinal mottle virus (PVMV) and TEV (Charron et al., 2008; Ruffel et al., 2006). Afterwards, several other variants of eIF4E/eIF(iso)4E were discovered in various hosts, conferring resistance to their corresponding viruses (Hwang et al., 2009; Rubio et al., 2009). Several other examples of recessive resistance genes are given in Table 5.2.

**TABLE 5.2** Recessive Resistance Genes to Develop Virus Resistant Plants Associated with Loss of Susceptibility Mutants and Translation Initiation

| Host Plant | Gene | Cause of Resistance* | Target Virus** | References |
|---|---|---|---|---|
| *Nicotiana* species | Tom1 | Decrease susceptibility by ethyl EMS mutagenesis. | YoMV | Ishikava et al. (1991) |
| *Capsicum annum* | Tom3 | Drop off susceptibility by EMS mutagenesis. | TMV | Yamanaka et al. (2000) |
| *A. thaliana* | Tom2a | Fast neutron mutagenesis deprive susceptibility. | YoMV | Ohshima et al. (1998) |
| *A. thaliana* | Tom2b | Fast neutron mutagenesis deprive susceptibility. | ToMV | Tsujimoto et al. (2003) |
| *A. thaliana* | ARL8 | Null mutation of ARL8 disappeared susceptibility. | ToMV | Nishikiori et al. (2011) |
| *A. thaliana* | EXA1 | Loss of susceptibility by EMS mutagenesis. | PlAMV | Hashimoto et al. (2016) |

**TABLE 5.2** *(Continued)*

| Host Plant | Gene | Cause of Resistance* | Target Virus** | References |
|---|---|---|---|---|
| *A. thaliana* | PCaP1 | T-DNA mutant overlook susceptibility. | TuMV | Vijayapalani et al. (2012) |
| *A. thaliana* | Sec24a | Drop off susceptibility by EMS mutagenesis. | TuMV | Jiang et al. (2015) |
| *A. thaliana* | RHD3 | T-DNA mutant overlook susceptibility. | TSWV | Feng et al. (2016) |
| Different species | IRE1 | T-DNA insertions squander susceptibility. | TuMV | Zhang et al. (2015) |
| *A. thaliana* | HAT1, HAT2, HAT3 | T-DNA insertions overlook susceptibility. | CMV | Zou et al. (2016) |
| Different plant species | VPS41 | cmv1 gene induce natural resistance. | CMV | Giner et al. (2017) |
| *Oryza glaberrima* | CPR5 | Natural resistance gene rymv1. | RYMV | Orjuela et al. (2013) |
| *Phaseolus vulgaris* | bc-3 | Deprivation of translation initiation factor eIF4E. | ClYVV | Hart & Griffiths (2013) |
| *Pisum sativum* | cyv1/ cyv2 | Disappearance of translation initiation factor eIF4E. | ClYVV | Choi et al. (2012) |
| *Solanum lycopersicum* | pvr1 | Dropping off translation initiation eIF4E. | TEV | Kang et al. (2007) |
| *Stererum hirsutum* | pot-1 | Mislaying of translation initiation eIF4E. | TEV, PVY | Ruffel et al. (2005) |
| *Capsicum annum* | pvr1/ pvr2 | Damage of translation initiation eIF4E. | PVY, TEV | Perez et al. (2012) |
| *Hordeum vulgare* | rym4 | Decline of translation initiation eIF4E. | BaMMV | Hofinger et al. (2011) |
| *Hordeum vulgare* | rym7 | Destruction of translation initiation eIF4E. | BaMMV | Yang et al. (2013) |
| *N. tabacum* | va | Depletion of translation initiation eIF4E. | PVY | Julio et al. (2015) |

* Ethyl methanesulfonate (EMS).

** Barley mild mosaic virus (BaMMV); clover yellow vein virus (ClYVV); cucumber mosaic virus (CMV); Plantago asiatica mosaic virus (PlAMV); potato virus Y (PVY); rice yellow mottle virus (RYMV); tobacco mosaic virus (TMV); turnip mosaic virus (TuMV); tomato spotted wilt virus (TSWV); tobacco etch virus (TEV); and Youcai mosaic virus (YoMV).

## 5.9 GENOME EDITING USING CRISPR/CAS9 SYSTEM

CRISPR/Cas9 (Clustered Regularly Interspaced Palindromic Repeats/ CRISPR associated 9) is the most revolutionary genome editing tool of the present age, with diverse applications including the development of resistance in plants against pathogens such as viruses. The system is simple, efficient, versatile, and reproducible. CRISPR/Cas9 is a system of molecular immunity used by prokaryotes (bacteria) to defend against invading phages or nucleic acids acquired through horizontal gene transfer. The DNA fragment incorporated into the bacterial genome is marked as a memory (Bolotin et al., 2005). Subsequent infection by the same virus or nucleic acid triggers the transcription of that short DNA fragment. The processed CRISPR RNA (crRNA) guides the Cas9 endonuclease to scan the invading viral or other nucleic acid based on sequence similarity. Cas9 cleaves the target segment of the genome at a site preceding the protospacer-associated motif (PAM) (Nunez et al., 2016; Wright et al., 2016), providing resistance to bacteria against invading phages or nucleic acids (Figure 5.2).

There are several studies demonstrating the use of CRISPR/Cas9 as a potential tool conferring resistance to plant viruses. Examples include the potential of this system against multiple geminiviruses tested in an experimental host, *Nicotiana benthamiana*. Resistance response was noted in *N. benthamiana* plants carrying the CRISPR/Cas9 apparatus against the infection of TYLCV, *Merremia mosaic virus* (MeMV), and *beet curly top virus* (BCTV) (Ali et al., 2015). Ji et al. (2015) prepared transgenic plants of *Arabidopsis* and *N. benthamiana* expressing single guide RNA (sgRNA) and Cas9 and challenged these plants with beet severe curly top virus (BSCTV). An inverse correlation was noted in the levels of Cas9 and virus titer, indicating virus suppression due to sgRNA and Cas9-created mutations in the target sequence. The transgenic plants were highly resistant to BSCTV infection. Similar results were obtained by Baltes et al. (2015), when they challenged their transgenic plants expressing the CRISPR/Cas system with *bean yellow dwarf virus* (BeYDV). In another study, Ali et al. (2016) targeted the conserved nonanucleotide sequence of begomoviruses to obtain broad-spectrum begomoviruses resistance. Transgenic plants of *N. benthamiana* expressing CRISPR/Cas9 with the sgRNA of a conserved nonanucleotide sequence displayed mild symptoms with reduced virus titer against multiple begomoviruses tested (Ali et al., 2016). The researchers further narrowed down their research to better understand the choice of viral target sequences. They targeted both coding and non-coding sequences and reached the conclusion that targeting non-coding sequences doesn't allow viruses to escape the CRISPR/Cas9 machinery and yields better resistance. Simultaneously targeting multiple regions of the viral genome ensured complete resistance to *cotton leaf curl Multan virus* (CLCuMuV) by expressing Cas9 with double gRNA in *N. benthamiana* (Yin et al., 2019). Kis et al. (2019) selected multiple regions of *wheat dwarf virus* (WDV) encompassing the movement protein (MP), coat protein (CP), and replication-associated protein (Rep) by designing sgRNA1, sgRNA2, sgRNA3, and sgRNA4, which did not have any *in silico* predicted off-target effects. Efficient resistance was obtained in the transgenic plants expressing the aforementioned sgRNAs/Cas9 constructs.

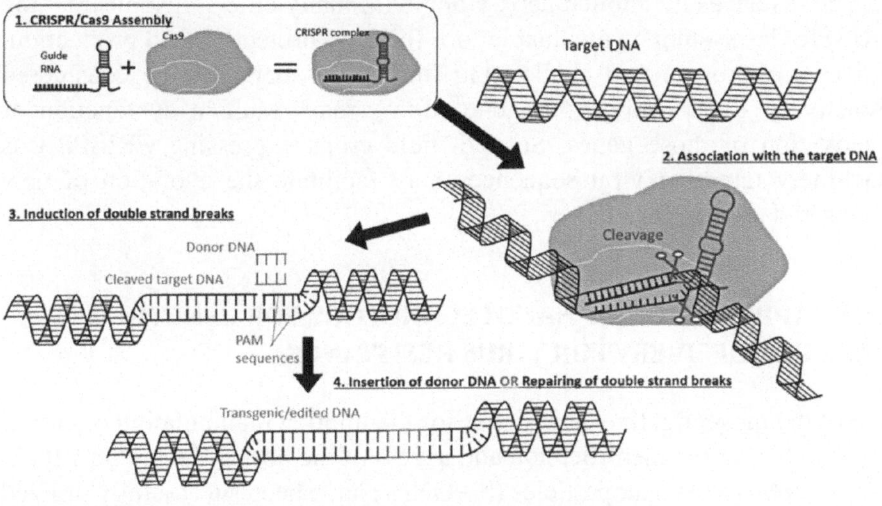


**FIGURE 5.2** The mechanism of CRISPR/Cas9 genome editing involves the attachment of a CRISPR complex, encompassing a sgRNA, to the target DNA, inducing double strand breaks. If donor DNA is provided, it is inserted into the break regions, resulting in the creation of transgenic DNA. In the absence of donor DNA, the host cell machinery repairs the breaks, leading to insertion or deletion, and the editing of the target DNA.

In the context of genome editing for resistance development in plants, instead of targeting viral sequences in RNA viruses, host plant genes responsible for recessive resistance, known as S Genes, have been used as targets. This is due to the limitation of the sgRNA-Cas9 system in working against RNA sequences. However, studies have proposed workable strategies for programming Cas9 to cleave RNA (O'Connell et al., 2014). Chandrasekaran et al. (2016), for the first time, developed resistance in cucumber plants by creating deletions and single nucleotide polymorphism (SNP) mutations in the eIF4E gene using CRISPR/Cas9 technology. The resulting non-transgenic mutant plants showed resistance to *cucumber vein yellowing virus* (CVYV), *zucchini yellow mosaic virus* (ZYMV), and *papaya ring spot mosaic virus* (PRSMV). The mutation of eIF4E did not have any undesired phenotypic effects in the mutant generations. Similarly, a point mutation created at the eIF(iso)4E locus in *A. thaliana* resulted in resistance against TuMV without affecting plant vigor (Pyott et al., 2016). Novel alleles of eIF4G were developed using CRISPR/Cas9 targeted mutagenesis to achieve resistance to RTSV (Macovei et al., 2018). Genetic resistance to PVMV was developed in cherry tomatoes by inactivating the eIF4E2 gene using CRISPR/Cas9-mediated genome editing (Kuroiwa et al., 2022).

The instances mentioned here, along with many others, demonstrate the CRISPR/Cas system as a valuable tool for crop protection and production. However, there are two challenging limitations that must be considered proactively. First, endogenous gene editing may result in dysfunction or inactivation of those genes. Second, field crops expressing CRISPR/Cas machinery targeting viral sequences may facilitate the evolution of new strains and species for viruses.

## 5.10 APPLICATION OF NANOTECHNOLOGY FOR GENETIC MATERIAL DELIVERY FOR VIRUS RESISTANCE

One of the most effective approaches for the implicit manipulation of antivirals is to hinder the identification and access of the host cell (Galdiero et al., 2011). Metal oxide nanoparticles (MeONPs) have been successfully utilized to recognize and interact with outer proteins, specifically viral glycoproteins and their receptors. In agriculture, MeNPs can be applied to restrict viruses within the seed coat (Figure 5.2), as virus transmission in plants often occurs through seeds. This transmission can happen either directly or indirectly. In indirect virus transmission, the seed coat or seed embryo becomes infected through plant gametes prior to fertilization. In direct transmission, the mother plant becomes infected through the embryonic tissue after fertilization (Cobos et al., 2019). MeNPs offer potential advantages as seed antiviruses due to their antiviral properties and their size, which may help confine virus particles within the seeds. Furthermore, MeNPs could be applied to seedlings to limit virus infection (Elbeshehy et al., 2015). However, *in vitro* propagation is considered the most powerful technique for virus elimination in seedlings (Wang et al., 2018). Recent research has shown that the virus suppression activity of MeNPs is particularly strong when applied prior to infection, enhancing their preventive effects (Cai et al., 2020) (Figure 5.2).

In general, the transmission of different viruses, particularly phloem-limited ones (such as ssDNA), for pathogenicity assays and viral vectors is achieved through Agrobacterium-mediated inoculation. However, there are situations where this method is not feasible due to the restricted host selection of the virus and/or viral vector, as well as limitations with Agrobacterium. As an alternative, a generic procedure for virus/vector transmission utilizes polyethylene glycol (PEG), which can be challenging when plant regeneration from protoplasts is not suitable for every host plant and tissue type. On the other hand, it has been discovered that NPs-mediated delivery systems using PEG-coated carbon nanotubes (CNTs), such as single-walled carbon nanotubes (SWCNTs), can overcome some of these obstacles (Ariga et al., 2020). PEG-coated SWCNTs (PEG-SWCNTs) provide a large attachment zone for genetic factors (RNA/DNA). These biocompatible carbon-based nanomaterials (NMs) can protect organic molecules (amino acids, proteins, polypeptides) from host cellular enzymes (DNA nucleases). Additionally, these NMs facilitate the alteration of RNA/DNA and the transfer of essential genetic material without integration into the plant cell genome (Derbalah & Elsharkawy, 2019). Interestingly, NMs have shown a high ability to deliver CRISPR components into undifferentiated or incompletely differentiated cells or germline tissues, regardless of species or genotype, both *in vivo* and *in vitro*. Therefore, they have the potential to significantly enhance gene editing in plants. Recently, SWCNTs have been applied for the successful transfer of synthetic siRNA within whole plant tissues in order to disrupt endogenous genes in plants (Demirer et al., 2020). Awan et al. (2022) reviewed that after infiltration in plants, SWCNTs-loaded siRNA displayed excessive plant cell endocytosis as well as inactive gene expression efficacy, which further authenticates the application of SWCNTs-organic molecule hybrids as a capable substitute for non-GMO (genetically modified organism) and tissue culture-free regeneration of plants. The in-planta transfer of NMs plated with miniature CRISPR-AsCas12f1, single-strand guide RNA (sgRNA)-expressing dsDNA construct, RNA template, or ubiquitously/heterogeneous nuclear ribonucleoproteins (RNPs) can be exploited to generate non-transgenic plants (Awan et al., 2022). Additionally, the high-accelerating transfer of plated NMs into pollens, besides implacably enhancing their effectiveness, can be used to produce edited pollens for pollination and anther culture to generate modified haploid or dihaploid production of plants.

## 5.11 CONCLUSION

To address the challenge of feeding the growing global population sustainably, agriculture faces the task of effectively managing both crop production and protection. Plant diseases, including viruses, significantly contribute to the decline in food, fiber, and ornamental crop yields. Plant viruses have been particularly challenging, causing epidemics and pandemics that result in substantial agricultural losses worldwide. To combat this menace, it is recommended to adopt an integrated management approach. Immunization and prophylaxis are two dynamic approaches used for virus disease management, either individually or in combination as part of an integrated disease management (IDM) plan.

Conventional breeding methods have traditionally been employed to develop plants capable of tolerating viral infections. MAS has revolutionized this process, significantly reducing the time required for variety development. Over the past two decades, advanced biotechnological tools, such as genome sequencing, genetic engineering, and tissue culture, have further accelerated the production of pathogen-resistant plants. Molecular biology advancements have revealed the innate mechanisms of plant tolerance, resistance, and immunity against phytopathogens at the sub-cellular level. Leveraging this knowledge, resistance against various pathogens, including viruses, has been successfully developed in plants.

In this regard, RNA silencing has been recognized as a primary innate defense strategy in plants against viral infections. It is a homology-dependent mechanism in which the viral genome, upon entering the plant cell, is converted into dsRNA and detected by the RNAi machinery of the host plant. The invading RNA molecule is ultimately broken down into siRNA fragments, which serve as a silencing signal for future invasions by the same virus. Resistance obtained through siRNA, miRNA, and DNA methylation is based on this mechanism, which has been extensively and successfully exploited against various viral diseases. dsRNA constructs are synthesized and cloned into artificially designed expression vectors for subsequent transformation in plants using Agrobacterium-mediated transformation or biolistic inoculation methods. More recently, the delivery and sustainability of these constructs have been further improved through the use of nanoparticle-based delivery systems.

A second layer of defense against viruses is provided by resistance genes, known as dominant and recessive resistance. HR and ER, induced by the interaction of R genes of the host and Avr genes of the pathogen, result in R gene-mediated resistance, known as dominant resistance. R genes from several monocots, dicot, and wild plants have been identified, isolated, and introduced in crop plants to make them resistance against viral infections. Contrary to the R genes, plant genome encompasses some susceptibility factors (or S genes) required for plant viruses to interact and to cause infection. Loss or mutation of such host genes required by the virus may induce resistance to that virus, termed as recessive resistance. eIFs and their isoforms of the plants are mainly found interacting with invading viruses, and the plants with mutant eIFs have shown promising resistance response. Recessive resistance is more frequently found against plant viruses as compared to other plant pathogens.

More recently, genome editing techniques have further facilitated resistance development protocols. CRISPR/Cas9 is the most revolutionary genome editing tool of the present age, with diverse applications including the development of resistance in plants against pathogens such as viruses. The system consists of a crRNA that serves as a guide for Cas9 endonuclease based on sequence similarity. Cas9 cleaves the target sequence at a specific site, and the host cell machinery repairs the break, resulting in mutations and edited DNA. Using this technique, transgenic plants containing target virus sequences as guide RNA (gRNA) and Cas9 are prepared. Upon challenge with the virus, these plants show resistance to viral infections. An alternate approach is the use of mutant, non-transgenic plants in which S Genes of the host are mutated, making them unable to interact with viruses. Currently, most of the research on resistance in plants against viral infections is based on this system. However, resistance cannot be sustained for long unless proper prophylactic measures are taken. These measures include: (i) regulatory control ensuring the implementation of quarantine laws, seed certification, and crop inspection systems; (ii) cultural control to reduce the virus inoculum including rogueing of volunteer plants, adopting sanitation methods, erecting yellow traps in the field, mulching of yellow polyethylene reflectants, and planting trap crops; and (iii) chemical control of insect vectors.

## KEYWORDS


- **chemical control**
- **CRISPR Cas9**
- **cross protection**
- **marker assisted selection**
- **quarantine**
- **recombinant DNA technology**
- **resistance development**
- **RNA interference**


## REFERENCES


Agrios, G., (2005). *Plant Pathology* (5th edn.). Elsevier Academic Press, Amsterdam, 26-27398-401.

Ahmad, N., Khan, M. A., Khan, N. A., Binyamin, R., & Khan, M. A., (2011). Identification of resistance source in potato germplasm against PVX and PVY. *Pak. J. Bot., 43*(6), 2745–2749.

Akhtar, S., (2014). *Characterization of Defective DNA-β Associated with Cotton Leaf Curl Disease and Their Use in Developing Enhanced Resistance*. PhD thesis, Quaid-I-Azam University Islamabad, Pakistan.

Akhtar, S., Tahir, M. N., Amin, I., & Mansoor, S., (2021). Amplicon-based RNAi construct targeting beta-C1 gene gives enhanced resistance against cotton leaf curl disease. *3 Biotech, 11*, 256. https://doi.org/10.1007/s13205-021-02816-6.

Ali, Z., Abulfaraj, A., Idris, A., Ali, S., Tashkandi, M., & Mahfouz, M. M., (2015). CRISPR/Cas9-mediated viral interference in plants. *Genome Biol., 16*, 238. doi: 10.1186/s13059-015-0799-6.

Ali, Z., Ali, S., Tashkandi, M., Zaidi, S. S., & Mahfouz, M. M., (2016). CRISPR/Cas9-mediated immunity to *Geminiviruses*: Differential interference and evasion. *Sci. Rep., 6*, 26912. doi: 10.1038/srep30223.

Anderson, P. K., Cunningham, A. A., Patel, N. G., Morales, F. J., Epstein, P. R., & Daszak, P., (2004). Emerging infectious diseases of plants: Pathogen pollution, climate change and agrotechnology drivers. *Trends Ecol. Evol., 19*, 535–544.

Ariga, H., Toki, S., & Ishibashi, K., (2020). *Potato virus X* vector-mediated DNA-free genome editing in plants. *Plant and Cell Physiology, 61*, 1946–1953.

Awan, M. J. A., Amin, I., & Mansoor, S., (2022). Mini CRISPR-Cas12f1: A new genome editing tool. *Trends Plant Sci., 27*, 110–112.

Baltes, N. J., Hummel, A. W., Konecna, E., Cegan, R., Bruns, A. N., Bisaro, D. M., et al., (2015). Conferring resistance to *Geminiviruses* with the CRISPR–Cas prokaryotic immune system. *Nat. Plants, 1*, 15145. doi: 10.1038/nplants.2015.145.

Bao, N., Lye, K. W., & Barton, M. K., (2004). MicroRNA binding sites in *Arabidopsis* class III HD-ZIP mRNAs are required for methylation of the template chromosome. *Developmental Cell, 7*(5), 653–662.

Bendahmane, A., Querci, M., Kanyuka, K., & Baulcombe, D. C., (2000). Agrobacterium transient expression system as a tool for the isolation of disease resistance genes: Application to the Rx2 locus in potato. *Plant J., 21*, 73–81. doi: 10.1046/j.1365-313x.2000.00654.x.

Bolotin, A., Quinquis, B., Sorokin, A., & Ehrlich, S. D., (2005). Clustered regularly interspaced short palindrome repeats (CRISPRs) have spacers of extrachromosomal origin. *Microbiology, 151*, 2551–2561. doi: 10.1099/mic.0.28048-0.

Breitbart, M., & Rohwer, F., (2005). Here a virus, there a virus, everywhere the same virus? *Trends in Microbiology, 13*(6), 278.

Broadbent, L., (1976). Epidemiology and control of tomato mosaic virus. *Annu. Rev. Phytopathol., 14*, 75–96. doi: 10.1146/annurev.py.14.090176.000451.

Brotman, Y., Normantovich, M., Goldenberg, Z., Zvirin, Z., Kovalski, I., Stovbun, N., et al., (2013). Dual resistance of melon to *Fusarium oxysporum* races 0 and 2 and to *Papaya ring spot virus* is controlled by a pair of head-to-head oriented NB-LRR genes of unusual architecture. *Mol. Plant, 6*, 235–238. doi: 10.1093/mp/sss121.

Cai, L., Jia, H., Liu, C., Wang, D., & Sun, X., (2020). Foliar exposure of $Fe_3O_4$ nanoparticles on *Nicotiana benthamiana*: Evidence for nanoparticles uptake, plant growth promoter and defense response elicitor against plant virus. *J. Hazard. Mater., 393*.

Chandrasekaran, J., Brumin, M., Wolf, D., Leibman, D., Klap, C., Pearlsman, M., et al., (2016). Development of broad virus resistance in non-transgenic cucumber using CRISPR/Cas9 technology. *Mol. Plant Pathol., 17*, 1140–1153. doi: 10.1111/mpp.12375.

Charron, C., Nicolaï, M., Gallois, J. L., Robaglia, C., Moury, B., Palloix, A., & Caranta, C., (2008). Natural variation and functional analyses provide evidence for co-evolution between plant eIF4E and potyviral VPg. *Plant J., 54*, 56–68.

Choi, S. H., Nakahara, K. S., Andrade, M., & Uyeda, I., (2012). Characterization of the recessive resistance gene cyv1 of *Pisum sativum* against *Clover yellow vein virus*. *J. Gen. Plant Pathol. 78*, 269–276.

Cobos, A., Montes, N., López-Herranz, M., Gil-Valle, M., & Pagán, I., (2019). Within-host multiplication and speed of colonization as infection traits associated with plant virus vertical transmission. *J. Virol., 93*, e01078–01019.

Collier, S. M., Hamel, L. P., & Moffett, P., (2011). Cell death mediated by the N-terminal domains of a unique and highly conserved class of NB-LRR protein. *Mol. Plant Microbe Interact., 24*, 918–931. doi: 10.1094/MPMI-03- 11-0050.

Cosson, P., Schurdi-Levraud, V., Le, Q. H., Sicard, O., Caballero, M., Roux, F., et al., (2012). The RTM resistance to potyviruses in *Arabidopsis thaliana*: Natural variation of the RTM genes and evidence for the implication of additional genes. *PLoS One, 7*, e39169. doi: 10.1371/journal.pone.0039169.

Delaney, T. P., Uknes, S., Vernooij, B., Friedrich, L., Weymann, K., Negrotto, D., et al., (1994). A central role of salicylic-acid in plant-disease resistance. *Science, 266*, 1247–1250. doi: 10.1126/science.266.5188.1247.

Demirer, G. S., Zhang, H., Goh, N. S., Pinals, R. L., Chang, R., & Landry, M. P., (2020). Carbon nanocarriers deliver siRNA to intact plant cells for efficient gene knockdown. *Science Advances,* 6.

Derbalah, A., & Elsharkawy, M., (2019). A new strategy to control *Cucumber mosaic virus* using fabricated NiO-nanostructures. *J. Biotechnol, 306*.

Diaz-Pendon, J. A., Truniger, V., Nieto, C., Garcia-Mas, J., Bendahmane, A., & Aranda, M. A., (2004). Advances in understanding recessive resistance to plant viruses. *Mol. Plant Pathol., 5*, 223–233.

Dimmock, N. J., Easton, A. J., & Leppard, K. N., (2007). *Introduction to Modern Virology* (6th edn.).

Dubcovsky, J., (2011). *Marker-Assisted Selection in Wheat.* Published online by the USDA National Institute of Food and Agriculture. http://maswheat.ucdavis.edu/ (accessed on 28 February 2024).

Elbeshehy, E. K. F., Elazzazy, A. M., & Aggelis, G., (2015). Silver nanoparticles synthesis mediated by new isolates of *Bacillus* spp., nanoparticle characterization and their activity against *Bean yellow mosaic virus* and human pathogens. *Frontiers in Microbiology, 6*, 453–453.

Feng, Z., Xue, F., Xu, M., Chen, X., Zhao, W., Garcia-Murria, M. J., et al., (2016). The ER-membrane transport system is critical for intercellular trafficking of the NSm movement protein and tomato spotted wilt *tospovirus. PLoS Pathog., 12*, e1005443. doi: 10.1371/journal.ppat.1005443.

Galdiero, S., Falanga, A., Vitiello, M., Cantisani, M., Marra, V., & Galdiero, M., (2011). Silver nanoparticles as potential antiviral agents. *Molecules, 16*, 8894–8918.

Giner, A., Pascual, L., Bourgeois, M., et al., (2017). A mutation in the melon vacuolar protein sorting 41prevents systemic infection of *Cucumber mosaic virus. Sci. Rep., 7*, 10471. https://doi.org/10.1038/s41598-017-10783-3.

Gupta, P. K., Langridge, P., & Mir, R. R., (2010). Marker-assisted wheat breeding: Present status and future possibilities. *Mol. Breed., 26*, 145–161.

Gururani, M. A., Venkatesh, J., Upadhyaya, C. P., Nookaraju, A., Pandey, S. K., & Park, S. W., (2012). Plant disease resistance genes: Current status and future directions. *Physiol. Mol. Plant, Pathol. 78*, 51–65. doi: 10.1016/j.pmpp.2012.01.002.

Hallwass, M., Silva Da, O. A., Dianese, E., Lohuis, D., Boiteux, L. S., InoueNagata, A. K., et al., (2014). The *Tomato spotted wilt virus* cell-to-cell movement protein (NSm) triggers a hypersensitive response in Sw-5 containing resistant tomato lines and *Nicotiana benthamiana* transformed with the functional Sw-5b resistance gene copy. *Mol. Plant Pathol.* doi: 10.1111/mpp.12144.

Hart, J. P., & Griffiths, P. D., (2013). A series of eIF4E alleles at the Bc-3 locus are associated with recessive resistance to *Clover yellow vein virus* in common bean. *Theor. Appl. Genet. 126*, 2849–2863.

Hashimoto, M., Neriya, Y., Yamaji, Y., & Namba, S., (2016). Recessive resistance to plant viruses: Potential resistance genes beyond translation initiation factors. *Front. Microbiol., 7*, 1695. doi: 10.3389/fmicb.2016.01695.

Hayes, A. J., Jeong, S. C., Gore, M. A., Yu, Y. G., Buss, G. R., Tolin, S. A., et al., (2004). Recombination within a nucleotide-binding-site/leucine-rich-repeat gene cluster produces new variants conditioning resistance to *Soybean mosaic virus* in soybeans. *Genetics, 166*, 493–503. doi: 10.1534/genetics.166.1.493.

Hewezi, T., Lane, T., Piya, S., Rambani, A., Rice, J. H., & Staton, M., (2017). Cyst nematode parasitism induces dynamic changes in the root epigenome. *Plant Physiol., 174*, 405–420. doi: 10.1104/pp.16.01948.

Hofinger, B. J., Russell, J. R., Bass, C. G., Baldwin, T., Dos, R. M., Hedley, P. E., et al., (2011). An exceptionally high nucleotide and haplotype diversity and a signature of positive selection for the eIF4E resistance gene in barley are revealed by allele mining and phylogenetic analyses of natural populations. *Mol. Ecol., 20*, 3653–3668.

Hull, R., (2002). *Mathews' Plant Virology*. London, UK: Elsevier.

Hwang, J., Li, J., Liu, W. Y., An, S. J., Cho, H., Her, N. H., Yeam, I., Kim, D., & Kang, B. C., (2009). Double mutations in eIF4E and eIFiso4E confer recessive resistance to *Chilli veinal mottle virus* in pepper. *Mol. Cells, 27*, 329–336.

Ishibashi, K., Masuda, K., Naito, S., Meshi, T., & Ishikawa, M., (2007). An inhibitor of viral RNA replication is encoded by a plant resistance gene. *Proc. Natl. Acad. Sci. U.S.A., 104*, 13833–13838. doi: 10.1073/pnas.07032 03104.

Ishibashi, K., Mawatari, N., Miyashita, S., Kishino, H., Meshi, T., & Ishikawa, M., (2012). Coevolution and hierarchical interactions of *Tomato mosaic virus* and the resistance gene Tm-1. *PLoS Pathog., 8*, e1002975. doi: 10.1371/journal.ppat.1002975.

Ishikawa, M., Naito, S., & Ohno, T., (1991). Isolation of mutants of *Arabidopsis thaliana* in which accumulation of *Tobacco mosaic virus* coat protein is reduced to low levels. *Mol. Gen. Genet., 230*, 33–38. doi: 10.1007/BF00290647.

Ji, X., Zhang, H., Zhang, Y., Wang, Y., & Gao, C., (2015). Establishing a CRISPRCas-like immune system conferring DNA virus resistance in plants. *Nat. Plants, 1*, 15144. doi: 10.1038/nplants.2015.144.

Jiang, J., Patarroyo, C., Garcia, C. D., Zheng, H., & Laliberté, J. F., (2015). The vesicle-forming 6K2 protein of *Turnip mosaic virus* interacts with the COPII coatomer Sec24a for viral systemic infection. *J. Virol., 89*, 6695–6710. doi: 10.1128/JVI.00503-15.

Jiang, L., Du, Z., Zhang, G., Wang, T., & Jin, G., (2022). Advances in RNA-silencing-related resistance against viruses in potato. *Genes, 13*, 731. https://doi.org/10.3390/ genes13050731.

Jin, M., Lee, S. S., Ke, L., Kim, J., Seo, M. S., Sohn, S. H., et al., (2014). Identification and mapping of a novel dominant resistance gene, TuRB07 to *Turnip mosaic virus* in *Brassica* Rapa. *Theor. Appl. Genet., 127*, 509–519. doi: 10.1007/s00122- 013-2237-z.

Jones, J. D. G., & Dangl, J. L., (2006). The plant immune system. *Nature, 444*, 323–329. doi: 10.1038/nature05286.

Jones, R. A. C., Boonham, N., Adams, I. P., & Fox, A., (2021). Historical virus isolate collections: An invaluable resource connecting plant virology's pre-sequencing and post-sequencing eras. *Plant Pathol., 70*, 235–248.

Julio, E., Cotucheau, J., Decorps, C., Volpatti, R., Sentenac, C., Candresse, T., et al., (2015). A eukaryotic translation initiation factor 4E (eIF4E) is responsible for the "VA" tobacco recessive resistance to potyviruses. *Plant Mol. Biol. Rep., 33*, 609–623.

Kaldis, A., Berbati, M., Melita, O., Reppa, C., Holeva, M., Otten, P., et al., (2017). Exogenously applied dsRNA molecules deriving from the *Zucchini yellow mosaic virus* (ZYMV) genome move systemically and protect cucurbits against ZYMV. Mol. *Plant Pathol., 19*, 883–895. doi: 10.1111/mpp.12572.

Kang, B. C., Yeam, I., Frantz, J. D., Murphy, J. F., & Jahn, M. M., (2005). The pvr1 locus in capsicum encodes a translation initiation factor eIF4E that interacts with *Tobacco etch virus* VPg. *The Plant Journal, 42*, 392–405.

Kang, B. C., Yeam, I., Li, H., Perez, K. W., & Jahn, M. M., (2007). Ectopic expression of a recessive resistance gene generates dominant potyvirus resistance in plants. *Plant Biotechnol. J., 5*, 526–536.

Kang, J. W., Shin, D., Cho, J. H., Lee, J. Y., Kwon, Y., Park, D. S., et al., (2019). Accelerated development of rice stripe virus-resistant, near-isogenic rice lines through marker-assisted backcrossing. *PLoS One, 14*(12), e0225974.

Katiyar-Agarwal, S., & Jin, H., (2010). Role of small RNAs in host-microbe interactions. *Annu. Rev. Phytopathol., 48*, 225–246.

Khan, A. J., Mansoor, S., & Briddon, R. W., (2014). Oman: A case for a sink of *begomovirus*es of geographically diverse origins. *Trends Plant Sci., 19*(2), 67–70. doi: 10.1016/j.tplants.2013.11.004. PMID: 24485409.

Kidulile, C. E., Ateka, E. M., Alakonya, A. E., & Ndunguru, J. C., (2018). Efficacy of chemotherapy and thermotherapy in elimination of East *African cassava mosaic virus* from Tanzanian cassava landrace. *J. Phytopathol., 166*, 739–745. https://doi.org/10.1111/jph.12725.

Kis, A., Hamar, É., Tholt, G., Bán, R., & Havelda, Z., (2019). Creating highly efficient resistance against *Wheat dwarf virus* in barley by employing CRISPR/Cas9 system. *Plant Biotechnol. J., 17*, 1004–1006.

Kørner, C. J., Klauser, D., Niehl, A., Domínguez-Ferreras, A., Chinchilla, D., Boller, T., et al., (2013). The immunity regulator BAK1 contributes to resistance against diverse RNA viruses. *Mol. Plant Microbe. Interact., 26*, 1271–1280. doi: 10.1094/MPMI-06-13-0179-R.

Kuroiwa, K., Thenault, C., Nogué, F., Perrot, L., Mazier, M., & Gallois, J. L., (2022). CRISPR-based knock-out of eIF4E2 in a cherry tomato background successfully recapitulates resistance to pepper veinal mottle virus. *Plant Science, 316*, 111160.

Liu, P. P., Bhattacharjee, S., Klessig, D. F., & Moffett, P., (2010). Systemic acquired resistance is induced by R gene-mediated responses independent of cell death. *Mol. Plant Pathol., 11*, 155–160. doi: 10.1111/j.1364-3703.2009.00564.x.

Ma, J. F., Hou, X. L., Xiao, D., Qi, L., Wang, F., Sun, F. F., et al., (2010). Cloning and characterization of the BcTuR3 gene related to resistance to *Turnip mosaic virus* (TuMV) from non-heading Chinese cabbage. *Plant Mol. Biol. Rep., 28*, 588–596. doi: 10.1007/s11105-010-0183-3.

Macovei, A., Sevilla, N., Cantos, C., Jonson, G., Slamet-Loedin, I., Èermák, T., et al., (2018). Novel alleles of rice eIF4G generated by CRISPR/Cas9-targeted mutagenesis confer resistance to *Rice tungro spherical virus. Plant Biotechnol. J., 16*, 1918–1927. doi: 10.1111/pbi.12927.

Maiti, S., Paul, S., & Pal, A., (2012). Isolation, characterization, and structure analysis of a non-TIR-NBS-LRR encoding candidate gene from MYMIV-resistant *Vigna mungo. Mol. Biotechnol. 52*, 217–233. doi: 10.1007/s12033-011-9488-1.

Maule, A. J., Caranta, C., & Boulton, M. I., (2007). Sources of natural resistance to plant viruses: Status and prospects. *Mol. Plant Pathol., 8*, 223–231.

Moffett, P., (2009). Mechanisms of recognition in dominant R gene mediated resistance. *Adv. Virus Res., 75*, 1–33. doi: 10.1016/S0065-3527(09)07501-0.

Moreno, A. B., & López-Moya, J. J., (2020). When viruses play team sports: Mixed infections in plants. *Phytopathology, 110*, 29–48. doi: 10.1094/PHYTO07-19-0250-F.

Mumford, R., Macarthur, R., & Boonham, N., (2016). The role and challenges of new diagnostic technology in plant biosecurity. *Food Secur., 8*, 103–109. doi: 10.1007/s12571-015-0533-y.

Nawaz-Ul-Rehman, M. S., & Fauquet, C. M., (2009). Evolution of *Geminiviruses* and their satellites. *FEBS Letters, 583*, 1825–1832.

Nishikiori, M., Mori, M., Dohi, K., Okamura, H., Katoh, E., Naito, S., et al., (2011). A host small GTP-binding protein ARL8 plays crucial roles in *Tobamovirus* RNA replication. *PLoS Pathog., 7*, e1002409. doi: 10.1371/journal.ppat.1002409.

Nunez, J. K., Harrington, L. B., & Doudna, J. A., (2016). Chemical and biophysical modulation of Cas9 for tunable genome engineering. *ACS Chem. Biol., 11*, 681–688. doi: 10.1021/acschembio.5b01019.

Nyaga, C., Gowda, M., Beyene, Y., Murithi, W. T., Burgueno, J., Toledo, F., Makumbi, D., et al., (2020). Hybrid breeding for MLN resistance: Heterosis, combining ability, and hybrid prediction. *Plants (Basel, Switzerland), 9*(4), 468. https://doi.org/10.3390/plants9040468.

O'Connell, M. R., Oakes, B. L., Sternberg, S. H., East-Seletsky, A., Kaplan, M., & Doudna, J. A., (2014). Programmable RNA recognition and cleavage by CRISPR/Cas9. *Nature, 516*, 263–266. doi: 10.1038/nature13769.

Ohshima, K., Taniyama, T., Yamanaka, T., Ishikawa, M., & Naito, S., (1998). Isolation of a mutant of *Arabidopsis thaliana* carrying two simultaneous mutations affecting *Tobacco mosaic virus* multiplication within a single cell. *Virology, 243*, 472–481. doi: 10.1006/viro.1998.9078.

Orjuela, J., Deless, E. F., Kolade, O., Chéron, S., Ghesquière, A., & Albar, L., (2013). A recessive resistance to rice yellow mottle virus is associated with a rice homolog of the CPR5 gene, a regulator of active defense mechanisms. *Mol. Plant Microbe Interact., 26*, 1455–1463. doi: 10.1094/MPMI-05-13-0127-R.

Perez, K., Yeam, I., Kang, B. C., Ripoll, D. R., Kim, J., Murphy, J. F., et al., (2012). *Tobacco etch virus* infectivity in *Capsicum* spp. is determined by a maximum of three amino acids in the viral virulence determinant VPg. *Mol. Plant-Microbe Interact., 25*. 1562–1573.

Pyott, D. E., Sheehan, E., & Molnar, A., (2016). Engineering of CRISPR/Cas9- mediated potyvirus resistance in transgene-free *Arabidopsis* plants. *Mol. Plant Pathol., 17*, 1276–1288. doi: 10.1111/mpp.12417.

Rahman, M., Hussain, D., & Zafar, Y., (2002). Estimation of genetic divergence among elite cotton cultivars–genotypes by DNA fingerprinting technology. *Crop Sci., 42*, 2137–2144.

Ronde, D., Butterbach, P., & Kormelink, R., (2014). Dominant resistance against plant viruses. *Frontiers in Plant Science*. doi: 10.3398/fpls.2014.00307.

Rubio, M., Nicolai, M., Caranta, C., & Palloix, A., (2009). Allele mining in the pepper gene pool provided new complementation effects between pvr2-eIF4E and pvr6- eIF(iso)4E alleles for resistance to *Pepper veinal mottle virus*. *J. Gen. Virol., 90*, 2808–2814.

Ruffel, S., Dussault, M. H., Palloix, A., Moury, B., Bendahmane, A., Robaglia, C., & Caranta, C., (2002). A natural recessive resistance gene against *Potato virus* Y in pepper corresponds to the eukaryotic initiation factor 4E (eIF4E). *Plant J., 32*, 1067–1075.

Ruffel, S., Gallois, J. L., Lesage, M. L., & Caranta, C., (2005). The recessive potyvirus resistance gene pot-1 is the tomato orthologue of the pepper pvr2-eIF4E gene. *Mol. Genet. Genomics 274*, 346–353.

Ruffel, S., Gallois, J. L., Moury, B., Robaglia, C., Palloix, A., & Caranta, C., (2006). Simultaneous mutations in translation initiation factors eIF4E and eIF(iso)4E are required to prevent *Pepper veinal mottle virus* infection of pepper. *J. Gen. Virol., 87*, 2089–2098.

Sastry, S. K., & Zitter, T. A., (2014). Management of virus and viroid diseases of crops in the tropics. In: Sastry, S. K., & Zitter, T. A., (eds.), *Plant Virus and Viroid Diseases in the Tropics, Epidemiology and Management* (Vol. 2, pp. 149–480). Springer: Dordrecht, The Netherlands.

Seemanpillai, M., Dry, I., Randles, J., & Rezaian, A., (2003). Transcriptional silencing of geminiviral promoter-driven transgenes following homologous virus infection. *Molecular Plant-Microbe Interactions, 16*, 429–438.

Shahriari, A. G., & Tahmasebi, A., (2018). MicroRNA-mediated resistance to plant viruses. *Journal of Cell and Molecular Research, 10*(1), 40–46.

Sharma, N., Sahu, P. P., Puranik, S., & Prasad, M., (2013). Recent advances in plant-virus interaction with emphasis on small interfering RNAs (siRNAs). *Mol. Biotechnol., 55*, 63–77. doi: 10.1007/s12033-012-9615-7.

Sijen, T., Fleenor, J., Simmer, F., Thijssen, K. L., Parrish, S., Timmons, L., Plasterk, R. H. A., & Fire, A., (2001). On the role of RNA amplification in dsRNA-triggered gene silencing. *Cell, 107*, 465–476.

Solomon-Blackburn, R. M., & Barker, H., (2001). Breeding virus resistant potatoes (*Solanum tuberosum*): A review of traditional and molecular approaches. *Heredity, 86*, 17–35. doi: 10.1046/j.1365-2540.2001.00799.x.

Soosaar, J. L., Burch-Smith, T. M., & Dinesh-Kumar, S. P., (2005). Mechanisms of plant resistance to viruses. *Nat. Rev. Microbiol., 3*, 789–798.

Syller, J., (2012). Facilitative and antagonistic interactions between plant viruses in mixed infections. *Mol. Plant Pathol., 13*, 204–216. doi: 10.1111/j.1364- 3703.2011.00734.x.

Tomita, R., Sekine, K. T., Mizumoto, H., Sakamoto, M., Murai, J., Kiba, A., et al., (2011). Genetic basis for the hierarchical interaction between *Tobamovirus* spp. and L resistance gene alleles from different pepper species. Mol. *Plant Microbe Interact., 24*, 108–117. doi: 10.1094/MPMI-06-10-0127.

Tsujimoto, Y., Numaga, T., Ohshima, K., Yano, M. A., Ohsawa, R., Goto, D. B., et al., (2003). *Arabidopsis* tobamovirus multiplication (TOM) 2 locus encodes a transmembrane protein that interacts with TOM1. *EMBO J., 22*, 335–343. doi: 10.1093/emboj/cdg034.

Vaucheret, H., (2008). Plant argonautes. *Trends Plant Sci., 13*, 350–358. doi: 10.1016/j.tplants.2008.04.007.

Verlaan, M. G., Hutton, S. F., Ibrahem, R. M., Kormelink, R., Visser, R. G., Scott, J. W., et al., (2013). The *Tomato yellow leaf curl virus* resistance genes Ty-1 and Ty-3 are allelic and code for DFDGD-Class RNA-dependent RNA polymerases. *PLoS Genet., 9*, e1003399. doi: 10.1371/journal.pgen.1003399.

Vijayapalani, P., Maeshima, M., Nagasaki-Takekuchi, N., & Miller, W. A., (2012). Interaction of the trans-frame potyvirus protein P3N-PIPO with host protein PCaP1 facilitates potyvirus movement. *PLoS Pathog., 8*, e1002639. doi: 10.1371/journal.ppat.1002639.

Villamor, D. E. V., Ho, T., Al Rwahnih, M., Martin, R. R., & Tzanetakis, I. E., (2019). High throughput sequencing for plant virus detection and discovery. *Phytopathology, 109*, 716–725. doi: 10.1094/PHYTO-07–18–0257-RVW.

Wang, C., Wang, C., Zou, J., Yang, Y., Li, Z., & Zhu, S., (2019). Epigenetics in the plant-virus interaction. *Plant Cell Rep., 38*(9), 1031–1038. doi: 10.1007/s00299-019-02414-0. Epub 2019. PMID: 31065780.

Wang, Z. Z., Zhan, L. Q., & Chen, X. X., (2018). Two types of lysozymes from the whitefly *Bemisia tabaci*: Molecular characterization and functional diversification. *Dev. Comp. Immunol., 81*, 252–261.

Waterhouse, P. M., Wang, M. B., & Lough, T., (2001). Gene silencing as an adaptive defense against viruses. *Nature, 411*, 834–842.

Wen, R. H., Khatabi, B., Ashfield, T., Saghai, M. M. A., & Hajimorad, M. R., (2013). The HC-Pro and P3 cistrons of an avirulent *Soybean mosaic virus* are recognized by different resistance genes at the complex Rsv1 locus. *Mol. Plant Microbe Interact., 26*, 203–215. doi: 10.1094/MPMI-06-12-0156-R.

Whitham, S., Dineshkumar, S. P., Choi, D., Hehl, R., Corr, C., & Baker, B., (1994). The product of the *Tobacco mosaic virus* resistance gene N: Similarity to toll and the interleukin-1 receptor. *Cell, 78*, 1101–1115. doi: 10.1016/0092- 8674(94)90283-6.

Wittmann, S., Chatel, H., Fortin, M. G., & Laliberte, J. F., (1997). Interaction of the viral protein genome linked of *Turnip mosaic potyvirus* with the translational eukaryotic initiation factor (iso) 4E of *Arabidopsis thaliana* using the yeast two-hybrid system. *Virology, 234*, 84–92.

Wright, A. V., Nunez, J. K., & Doudna, J. A., (2016). Biology and applications of CRISPR systems: Harnessing nature's toolbox for genome engineering. *Cell, 164*, 29–44. doi: 10.1016/j.cell.2015.12.035.

Xiao, D., Zhou, K., Yang, X., Yang, Y., Ma, Y., & Wang, Y., (2021). Crosstalk of DNA methylation triggered by pathogen in poplars with different resistances. *Front. Microbiol., 12*, 750089. doi: 10.3389/fmicb.2021.750089.

Yamanaka, T., Ohta, T., Takahashi, M., Meshi, T., Schmidt, R., Dean, C., et al., (2000). TOM1, an *Arabidopsis* gene required for efficient multiplication of a tobamovirus, encodes a putative transmembrane protein. *Proc. Natl. Acad. Sci. U.S.A. 97*, 10107–10112. doi: 10.1073/pnas.170295097.

Yang, P., Perovic, D., Habekuß, A., Zhou, R., Graner, A., Ordon, F., et al., (2013). Gene-based high-density mapping of the gene rym7 conferring resistance to *Barley mild mosaic virus* (BaMMV). *Mol. Breed., 32*, 27–37.

Yin, K., Han, T., Xie, K., et al., (2019). Engineer complete resistance to *Cotton leaf curl multan virus* by the CRISPR/Cas9 system in *Nicotiana benthamiana*. *Phytopathol. Res., 1*, 9. https://doi.org/10.1186/s42483-019-0017-7.

Zhang, L., Chen, H., Brandizzi, F., Verchot, J., & Wang, A., (2015). The UPR branch IRE1-bZIP60 in plants plays an essential role in viral infection and is complementary to the only UPR pathway in yeast. *PLoS Genet., 11*, e1005164. doi: 10.1371/journal.pgen.1005164.

Zou, L. J., Deng, X. G., Han, X. Y., Tan, W. R., Zhu, L. J., Xi, D. H., et al., (2016). Role of transcription factor HAT1 in modulating *Arabidopsis thaliana* response to *Cucumber mosaic virus*. *Plant Cell Physiol., 57*, 1879–1889. doi: 10.1093/pcp/pcw109.

Zvereva, A. S., & Pooggin, M. M., (2012). Silencing and innate immunity in plant defense against viral and non-viral pathogens. *Viruses, 4*, 2578–2597. doi: 10.3390/v4112578.

# PART II

# Recent Trends in Viral Disease Control

CHAPTER 6

# Plant Virus Management: Account of Conventional and Next-Generation Approaches

PARNEETA MISHRA, RASHMI PANDEY, RANJANA CHAUHAN, ASHISH VISHWAKARAMA, MANISHA MISHRA, ANOOP K. SHUKLA, PRADHYUMNA K. SINGH, and SHARAD SAURABH

*Insect Defense Laboratory, Molecular Biology and Biotechnology Division, CSIR–National Botanical Research Institute, Rana Pratap Marg, Lucknow, Uttar Pradesh, India*

## ABSTRACT

Viruses pose a critical threat to plant growth, often hindering agricultural productivity worldwide and jeopardizing food security. Viral epidemics and pandemics have led to catastrophic consequences, emphasizing the urgent need for effective and advanced virus management. Managing plant viruses and associated diseases is challenging due to their ability to be transmitted by multiple insects. Conventional strategies for management involve prophylactic measures such as chemical control of vectors, eradication of infected plants, and the use of resistant cultivars. While these approaches are effective to some extent, they have adverse effects on the environment and come with certain limitations. The application of modern technologies, including genetic engineering, RNA interference (RNAi), genome editing, and nanotechnology, offers promising potential for controlling plant viral diseases. These next-generation approaches are more effective than conventional strategies and are relatively specific to targeted viruses. In this chapter, we discuss both conventional and next-generation methods developed thus far for combating plant viral infections. We also shed light on non-transgenic

Detection and Management of New and Emerging Mystery Plant Virus Sources.
Alireza Golnaraghi, Rajarshi Kumar Gaur, and Rakesh Kumar Verma (Eds.)

strategies that can avoid complexities associated with genetically modified crops, thereby addressing concerns related to the use of foreign DNA.

## 6.1 INTRODUCTION

Plant viruses represent the most invasive group of pathogens that pose a constant constraint to sustainable agriculture. Viral infections lead to drastic effects on plant health, causing boundless loss to agriculture and declining global crop production every year (Mumford, Macarthur, & Boonham, 2016; Rubio, Galipienso, & Ferriol, 2020). There are 1,901 plant viruses comprising 6 orders, 32 families, and 141 genera, as per the International Committee on Taxonomy of Viruses (Hančinský et al., 2020). Nevertheless, in the current scenario, plant virus populations are speculated to be more complex owing to the high evolvability of the viral genome (Rubio, Galipienso, & Ferriol, 2020).

Plant viral infections are inevitable due to the availability of diverse pest vectors, lack of curable methods, and growing international trade. The increased occurrence of plant viral diseases is a significant threat to food security and needs thorough addressal (Rubio, Galipienso, & Ferriol, 2020; Jones, 2021). Though a multitude of prophylactic measures are practiced in different parts of the world to counteract virus attacks, the growing of resistant cultivars together with the application of pesticides is the only strategy that has significantly contributed to circumventing plant viral infections (Kumar, Singh, & Kianian, 2017). However, the latter is associated with certain drawbacks that limit their efficacious utilization. Given that viral diseases lay an immense setback to the world economy and the pitfalls associated with conventional methods, the adoption of next-generation approaches is an urging demand. In the era of today when biotechnology has flourished on the horizon, engineered virus resistance is the most reliable practice for managing plant viral diseases. Amongst varied approaches, the concept of pathogen-derived resistance (PDR) and RNA interference (RNAi) have been extensively exploited for generating viral resistance (Gandhi et al., 2021). Further rapid advances in the CRISPR/CAS system and nanotechnology have added new dimensions for combating plant viral infections (Liu et al., 2018; Worrall et al., 2018).

This chapter provides an account of the various approaches that have been utilized thus far to achieve virus resistance. We evaluate different paradigms on a case-by-case basis and offer insights into the advantages and disadvantages of each strategy. Additionally, we provide a glimpse of emerging approaches that may make a significant impact in the near future. Our goal is to present a comprehensive overview that will assist readers from diverse backgrounds in understanding the latest advancements in plant virus management.

## 6.2 CONVENTIONAL METHODS OF MANAGING PLANT VIRAL INFECTION

Diverse conventional approaches are practiced worldwide to protect plants from viral infections. Eradicating viral infections after transmission is unlikely feasible due to the highly evolved and complex viral machinery. Most of these measures are prophylactic, including strict quarantine and certification of plants and plant parts, control of natural vectors, application of chemicals, along with timely detection and removal of infected plants on a regular basis (Rubio, Galipienso, & Ferriol, 2020). The reliability of available diagnostic methods is crucial for managing viral diseases in the fields, as some preventive methods rely on early and accurate detection (Rubio, Galipienso, & Ferriol, 2020). Spraying chemicals is a commonly followed method to restrain virus growth, which has proven effective in preventing viral threats to some extent. The presence of viruses as intracellular particles in the host plant makes the direct application of chemicals ineffective in combating viral infection, but it proves effective in killing viral vectors, thereby restricting virus transmission (Castle, Palumbo, & Prabhaker, 2009). Moreover, the widespread use of chemicals is not only hazardous to the ecosystem but may also affect the fitness of host plants, limiting their productivity (Sharma et al., 2019; Tudi et al., 2021).

The use of genetically resistant plants is one of the most ancient conventionally used strategies for controlling viral infections in the field. These strategies rely on the host plants' ability to recognize pathogens and activate a hypersensitive cell death response (HR) (Kumar, Singh, & Kianian, 2017). This phenomenon is known as host plant resistance (HPR). There is a vast amount of literature supporting the importance of HPR in combating viral infections in various crops. Wild relatives of domesticated crops harbor several resistance genes that can be effectively utilized for antiviral breeding (Table 6.1). One notable example is the application of HPR to control *tomato yellow leaf curl virus* (TYLCV). While domesticated tomato varieties are naturally susceptible to TYLCV, closely related wild species exhibit a high level of resistance to the virus (Verlaan et al., 2013). Studies have shown that TYLCV-resistant cultivars may still contain a detectable amount of the virus, but the viral load remains below 10% of that found in susceptible cultivars due to the presence of *Ty* genes (*Ty1–Ty6*). *Ty-1*, *Ty-3*, *Ty-4*, and *Ty-6* are derived from *Solanum chilense*, while *Ty-2* and *Ty-5* are derived from *Solanum habrochaites* and *Solanum esculentum,* respectively (Hanson, Green, & Kuo, 2006). Gill et al. (2019) demonstrated the effectiveness of Ty-6 alone and in combination with *Ty-3* and *Ty-5* through introgression. They found that Ty-6 alone can provide protection against TYLCV but combining *Ty-6* with *Ty-3* and *Ty-5* separately can enhance the level of resistance (Gill et al., 2019). Recently, Voorburg et al. (2020) discovered that *Ty-1* introgressed lines of tomato also exhibit resistance to various TYLCV-like viral species, such as *tomato yellow leaf curl Sardinia virus* (TYLCSV) and BCTV, suggesting that *Ty-1* can confer broad-spectrum viral resistance.

One of the *Ty* genes, *Ty6*, can also provide potential protection against *tomato mottled virus* (ToMoV), as revealed by the study conducted by Gill et al. (2019). They found that *Ty6* effectively reduces the severity of ToMoV infection. Additionally, *Tm-1* from *Solanum habrochites,* and *Tm-2* and *Tm2*$^2$ from *Solanum peruvianum*, have been reported to confer protection against ToMV (Foolad, 2007; Lanfermeijer et al., 2003). Another set of natural resistance genes from wild tomato species (*Lycopersicon peruvianum, Lycopersicon chilense, Lycopersicon pimpinellifolium*), designated as "*Sw-1 to Sw-7*," have been found to provide defense against wilt disease caused by *tomato spotted wilt virus* (TSWV) (Finlay, 1953). Among these, "*Sw-5*" imparts strong resistance and is also present in many commercially available TSWV-resistant tomato cultivars such as Tycoon, Tous 91, Talladega, Red Defender, BHN 444, and Nico (Riley et al., 2011).

There are several reports that highlight the importance of HPR in controlling soybean mosaic diseases caused by *soybean mosaic virus* (SMV) and *bean common mosaic virus* (BCMV). Genetic loci conferring resistance against SMV strains have been identified in various soybean germplasms. The V94-5152 variety carries the *Rsv4* locus, which provides resistance against all seven SMV-G strains. Similarly, the *Rsv3* locus in the soybean cultivar "L29" also confers strong resistance against SMV isolates (Seo, Lee, & Kim, 2009). Liu et al. (2021) crossed "V94-5152" with the susceptible cultivar "Williams 82" and found that the resulting hybrids were resistant to both BCMV and SMV. Wu et al. (2017) evaluated the performance of SMV-resistant soybean genotypes against BCMV isolates and found that only some of the genotypes conferred resistance to all tested BCMV strains, likely due to the presence of *Rsv1*, *Rsv3*, *Rsv4*, and *Rsv5*.

**TABLE 6.1** List of Antiviral Resistance Genes

| Resistance Gene | Accession/Cultivar/Line/Ecotype | Linkage Group | Source of Resistance Gene | Virus(s) Targeted by the Resistance Gene | References |
|---|---|---|---|---|---|
| Ty-1 | LA1969 | Chromosome 11 | *Solanum chilense* | Tomato yellow leaf curl virus (TYLCV), tomato yellow leaf curl Sardinia virus (TYLCSV), tomato mottle virus (ToMoV) | Verlaan et al. (2013) |
| Ty-3 | LA1932/LA2779/LA1938, | Chromosome 6 | *Solanum chilense* | Tomato yellow leaf curl virus (TYLCV) | Gill et al. (2019) |
| Ty-2 | B6013 | Chromosome 6 | *Solanum habrochaites* | Tomato yellow leaf curl virus (TYLCV) | Shen et al. (2020); Yamaguchi et al. (2018) |
| Ty-4 | LA1932 | Chromosome 3 | *Solanum chilense* | Tomato yellow leaf curl virus (TYLCV) | Ji, Schuster, & Scott (2007) |
| Ty-5 | TY 172 | Chromosome 4 | *Solanum esculentum* "tyking" | Tomato yellow leaf curl virus (TYLCV) | Lapidot et al. (2015) |
| Ty-6 | LA2779 | Chromosome 10 | *Solanum chilense* | Tomato yellow leaf curl virus (TYLCV), tomato mottle virus (ToMoV) | Gill et al. (2019); Hutton, Ji, & Scott (2015) |
| Sw-7 | LA 1938 | Chromosome 12 | *Solanum chilense* | Tomato spotted wilt virus (TSWV) | Canady et al. (2001) |
| Sw-5 | NC 127S, Stevens | Chromosome 9 | *Solanum peruvianum* | Tomato spotted wilt virus (TSWV), tomato chlorotic spot virus (TCSV) | Boiteux & De Giordano (1993); Gardner & Panthee (2012) |
| Sw-7 | LA 1938 | Chromosome 12 | *Solanum chilense* | Tomato spotted wilt virus (TSWV) | Saidi & Warade (2008) |
| Sw-1a, sw-2, sw-3, sw-4 | Porter's, Pearl Harbor | – | *Solanum pimpinellifolium* | Tomato spotted wilt virus (TSWV) | Finlay (1953) |
| Sw-6 | PE-281/UPV-32 | – | *Solanum peruvianum* | Tomato spotted wilt virus (TSWV) | Roselló et al. (1997) |
| Rx1 | Cara, Atlantic | Chromosome 11 | *Solanum andigena* | Potato virus X (PVX) | Bendahmane, Kanyuka, & Baulcombe (1997) |
| Rx2 | – | Chromosome V | *Solanum acuale* | Potato virus X (PVX) | Bendahmane, Kanyuka, & Baulcombe (1997) |

**TABLE 6.1** *(Continued)*

| Resistance Gene | Accession/Cultivar/ Line/Ecotype | Linkage Group | Source of Resistance Gene | Virus(s) Targeted by the Resistance Gene | References |
|---|---|---|---|---|---|
| Rysto | Payette russet, castle russet | Chromosome 12 | *Solanum stoloniferum* | Potato virus Y (PVY) | Herrera et al. (2018); Grech-Baran et al. (2020) |
| Rychc | DXY.7 | Chromosome 9 | *Solanum chacoense* | Potato virus Y (PVY) | Herrera et al. (2018); Grech-Baran et al. (2020) |
| Ryadg | Uruguayan | Chromosome 11 | *Solanum tuberosum* (group Andigena) | Potato virus Y (PVY) | Herrera et al. (2018); Grech-Baran et al. (2020) |
| Tm-1 | DJ344478, DJ344481, DJ344505 | Chromosome 2,5 | *Solanum habrochites* | Tomato mosaic virus (ToMV) strains 0 and 2 | Ohmori, Murata, & Motoyoshi (1996) |
| Tm-2, Tm-22 | AF536199, AF536200, AF536201 | Chromosome 9 | *Solanum peruvianum* | Tomato mosaic virus (ToMV) strains 0 and 1 | Lanfermeijer, Warmink, & Hille (2005) |
| RCY1 | C24 | Chromosome 5 | *Arabidopsis thaliana* | Cucumber mosaic virus (CMV-Y) | Takahashi et al. (2002) |
| RPP8 | Landsberg erecta | Chromosome 5 | *Arabidopsis thaliana* | Cucumber mosaic virus (CMV-Y) | Takahashi et al. (2002) |
| HRT | Dijon-17 | Chromosome 5 | *Arabidopsis thaliana* | Turnip crinkle virus (TCV) | Takahashi et al. (2002) |

A few potato varieties har*bor* loci that provide natural resistance to viral diseases in potatoes. The wild potato species, *Solanum chacoense, Solanum tuberosum*, and *Solanum stoloniferum*, possess the "*Ry*" genes that can confer exceptional resistance to *potato virus Y* (PVY) and *potato virus X* (PVX) (Herrera et al., 2018). The $Ry_{sto}$ gene from *Solanum stoloniferum* has been introgressed into many commercial American and European potato varieties, including Payette Russet, Castle Russet, Alicja, White Lady, and Pirola. A few more loci, the "*Ny-1*" from *Solanum tuberosum* cv. Rywal, "*Rx1*" and "*Rx2*" from *Solanum andigen* and *Solanum acuale*, respectively, have also been observed to combat PVY and PVX (Grech-Baran et al., 2020).

Additionally, there may be crop varieties that exhibit a natural ability to restrict viral diseases but have not been investigated in detail. For example, one of the potato cultivars, "Saturna," exhibits an extraordinary resistance response against *tobacco rattle virus* (TRV) (Ghazala & Varrelmann, 2007). It is speculated that this cultivar harbors a gene that could offer protection against TRV, which needs to be identified for imparting resistance in susceptible varieties.

## 6.3 RNA-MEDIATED SILENCING STRATEGIES OF PLANT VIRUS RESISTANCE: RNA INTERFERENCE (RNAI)

Transgenic resistance against plant viruses has been the earliest biotechnological approach for managing plant viral infections and was initially based on PDR. The concept of PDR was introduced by Sanford and Johnston in 1985 (Sanford & Johnston, 1985), which suggests that the expression of genetic elements derived from pathogens as transgenes can provide significant resistance in host plants (Sanford & Johnston, 1985). The initial study that demonstrated the feasibility of PDR for stimulating antiviral defense was conducted by Abel et al. (1986), while they were attempting to develop resistance against *tobacco mosaic virus* (TMV) by expressing the TMV coat protein (CP) in *Nicotiana tobaccum* (Abel et al., 1986). Subsequently, the applicability of this approach was established in numerous other plants against important phytoviruses by expressing viral proteins such as CP, replicase, and movement proteins (MPs) (Baulcombe, 1996; Gottula & Fuchs, 2009; Wilson, 1993). Despite considerable efforts in the PDR-based transgenic approach, only a few virus-resistant transgenic crops have undergone regulatory trials and been released for commercialization. The ZYMV and *watermelon mosaic virus* (WMV) resistant transgenic Squash was the first virus-resistant transgenic crop that was released in 1996 (Fuchs & Gonsalves, 1995; Tricoll et al., 1995). Thereafter, six more PDR-based virus-resistant transgenic crops were approved for commercialization (Table 6.2).

**TABLE 6.2** List of PDR-based Virus Resistant Transgenic Crops Approved for Commercialization

| Crop | Targeted Virus(s) | Introduced Viral Gene(s) | Details of Approval | References |
|---|---|---|---|---|
| Squash | Cucumber mosaic virus (CMV), watermelon mosaic virus (WMV), zucchini yellow mosaic virus (ZYMV) | Coat protein of CMV, WMV, ZYMV | 1996 in USA | Tricoll et al. (1995) |
| Papaya | Papaya ringspot virus (PRSV) | Coat protein of PRSV | 1996 in USA, 2006 in China, 2011 in Japan | Gonsalves (1998, 2006) |
| Potato | Potato leaf roll virus (PLRV) | Replicase and helicase of PLRV | 1998 in USA | Kaniewski & Thomas (2004); Thomas, Kaniewski, & Lawson (1997) |
| Potato | Potato virus Y (PVY) | Coat protein of PVY | 1998 in USA, 1999 in Canada | Kaniewski & Thomas (2004); Newell et al. (1991) |
| Sweet pepper | Cucumber mosaic virus (CMV) | Coat protein of CMV | 1998 in China | Zhu et al. (1996) |
| Tomato | Cucumber mosaic virus (CMV) | Coat protein of CMV | 1999 in China | Yang et al. (1995) |
| Plum | Plum pox virus (PPV) | Coat protein of PPV | 2007 in USA | Ilardi & Tavazza (2015); Scorza et al. (1994) |

The progression of PDR-based transgenic approach was discouraged due to a lack of clarity regarding the underlying mechanisms causing viral inhibition. Initially, constructs harboring genes for viral proteins were introduced with the assumption that transgenic expression of viral protein would induce molecular immunity in the genetically engineered plant. However, the hypothesis did not hold true in every case. It was found that the observed resistance in many cases was due to the gene silencing activity caused by viral RNA sequence, rather than the activity of the viral protein. In fact, in a few instances, the displayed resistance could be partly directed by both viral RNA and viral protein (Baulcombe, 1996; Palukaitis, 2011; Prins et al., 2008).

The RNA-based silencing approaches are referred to as RNA interference (RNAi), as they are attributed to hindering the structure and function of target RNA molecules. RNAi is a conserved gene regulatory mechanism in eukaryotes, which involves a non-coding, regulatory RNA that acts in a sequence-specific manner to downregulates the target gene expression (Voinnet, 2005). Small RNAs, particularly microRNA (miRNA) and short interfering RNAs ranging from 20–24 nucleotides, are the major entities that, in association with Dicer-like (DCL) and argonaute (AGO) family of enzymes, mediate RNAi phenomena in plants. The DCL family of enzymes possesses endoribonuclease activity, which is necessary for the synthesis of regulatory small RNA(s) from the precursor transcripts, while the AGO family proteins are part of the RISC (RNA-induced silencing complex) that mediates gene silencing (Voinnet, 2005; Bouché et al., 2006; Ding & Voinnet, 2007).

Exfoliation of the mechanisms underlying PDR, along with advances in genetic engineering, has prompted virologists to exploit the naturally occurring RNAi tool for plant virus management. The theoretical approach involves introducing an active RNAi module against a specific virus or group of viruses, which would hinder the function of viral RNA(s) soon after the plant detects a viral threat (Kuo & Falk, 2020). This is known as host-induced gene silencing and can be achieved through the transgenic expression of RNAi precursors such as hairpin RNA (hpRNA), double-stranded RNA (dsRNA), sense/antisense RNA, and artificial miRNA precursors (Duan, Wang, & Guo, 2012). There are numerous examples demonstrating the successful application of this tool against various DNA and RNA viruses in diverse host systems (Table 6.3).

In addition to targeting viral sequences, engineered virus resistance can also be achieved through small RNA-directed silencing of host genes. One notable example worth discussing is the modulation of translation initiation factors in plants. Wang et al. (2013) manipulated the expression of *eIF(iso)4E* in *Prunus dometica* through RNAi and found that transgenic lines with a loss of *eIF(iso)4E* were resistant to *plum pox virus* (PPV). Later, Gao et al. (2020) reported that transgenic expression of hpRNA generating siRNAs targeting the *eIF4E* transcript in Soybean could generate broad-spectrum resistance against SMV (Gao et al., 2020). Coilin proteins, which serve as scaffolds in Cajal bodies (CBs), are known to influence the interaction between plants and different viruses. Shaw et al. (2015) demonstrated that RNA silencing of the gene encoding coilin through transgenic expression of coilin hpRNA reduced the susceptibility of transgenic plants *to* PVY and *turnip vein clearing virus* (TVCV) (Shaw et al., 2014).

**TABLE 6.3** List of Transgenic Antiviral RNAi in Diverse Plants

| Virus(s) | Viral Sequence Targeted | Host Plant | Outcome | References |
|---|---|---|---|---|
| Chickpea chlorotic dwarf Pakistan virus (CpCDPKV) | N-terminus of replicase, long intergenic region and N-terminus of movement protein of CpCDPKV. | *Nicotiana benthamiana* | No viral symptoms post CpCDPKV challenge. | Nahid et al. (2011) |
| Rice gall dwarf virus (RGDV) | Pns9 (viroplasm matrix protein) of RGDV | Rice | Robust resistance due to hindered viral replication. | Shimizu et al. (2012) |
| Watermelon silver mottle virus (WSMoV) | L (replicase) of WSMoV | *Nicotiana benthamiana* | Significant resistance against WSMoV | Kung et al. (2012) |
| Cassava brown streak virus (CBSV) and Ugandan cassava brown streak virus (UCBSV) | Coat protein | Cassava | Delayed disease development; resistance against cassava brown streak disease (CBSD). | Odipio et al. (2014); Ogwok et al. (2012) |
| Rice grassy stunt virus (RGSV) | Nucleocapsid or movement proteins | Rice | Limited RGSV multiplication laying strong resistance against RGSV. | Shimizu et al. (2013) |
| Citrus tristeza virus (CTV) | Silencing suppressors (p25, p20, and p23) of CTV | Mexican lime (*Citrus aurantifolia*) | Effective resistance against CTV. | Soler et al. (2012) |
| Cucumber mosaic virus (CMV) | Replicase | Potato | Resistance against CMV | Ntui et al. (2013) |
| Chili-infecting begomoviruses (CIBs) | Viral ORFs AC1/AC2/βC1 | *Nicotiana benthamiana* | Up to 90% decline in viral DNA; resistance against chili leaf curl disease (ChiLCD). | Sharma, Basu, & Chakraborty (2015) |
| Sri Lankan cassava mosaic virus (SLCMV) | Region between the AV2 and AV1 of SLCMV | Cassava (*Manihot esculenta*) | Strong resistance to SLCMV. | Ntui et al. (2015) |
| Soybean mosaic virus (SMV) | Helper component-protease (HC-Pro) of SMV. | Soybean | Vigorous resistance against SMV | Gao et al. (2015) |

**TABLE 6.3** *(Continued)*

| Virus(s) | Viral Sequence Targeted | Host Plant | Outcome | References |
|---|---|---|---|---|
| Sorghum mosaic virus (SrMV) | Coat protein | Sugarcane | Resistance against SrMV | Guo et al. (2015) |
| Rice black-streaked dwarf virus (RBSDV) | S1 (RNA-dependent RNA polymerase), S2 (putative core protein), S6 (RNA silencing suppressor), S10 (outer capsid protein) of RBSDV. | Rice | Stable resistance due to targeting of multiple loci. | Wang et al. (2016) |
| Cotton leaf curl Kokhran virus-Burewala strain (CLCuKoV-Bur) | V2 gene | Cotton | Hindered viral replication resulting in low virus titer. | Yasmeen et al. (2016) |
| Multiple potyvirids | Replicase (NIb) | Soybean (*Glycine max*) | Transgenic lines displayed significantly enhanced resistance to SMV field conditions for consecutively three generations. | Yang et al. (2017) |
| Cotton leaf curl Kokhran virus-Burewala (CLCuKoV-Bu) and cotton leaf curl Multan betasatellite (CLCuMB). | AC1 of CLCuKoV-Bu and the βC1 and satellite conserved region CLCuMB. | Cotton | Resistance against cotton leaf curl virus disease (CLCuD). | Ahmad et al. (2017) |
| Potato virus X (PVX), potato virus Y (PVY), potato virus S (PVS). | Coat protein | *Solanum tuberosum* | Induction of stable antiviral immunity. | Hameed et al. (2017) |
| Sugarcane mosaic virus (SCMV) | Coat protein | Sugarcane | Resistance to SCMV | Aslam et al. (2018) |
| Rice black-streaked dwarf virus (RBSDV) | S7-2 or S8 of RBSDV | Rice | Enhanced resistance against RBSDV. | Ahmed et al. (2017) |
| Soybean mosaic virus (SMV) | P3 cistron of SMV | *Glycine max* | Reduced viral accumulation of seven potyvirus strains; resistance against multiple strains of potyvirus, BCMV, and WMV. | Yang et al. (2018) |

**TABLE 6.3** *(Continued)*

| Virus(s) | Viral Sequence Targeted | Host Plant | Outcome | References |
|---|---|---|---|---|
| Tobacco streak virus (TSV) | Coat protein | *Arachis hypogea* | Resistance to TSV could be induced. | Senthilraja et al. (2018) |
| Rice tungro bacilliform virus (RTBV) and rice tungro spherical virus (RTSV) | 300 bp of RTBV DNA and 300 bp of RTSV cDNA. | Rice | Up to 500-fold reduction in accumulation of RTBV and RTSV. | Sharma, Kumar, & Dasgupta (2018) |
| South African cassava mosaic virus (SACMV) | Region overlapping AC1 and AC4 of SACMV. | Cassava | Tolerance against cassava mosaic disease (CMD). | Walsh et al. (2019) |
| Rice tungro bacilliform virus (RTBV), rice tungro spherical virus (RTSV) | Highly conserved partial sequence of CP3 of RTSV. | *Oryza sativa* | Tungro disease resistance in rice. | Malathi et al. (2019) |
| Wheat streak mosaic virus (WSMV) and Triticum mosaic virus (TriMV). | Nib (replicase) of WSMV and TriMV | Wheat | Dual resistance against WSMV and TriMV. | Tatineni et al. (2020) |
| Potato leafroll virus (PLRV) | ATPase fold region of movement protein. | *Solanum tuberosum* | No rolling symptoms post PLRV infection. | Kumari et al. (2020) |
| Rice black-streaked dwarf virus (RBSDV) | Fragments of S1, S2 and S6 genes | *Oryza japonica* | Complete resistance conferred by RNAi of S6, while partial resistance due to RNAI of S1 and S2. | Feng et al. (2021) |
| Sugarcane mosaic virus (SCMV) | Coat protein | Sugarcane | 57.69% and 82.35% resistant transgenic lines with HpSCMVCp-CaMV and HpSCMVCp-Ubi0. | Widyaningrum et al. (2021) |
| Cucumber green mottle mosaic virus (CGMMV) | Conserved regions of different strains of CGMMV genome. | Cucumber (*Cucumis sativus*) | Hindered accumulation of CGMMV. | Miao et al. (2021) |

To address concerns regarding the transgenic approach for engineering viral resistance, researchers have also explored non-transgenic methods for delivering antiviral RNAi cassettes in various host systems (Shen et al., 2014; Sun et al., 2010; Tenllado et al., 2003). In this approach, *in-vitro* synthesized dsRNA/hpRNA is either mechanically introduced into plant cells or delivered through agrobacterium-mediated transient expression. The first study demonstrating the effectiveness of antiviral RNAi without transgenesis was conducted by Tenllado & Díaz-Ruíz (2001). The authors showed that both mechanical and agro-inoculation of dsRNAs derived from *pepper mild mottle virus* (PMMoV), *tobacco etch virus* (TEV), and *alfalfa mosaic virus* (AMV) could specifically interfere with the targeted viruses (Tenllado & Dıaz-Ruız, 2001). However, this approach had limitations that hindered its further application. Mechanical inoculation of dsRNA could not be scaled up beyond laboratory trials due to the high cost of *in-vitro* dsRNA synthesis (Tenllado et al., 2003). While transient expression via agrobacterium is more cost-effective, there were concerns about agrobacterium mobility from the site of inoculation to other parts of the plant (Tenllado et al., 2003).

Thus, the potential of bacterial systems for *in vivo* expression of antiviral dsRNA was evaluated, which circumvented the issue of longevity and systemic delivery to a significant extent. According to studies, RNAse III deficient bacterial systems delivered in the form of crude bacterial extracts can maintain expression of dsRNA/hpRNA to stimulate RNA silencing of viral sequences. It has also been observed that dsRNA delivered through this method was systematically transmitted and remained stable for several days post-inoculation (Sun et al., 2010; Tenllado et al., 2003; Gan et al., 2010; Yin et al., 2009). The applicability of this method has been witnessed against a few plant viruses, namely TMV (Yin et al., 2009), PMMoV (Tenllado et al., 2003), PPV (Tenllado et al., 2003), *sugarcane mosaic virus* (SCMV) (Gan et al., 2010), PVY (Sun et al., 2010), and *papaya ringspot virus* (PRSV) (Shen et al., 2014), suggesting that it could be an economical, safe, and expeditious method of plant virus management.

## 6.4 APPLICATION OF NANOMATERIAL FOR COMBATING PLANT VIRAL INFECTION

Nanotechnology, the most innovative technology in the present era, has gained unprecedented recognition in diverse spheres of science, including agriculture. It has provided novel options for overcoming versatile challenges involved in the regulation of plant disease management (Worrall et al., 2018; Usman et al., 2020). The initial study highlighting the significance of nanoparticles (NPs) as an antiviral material was performed by Aref et al. in 2012 (Aref et al., 2012) against *barley yellow mosaic virus* (BaYMV). Thereafter, many studies have demonstrated the efficacy of nanomaterial-based agrochemicals in regulating phytoviruses. The growing branch of "Nanophytovirology" has been viewed as the most promising strategy in the direction of plant virus management (Farooq et al., 2021). NPs are highly efficient in penetrating the cell wall and plasma membrane due to their nano-scale size (1–100 nm), large surface area, and high reactivity. These attributes render nanomaterial much superior to their macro-scale counterparts. Thus, antiviral nanomaterial represents a new set of weapons in the war against plant viruses (Rai & Ingle, 2012).

An array of nanomaterials (NMs) has been exploited for combating plant viruses both *in vitro* and *in vivo*. However, the majority of these materials are based on metals or metal oxides. In addition, several non-metals and carbon nanomaterials (CNMs) have also been evaluated for their significance in generating plant antiviral defense (Table 6.4). The antiviral defense directed by NMs may possibly be mediated by a dual mode. There is substantial literature confirming that NPs may exhibit direct toxicity against virus particles and cause viral mortality by preventing viral replication. At the same time, NPs are suggested to boost the innate plant defense system and positively modulate plant fitness, thereby restricting viral growth (Cai et al., 2019, 2020).

A few studies discussing the application of NMs in plant virus resistance are presented here. Elazzazy et al. (2017) developed a composite (GO-AgNPs) by adsorbing silver nanoparticles (AgNPs) onto graphene oxide (GO) sheets to test its effectiveness against *tomato bushy stunt virus* (TBSV). They found that GO-AgNPs exhibited strong activity against TBSV. Noha et al. (2018) demonstrated the antiviral properties of AgNPs against *tomato mosaic virus* (ToMV) and PVY. They reported that AgNPs induced systemic acquired resistance (SAR) in tomatoes, reducing ToMV and PVY infection by 67% and 74% respectively. In China, a group of researchers tested the efficacy of two engineered nanomaterials (ENMs), $Fe_2O_3$ and $TiO_2$, along with two CNMs, multi-walled carbon nanotubes (MWCNTs) and graphene ($C_{60}$), against TuMV in *Nicotiana benthamiana*. The study concluded that both ENMs and MWCNTs could generate resistance against TuMV, as evidenced by significantly hindered viral proliferation on newly emerged leaves. Exposure to $Fe_2O_3$ and $TiO_2$ resulted in a 40% and 20% decrease in the level of TuMV coat protein (CP), respectively, while MWCNTs treatment led to a

**TABLE 6.4** List of Potential Antiviral Nanomaterial Evaluated in Plants

| Nanomaterial | Host Plant | Targeted Virus(s) | Method of Application | Outcome | References |
|---|---|---|---|---|---|
| Gold nanoparticles (AuNPs) | *Hordeum vulgare* | Barley yellow mosaic virus (BaYMV) | Foliar application | Led to dissociation of viral particles. | Aref et al. (2012) |
| Silver nanoparticles (AgNPs) | *Vicia faba* | Bean yellow mosaic virus (BYMV) | Foliar application | Decreased viral titer, along with reduced disease severity. | Jain & Kothari (2014) |
| Silica nanoparticles ($SiO_2$NPs) | *Cucumis sativus* | Papaya ring spot virus (PRSV) | Soil drench method | Reduced virus accumulation | Elsharkawy & Mousa (2015) |
| Silver nanoparticles (AgNPs) | *Nicotiana tabacum* | Tobacco mosaic virus (TMV) | Foliar spray | Enhanced plant immunity leading to increased virus resistance. | Wang et al. (2016) |
| Cerium dioxide nanoparticles ($CeO_2$NPs) | *Nicotiana tabacum*, *Datura stramonium* | Tobacco mosaic virus (TMV) | Leaf disc assay | Reduced viral symptoms | Eugene & Zholobak (2016) |
| Gold nanoparticles (AuNPs) | *Hordeum vulgare* | Barley yellow dwarf virus (BYDV-PAV) | Foliar spray | Deterioration of viral particles | Alkubaisi & Aref (2017) |
| Zinc nanoparticles (ZnONPs) | *Solanum melongena* | Cucumber mosaic virus (CMV) | Foliar application | Substantial reduction in disease incidence and severity. | El-Sawy et al. (2017) |
| Graphene-based silver composite nanoparticles | Lettuce | Tomato bushy stunt virus (TBSV) | Foliar application | Potent activity against TBSV | Elazzazy, Elbeshehy, & Betiha (2017) |
| Silica nanoparticles ($SiO_2$NPs) | *Solanum lycopersicum* | Tomato yellow leaf curl virus (TYLCV) | Foliar spray | Decrease in virus titer, reduced disease severity. | El-Sawy et al. (2018) |
| Silver nanoparticles (AgNPs) | *Solanum tuberosum* | Tomato spotted wilt virus (TSWV) | Foliar application | Reduced TSWV infection and inhibited local lesions. | Shafie, Salama, & Farroh (2018) |
| Iron nanoparticles ($Fe_2O_3$) | *Nicotiana benthamiana* | Turnip mosaic virus (TuMV) | Foliar spray | Efficient reduction in virus replication and infection. | Hao et al. (2018) |

**TABLE 6.4** *(Continued)*

| Nanomaterial | Host Plant | Targeted Virus(s) | Method of Application | Outcome | References |
|---|---|---|---|---|---|
| Silver nanoparticles (AgNPs) | Tomato | Tomato mosaic virus (ToMV) and potato virus Y (PVY). | Foliar application of AgNPs | Decreased virus concentration and disease severity. | Noha, Bondok, & El-Dougdoug (2018) |
| Titanium Dioxide ($TiO_2$NPs) | *Vicia faba* | Broad bean strain virus (BBSV) | Foliar spray and soil drench treatment. | Reduced disease severity | Elsharkawy & Derbalah (2019) |
| Fabricated nickel oxide nanoparticles (NiO) | *Cucumis sativus* | Cucumber mosaic virus (CMV) | Foliar spray | Significant reduction in virus titer, decreased disease severity. | Derbalah & Elsharkawy (2019) |
| Zinc oxide nanoparticles (ZnONPs) and silica nanoparticles ($SiO_2$NPs). | *Nicotiana benthamiana* | Tobacco mosaic virus (TMV) | Foliar application | Inhibited viral replication and accumulation. | Cai et al. (2019) |
| Silver nanoparticles (AgNPs) | Banana plant | Banana bunchy top virus (BBTV) | Foliar application | Reduced disease symptoms | Mahfouze, El-Dougdoug, & Mahfouze (2020) |
| Iron nanoparticles ($Fe_3O_4$) | *Nicotiana benthamiana* | Tobacco mosaic virus (TMV) | Foliar application | Enhanced plant resistance against TMV through upregulation of salicylic acid genes and activation of antioxidants pathways. | Cai et al. (2020) |
| Carbon nanotubes (CNTs) | *Nicotiana benthamiana* | Tobacco mosaic virus (TMV) | Foliar application | Inhibited TMV replication and movement. | Adeel et al. (2021) |
| Quercetin nanoliposomes | *Nicotiana benthamiana* | Tobacco mosaic virus (TMV) | Foliar application | Effective viral control | Wang et al. (2022) |
| Chitosan nanoparticles | Vicia faba | Bean yellow mosaic virus (BYMV) | Foliar application | Inhibited viral disease after 48 hours of spray. | El Gamal et al. (2022) |

70% reduction compared to the control. Graphene did not have a significant effect (Hao et al., 2018).

Zinc oxide (ZnO) and silica oxide ($SiO_2$) NPs also exhibit potent antiviral properties, as concluded from the significantly reduced colonization of TMV in *Nicotiana benthamiana* post-treatment. It has been reported that ZnO and $SiO_2$ NPs suppress TMV infection by deactivating TMV and enhancing plant immunity (Cai et al., 2019). The application of ZnONP and $SiO_2$NPs leads to an increase in catalase activity, peroxidase activity, and the level of reactive oxygen species. Furthermore, both ZnONPs and $SiO_2$NPs increase the expression of genes related to plant systemic resistance. ZnONPs also enhance the level of phytohormones, SA, and abscisic acid by 162% and 517%, respectively (Cai et al., 2019). Elsarkhawy et al. (2019) demonstrated the ability of fabricated nickel NPs in combating *cucumber mosaic viruses* (CMV). They found that the use of fabricated nickel NPs at a concentration of 150 μg/L caused a reduction in CMV accumulation along with the over-expression of some defense-related genes. In a study conducted by Muhammad Adeel et al. (2021), various metal NPs (NiO, Ce, Ag, $TiO_2$, Cu, Si, Zn, Fe), along with carbon nanotubes (CNTs) and graphene ($C_{60}$), were tested against TMV on *Nicotiana benthamiana.* The exposure to metal NPs did not significantly control virus infection, whereas CNTs and graphene exposure caused an 81% and 74% reduction in TMV CP transcript, respectively (Adeel et al., 2021).

Moreover, NPs can serve as potent carriers for the delivery of other antiviral molecules. The application of NPs as carriers adds beneficial attributes, making the antiviral molecule more stable and efficacious compared to the naked antiviral molecule. Recently, a few studies have demonstrated that layered double hydroxide (LDH) NPs can serve as efficient carriers of dsRNA targeting viral sequences, rendering consistent RNAi by protecting the dsRNA from the action of nucleases residing in the cell (Mitter et al., 2017; Worrall et al., 2019). Mitter et al. (2017) examined the potential of dsRNA loaded on LDH (BioClay) for protection against *Cucumovirus* CMV and PMMoV. They reported that one-time BioClay spray was sufficient for imparting viral resistance for a period of 20 days when the virus was mechanically inoculated on both BioClay-treated and non-treated leaves, while the naked dsRNA was effective only for 5 days (Mitter et al., 2017). A somewhat similar study was performed by Worrall et al. (2019), in which dsRNAs (BCMVCP-dsRNA) targeting the coding region of the CP of *BCMV* were loaded onto the LDH NPs, and the cargo (BCMVCP-BioClay) was foliarly applied to *Nicotiana benthamiana* and *Vigna unguiculata*. In this study, it was observed that topical application of BCMVCP-BioClay could be effective against aphid-mediated transmission of BCMV in both host plants (Worrall et al., 2019).

Although nanotechnology has emerged as a blessing for mankind, offering potential protection against plant viral infections, its scope is still limited to laboratories and in its early stages. One major challenge in developing nanomaterial-based pesticides is the evaluation of their toxicity (Ray, Yu, & Fu, 2009). Therefore, it is crucial for the scientific community and government authorities to focus on the responsible use of nanotechnology. Additionally, it is important to understand how potential antiviral NPs function after being absorbed by plants, as the detailed mechanism of action remains largely unknown in most cases. Unraveling these underlying mechanisms would likely address existing concerns and facilitate timely adoption.

## 6.5 GENOME EDITING TECHNOLOGY: A POTENTIAL TOOL FOR DEVELOPING VIRAL RESISTANCE

In the past two decades, numerous studies have demonstrated the application of site-directed nucleases for targeted editing of desired nucleic acid segments with the aim of generating viral resistance in plants. In the present scenario, the clustered regularly interspaced short palindromic repeat and associated protein (CRISPR/Cas) system is the most reliable genome editing tool due to its precision, simplicity, and highly efficient targeting, and has been extensively exploited in diverse plant species for generating viral resistance. This system includes an easily customized guide RNA (gRNA) and a Cas effector protein, which can effectively focus on nearly any genomic location of choice by employing a brief gRNA that targets the loci of interest (Liu et al., 2018; Jiao et al., 2022; Kis et al., 2019; Mubarik et al., 2021; Tashkandi et al., 2018; Tripathi et al., 2019) (Figures 6.1 and 6.2).

Before the establishment of the CRISPR/Cas system, zinc finger nucleases (ZFNs) and transcription activator-like effector nucleases (TALENs) were also used to generate resistance against plant viruses (Table 6.5). ZFNs and TALENs-based approaches were costly, difficult to handle, and time-consuming, which restricted their broad application. Furthermore, the specificity of ZFNs and TALENs is very low compared to the CRISPR/Cas system and designing new proteins for every further DNA alteration is cumbersome (Ceasar et al., 2016; Gaj & Gersbach, 2013; Gupta & Musunuru, 2014).

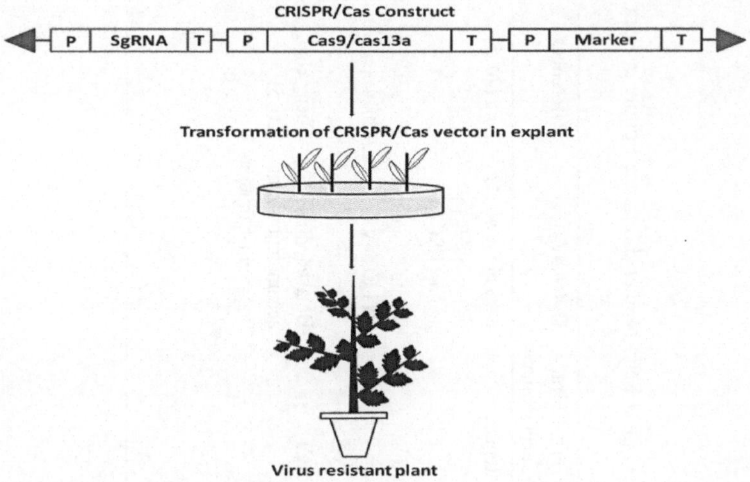


**FIGURE 6.1** Schematic representation showing generation of virus resistant plant. Construction of CRISPR/Cas cassette targeting viral genome and host susceptibility factor, followed by transformation into explants and generation of virus resistant plants.

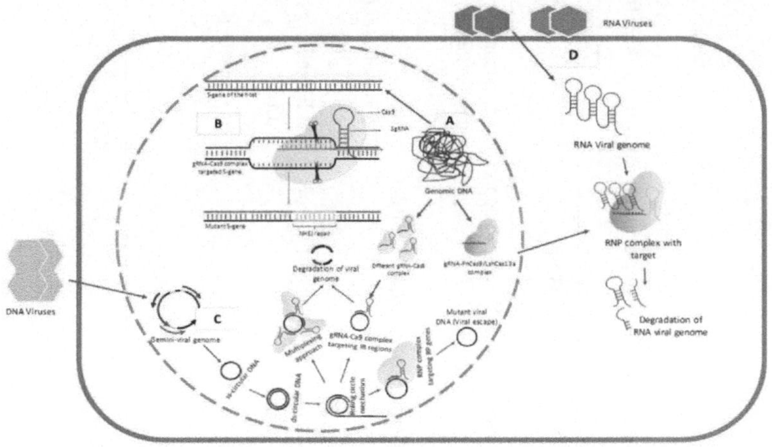


**FIGURE 6.2** Schematic overview of wide spectrum virus resistance in plants through genome editing of host plants as well as viral genome (both DNA and RNA genome).

*Note:* (A) shows integration of CRISPR/Cas cassettes in the host genome; (B) shows targeting of the S-gene (host susceptibility factors) creating deleterious point mutations; (C) represents general organization of Gemini-viral genome and its replication; targeting viral ORF (RP) and non-ORF intergenic region (IR) region through CRISPR/Cas9 system may confer resistance against the virus, but the mutant virus can evade the CRISPR/Cas system if the repaired ORF (RP gene) maintained its function. However, no escape happens when non-ORF intergenic region (IR) of virus is targeted and ultimately virus is degraded within host cells. Another approach is multiplexing in which multiple gRNA targets viral genome, and the repaired viral genome could not multiply; and (D) represents gRNA-FnCas9/LshCas13a-mediated RNA virus interference following degradation of the RNA viral genome in the cytoplasm of plant cells.

**TABLE 6.5** Zinc Finger Nucleases (ZFNs) and Transcription Activator-Like Effector Nucleases (TALENs)-Mediated Genome Editing in Plant Viruses

| Virus(s) | Viral Genome | Family | Vector | Effector Protein | Host Plant | Target Viral Gene/Sequence | Outcome | References |
|---|---|---|---|---|---|---|---|---|
| Beet severe curly top virus (BSCTV) | ss-DNA | Geminiviridae | Beet leafhopper (*Circulifer tenellus*) | Artificial zinc-finger proteins (AZPs) | *Arabidopsis thaliana* | Replicase | 84% of the plants developed resistance against BSCTV. | Sera (2005) |
| Tomato yellow leaf curl China virus (TYLCCNV) | ss-DNA | Geminiviridae | *Bemisia tabaci* | Artificial zinc finger nuclease ZFN (AZFN) | *Nicotiana benthamiana* | AC1 | Immunity against TYLCCNV | Chen et al. (2014) |
| Tobacco curly shoot virus (TbCSV)<br>Tomato yellow leaf curl China virus (TYLCCNV)<br>Tomato leaf curl Yunnan virus (TLCYnV) | ss-DNA | Geminiviridae | *Bemisia tabaci* | TALE1 and TALE2 | *Nicotiana benthamiana* | AC1 | Delayed virus accumulation with very mild symptoms | Cheng et al. (2015) |

The CRISPR/Cas system primarily exists as an adaptive immune system in bacteria, providing protection against phages and promiscuous plasmids. Based on the structure and function of Cas proteins, the CRISPR/Cas system can be classified into two classes: class I and class II. Class I (type I, III, and IV) utilizes a multi-protein effector complex, while class II (type II, V, and VI) relies on the activity of a single effector protein. Cas proteins of type II and V have the ability to edit DNA target sequences, whereas type VI Cas proteins can edit RNA sequences (Makarova et al., 2015; Makarova, Zhang, & Koonin, 2017). The most commonly used CRISPR system is based on the Cas9 endonuclease from *Streptomyces pyogenes*, which cleaves the target sequence upstream of the protospacer sequence, NGG (Wu, Kriz, & Sharp, 2014).

### 6.5.1 CRISPR/CAS-MEDIATED VIRAL GENOME MODIFICATIONS

Several groups have demonstrated the efficiency of the CRISPR/Cas9 system in targeting essential genes of DNA viruses, such as the replicase protein, movement proteins (MPs), and CPs (Table 6.6). One remarkable feature of the CRISPR/Cas system is its ability to target multiple loci in the same or different viral genomes using multiple gRNAs. Tashkandi et al. (2018) achieved immunity against TYLCV in *Nicotiana benthamiana* and tomato by expressing Cas9 along with gRNA targeting the CP and Rep regions of the TYLCV genome for multiple generations. Yin et al. (2019) attempted to generate resistance against the *cotton leaf curl Multan virus* (CLCuMuV) in *Nicotiana benthamiana* by targeting two different regions of the CLCuMuV genome. They observed that *Nicotiana* plants expressing Cas9 along with dual gRNAs targeting the Rep and intergenic region (IR) of CLCuMuV could confer resistance without any symptoms of viral infection (Yin et al., 2019). Kis et al. (2019) demonstrated an excellent example of multiplex viral genome editing by simultaneously disrupting four different regions of the wheat dwarf viral (WDV) genome. Expression of four WDV-specific sgRNAs along with Cas9 was found to develop effective resistance against WDV in barley (Kis et al., 2019). Tripathi et al. (2019) individually targeted three conserved regions of the endogenous *Banana streak virus* (eBSV) genome, resulting in its inactivation in the plant genome. It was observed that 75% of the edited events successfully generated resistance against eBSV under conditions of water stress (Tripathi et al., 2019). Recently, Mubarik et al. (2021) evaluated a multiplexing approach against *Cotton leaf curl viruses* (CLCuVs) in *Nicotiana benthamiana* plants and cotton seedlings. They concluded that events expressing a multiplex CRISPR-Cas9 construct containing gRNAs targeting six genes of CLCuV were able to suppress CLCuV infection more effectively compared to those targeting single regions individually (Mubarik et al., 2021).

Furthermore, several reprogrammed CRISPR-associated nucleases, such as Cas9 from *Francisella novicida* (FnCas9) and Cas13a from *Leptotrichia shahii* (LshCas13a), have been discovered to effectively cleave RNA target sites, leading to successful interference with RNA viral genomes (Table 6.7). Zhang et al. (2018) demonstrated the efficacy of the *Francisella novicida* CRISPR/Cas9 system in generating immunity against the RNA genomes of CMV or TMV in *Arabidopsis* and *Nicotiana*. This immunity was stable and transmitted to the next generation, as evidenced by reduced viral load in the progeny of transgenic plants (Zhang et al., 2018). Zhang et al. (2019) utilized the CRISPR/Cas13a system to impart protection against PVY in the crop plant potato. In this study, the authors were able to generate broad-range resistance by targeting the conserved region of three PVY strains. They also demonstrated that the anti-PVY CRISPR/Cas13a specifically targeted PVY genomes and no other related RNA viral genome sequences such as PVA and PVS (Zhang et al., 2019). Jiao et al. (2022) generated *grapevine leafroll-associated virus 3* (GLRaV-3) resistant grapevine by harnessing the RNA editing ability of both FnCas9 and LshCas13. They found that both FnCas9 and LshCas13a activity restricted GLRaV-3 accumulation, but LshCas13a activity caused more viral inhibition compared to FnCas9 (Jiao et al., 2022).

### 6.5.2 CRISPR/CAS-MEDIATED HOST PLANT GENOME EDITING

Over the past few decades, studies dealing with the phenomena of plant-virus interaction have suggested the critical significance of many plant genes in the establishment of viral infection. The CRISPR/Cas system can be used to generate antiviral immunity by disrupting key genes in the host plant genome that are vital for the infection process of viruses (Table 6.8). One of the best candidates for CRISPR-based plant-driven antiviral defense are the host plant susceptibility genes "S" that aid viral proliferation. The conceptual idea is that the incorporation of deleterious indels in "S" genes limits viral growth by hindering the affinity between the host plant and the virus (Garcia-Ruiz, 2018).

Several studies have demonstrated the proof of concept using the CRISPR/Cas9 module by targeting *eukaryotic translation initiation factor 4E proteins* (*eIF4Es*) and their isoforms. The eIF4Es and their isoforms play an important

**TABLE 6.6** CRISPR/CAS-Mediated DNA Viral Genome Editing

| Virus | Viral Genome | Family | Vector | Effector Protein | Viral Gene/Sequence Targeted | Host Plant | Outcome | References |
|---|---|---|---|---|---|---|---|---|
| *Beet severe curly top virus* (BSCTV) | ss-DNA | Geminiviridae | Beet leafhopper (*Circulifer tenellus*) | *Streptomyces pyogenes* Cas9 (SpCas9) | Intergenic region, coat protein, replicase | *Nicotiana benthamiana* and *Arabidopsis* | Reduced accumulation of BSCTV in the leaves. | Ji et al. (2015) |
| *Merremia mosaic virus* (MeMV) | ss-DNA | Geminiviridae | *Bemisia tabaci* | SpCas9 | Intergenic region, coat protein, replicase. | *Nicotiana benthamiana* | Resistance against MeMV | Ali et al. (2016) |
| *Tomato yellow leaf curl Sardinian virus* (TYLCSV) | ss-DNA | Geminiviridae | *Bemisia tabaci* | SpCas9 | Intergenic region, coat protein, replicase. | *Nicotiana benthamiana* | Resistance against TYLCSV | Ali et al. (2016) |
| *Tomato yellow leaf curl virus* (TYLCV) | ss-DNA | Geminiviridae | *Bemisia tabaci* | SpCas9 | Coat protein, replicase | *Solanum lycopersicum* and *Nicotiana benthamiana* | Low accumulation of TYLCV | Tashkandi et al. (2018) |
| *Cauliflower mosaic virus* (CaMV) | ds-DNA | Caulimoviridae | Aphids | SpCas9 | Coat protein | *Arabidopsis* | Resistance against CaMV. | Liu et al. (2018) |
| *Banana streak virus* (BSV) | ds-DNA | Caulimoviridae | Mealybug | SpCas9 | *ORF-1,2* and *3* of BSV | *Musa* spp. | 75% decrease in virus titer; loss of viral symptoms. | Tripathi et al. (2019) |
| *Wheat dwarf virus* (WDV) | ss-DNA | Geminiviridae | Leafhopper (*Psammotettix alienus* Dahlb.) | SpCas9 | Overlapping region of the movement protein and coat protein, replicase, and long intergenic region. | *Hordeum vulgare* | Resistance against WDV | Kis et al. (2019) |

**TABLE 6.6** *(Continued)*

| Virus | Viral Genome | Family | Vector | Effector Protein | Viral Gene/Sequence Targeted | Host Plant | Outcome | References |
|---|---|---|---|---|---|---|---|---|
| *Cotton leaf curl Multan virus* (CLCuMuV) | ss-DNA | Geminiviridae | *Bemisia tabaci* | SpCas9 | Replicase and intergenic region. | *Nicotiana benthamiana* | Complete resistance to CLCuMuV. | Yin et al. (2019) |
| *Cotton leaf curl virus* (CLCuV) | ss-DNA | Geminiviridae | *Bemisia tabaci* | SpCas9 | Overlapping region of movement protein/coat protein, transcriptional activator protein/replication enhancer protein and replication associate protein/silencing protein. | *Nicotiana benthamiana* | Low virus titer | Mubarik et al. (2021) |

**TABLE 6.7** CRISPR/CAS-Mediated RNA Viral Genome Editing

| Virus(s) | Genome | Family | Vector | Effector Protein | Targeted Viral Sequence | Host Plant | Outcome | References |
|---|---|---|---|---|---|---|---|---|
| Turnip mosaic virus (TuMV) | +ve sense RNA | Potyviridae | Aphids | LshCas13a | Coat protein | *Nicotiana benthamiana* | Resistance to leaf curl disease. | Aman et al. (2018) |
| Cucumber mosaic virus (CMV) | +ve sense RNA | Bromoviridae | Aphids | FnCas9 | ORF1a, ORF 3a and 3UTR of CMV | *Nicotiana benthamian* and *Arabidopsis* | Resistance against CMV. | Zhang et al. (2018) |
| Tobacco mosaic virus (TMV) | +ve sense ss-RNA | Virgaviridae | – | LshCas13a | Five gRNAs targeting genome of TMV; three gRNAs targeting genome of SRBSDV. | *Nicotiana benthamiana* and *Oryza sativa* | Resistance against TMV in *Nicotiana benthamiana*, mild symptom in *Oryza sativa* | Zhang et al. (2019) |
| Southern rice black streaked dwarf virus (SRBSDV) | ds RNA | Reoviridae | Whitebacked planthopper (*Sogatella furcifera*) | | | | | |
| Rice stripe mosaic virus (RSMV) | –ve sense RNA | Rhabdoviridae | Leafhopper (*Recilia dorsalis*) | LshCas13a | gRNA targeting single strand RNA genome. | *Oryza sativa* | Resistance against RSMV | Zhang et al. (2019) |
| Potato virus Y (PVY) | +ve sense RNA | Potyviridae | Aphids | LshCas13a | Viral replicase (NIb), coat protein, poty viral membrane protein (P3), CI protein. | *Solanum tuberosum* L. | Resistance against PVY | Zhan et al. (2019) |
| Grapevine leafroll-associated virus 3 (GLRaV-3) | +ve sense RNA | Closteroviridae | Mealybugs | FnCas9 and LshCas13a | Movement protein (p5), heat shock protein 70 homolog (Hsp70h), heat shock protein 90 homolog (Hsp90h), coat protein, coat protein minor (CPm). | Grapevine | Inhibition of GLRaV-3 titer | Jiao et al. (2022) |

**TABLE 6.8** CRISPR/CAS-Mediated Host Gene Modifications

| Virus(s) | Genome | Family | Vector | Effector Protein | Host Plant | Host Gene Targeted | Outcome | References |
|---|---|---|---|---|---|---|---|---|
| *Cucumber vein yellowing virus* (CVYV) | +ve sense RNA | Potyviridae | *Bemisia tabaci* | SpCas9 | *Cucumis sativus* L. | Eukaryotic translation initiation factor 4E proteins (eIF4E) | Broad virus resistance | Chandrasekaran et al. (2016) |
| *Zucchini yellow mosaic virus* (ZYMV) | | | Aphids | | | | | |
| *Papaya ring spot mosaic virus-W* (PRSMV-type-W) | | | – | | | | | |
| *Rice tungro spherical virus* (RTSV) | +ve sense RNA | Secoviridae | Leafhopper (*Nephotettix virescens*) | SpCas9 | *Oryza sativa* | Translation initiation factor 4 gamma (eIF4G) | Resistance to RTSV | Macovei et al. (2018) |
| *Cassava brown streak virus* (CBSV) | +ve sense RNA | Potyviridae | *Bemisia tabaci* | SpCas9 | Cassava | nCBP-1 and nCBP-2 | Resistance to CBSV | Gomez et al. (2019) |
| *Soya bean mosaic virus* (SMV) | +ve sense RNA | Potyviridae | Aphids | Spcas9 | *Glycine max* (L.) | Flavanone-3-hydroxylase (F3H1, F3H2) and flavone synthase II (FNSII-1) | Resistance to SMV | Zhang et al. (2020) |
| *Tomato yellow leaf curl virus* (TYLCV) | ss-DNA | Geminiviridae | *Bemisia tabaci* | Spcas9 | *Solanum lycopersicum* | *SlPelo* locus (carries susceptible factor) | TYLCV resistance, restricted viral DNA proliferation. | Pramanik et al. (2021) |

role in facilitating the infection of certain RNA viruses. Pyott et al. (2016) targeted the *eIF(iso)4E* locus in the *Arabidopsis thaliana* genome to generate deleterious point mutations, and they observed complete resistance against TuMV which remained stable until the T3 generation even after the segregation of the *CRISPR/Cas9* transgene. It was also observed that homozygous *eIF(iso)4E* mutants were very similar to the wild type without any decline in plant vigor. Chandrasekaran et al. (2016) demonstrated the effectiveness of plant-driven resistance by targeting *eIF4E* in Cucumber to develop resistance against *cucumber vein yellowing virus* (CMV), ZYMV, and *Papaya ringspot mosaic virus* (PMV). Macovei et al. (2018) mutated the *eIF4G* of the RTSV-susceptible variety IR64 (Rice) to generate a novel allele for achieving resistance against RTSV. Furthermore, Gomez et al. (2019) reported that simultaneous editing of the *nCBP-1* and *nCBP-2* genes in Cassava, the two isoforms of the *eIF4E* gene required for interaction with viral genome-linked protein (VPg), resulted in tolerance against cassava brown streak disease (CBSD).

More recently, Zhang et al. (2020) employed a multiplexed genome editing strategy to combat SMV in *Glycine max.* Simultaneous targeting (knocking out) of three genes of the isoflavone competition pathway (*GmF3H1*, *GmF3H2*, and *GmFNSII-1*) caused higher isoflavone content and increased resistance against SMV. Pramanik et al. (2021) attempted to target the *Ty-5* locus carrying the *SlPelo* gene as a susceptibility factor in the tomato cultivar BN-86, where it was found that *SlPelo* knockout homozygous mutants had a lower accumulation of *tomato yellow leaf curl virus* (TYLCV).

## 6.6 CONCLUSION AND PERSPECTIVE

The rapidly increasing world population necessitates an enhancement in crop production. The rigorous practice of traditional strategies has gained limited success in regulating plant viral diseases. Natural defense based on HPR offers a sustainable way to produce virus-resistant varieties with high genetic stability. In spite of tremendous efforts spanning several decades, antiviral breeding could not progress to the level needed for successful plant virus management, due to constraints of hybrid incompatibility, outbreeding depression, heavy cost, and laborious efforts. Thus, biotechnological approaches collectively constitute a set of armor that is anticipated to meet the challenge. The adoption of antiviral biotechnological methods is impeded due to biosafety concerns pertaining to genetically modified plants. So far, only a handful of virus-resistant transgenic crops have been deregulated for commercial cultivation.

The CRISPR/Cas directed genome editing system is considered superior to other biotechnological approaches due to its potential as a transgene-free method for developing antiviral immunity. By stably silencing host genes that promote viral replication, this system can be utilized to develop protection against both DNA and RNA viruses across generations. Additionally, genome editing of host plants offers the benefit of developing functionally resistant virus-resistant plants, even after the segregation of CRISPR/Cas transgenes. However, it is important to note that these genome-edited virus-resistant plants may eventually lose immunity against the targeted virus(es) over time, due to the rapid rate of mutations among viruses. Nevertheless, the genome editing technology holds great promise for plant virologists, as it can effectively target both viral and host genomes to ensure robust resistance. It is speculated that the simultaneous exploitation of both approaches may be highly beneficial in generating broad spectrum resistance against a multitude of RNA and DNA viruses. However, further analysis is required to determine the feasibility and extent of applying both approaches simultaneously, taking into account potential off-target effects.

While non-transgenic antiviral strategies, such as those based on the application of NMs, *in-vitro* delivery of RNAi, and the use of prokaryotic systems to express dsRNA/hpRNA, offer innovative and cost-effective methods for restricting viral threats, these approaches are still in their infancy and limited to laboratory settings. Therefore, it may take a considerable amount of time before they can be implemented in practical applications.

## KEYWORDS

- **CRISPR/CAS**
- **genetic engineering**
- **genome editing**
- **host plant resistance**
- **nanoparticles**
- **pathogen-derived resistance**
- **plant virus**
- **RNA interference**
- **viral diseases**

## REFERENCES

Abel, P. P., et al., (1986). Delay of disease development in transgenic plants that express the *Tobacco mosaic virus* coat protein gene. *Science, 232*(4751), 738–743.

Adeel, M., et al., (2021). Carbon-based nanomaterials suppress *Tobacco mosaic virus* (TMV) infection and induce resistance in *Nicotiana benthamiana*. *Journal of Hazardous Materials, 404*, 124167.

Ahmad, A., et al., (2017). Engineered disease resistance in cotton using RNA-interference to knock down *Cotton leaf curl Kokhran virus*-burewala and *Cotton leaf curl Multan* betasatellite expression. *Viruses, 9*(9), 257.

Ahmed, M., et al., (2017). RNAi-mediated resistance to *Rice black-streaked dwarf virus* in transgenic rice. *Transgenic Research, 26*(2), 197–207.

Ali, Z., et al., (2016). CRISPR/Cas9-mediated immunity to *Geminiviruses*: Differential interference and evasion. *Scientific Reports, 6*(1), 1–13.

Alkubaisi, N. A., & Aref, N., (2017). Dispersed gold nanoparticles potentially ruin gold *Barley yellow dwarf virus* and eliminate virus infectivity hazards. *Applied Nanoscience, 7*(1), 31–40.

Aman, R., et al., (2018). RNA virus interference via CRISPR/Cas13a system in plants. *Genome Biology, 19*(1), 1–9.

Aref, N., et al., (2012). Multi-functional effects of gold nano-particles inducing plant virus resistance crops. In: *Proceedings of the 5th Annual World Congress of Industrial Biotechnology-2012*. Xi'an, China.

Aslam, U., et al., (2018). A virus-derived short hairpin RNA confers resistance against *Sugarcane mosaic virus* in transgenic sugarcane. *Transgenic Research, 27*(2), 203–210.

Baulcombe, D. C., (1996). Mechanisms of pathogen-derived resistance to viruses in transgenic plants. *The Plant Cell, 8*(10), 1833.

Bendahmane, A., Kanyuka, K., & Baulcombe, D., (1997). High-resolution genetical and physical mapping of the Rx gene for extreme resistance to *Potato virus X* in tetraploid potato. *Theoretical and Applied Genetics, 95*(1), 153–162.

Boiteux, L., & De Giordano, L. B., (1993). Genetic basis of resistance against two *Tospovirus* species in tomato (*Lycopersicon esculentum*). *Euphytica, 71*(1), 151–154.

Bouché, N., et al., (2006). An antagonistic function for *Arabidopsis* DCL2 in development and a new function for DCL4 in generating viral siRNAs. *The EMBO Journal, 25*(14), 3347–3356.

Cai, L., et al., (2019). Preventing viral disease by ZnONPs through directly deactivating TMV and activating plant immunity in *Nicotiana benthamiana*. *Environmental Science: Nano, 6*(12), 3653–3669.

Cai, L., et al., (2020). Foliar exposure of $Fe_3O_4$ nanoparticles on *Nicotiana benthamiana*: Evidence for nanoparticles uptake, plant growth promoter and defense response elicitor against plant virus. *Journal of Hazardous Materials, 393*, 122415.

Canady, M., et al., (2001). *Tomato spotted wilt virus* (TSWV) resistance in tomato derived from *Lycopersicon chilense* Dun. LA 1938. *Euphytica, 117*(1), 19–25.

Castle, S., Palumbo, J., & Prabhaker, N., (2009). Newer insecticides for plant virus disease management. *Virus Research, 141*(2), 131–139.

Ceasar, S. A., et al., (2016). Insert, remove or replace: A highly advanced genome editing system using CRISPR/Cas9. *Biochimica et Biophysica Acta (BBA)-Molecular Cell Research, 1863*(9), 2333–2344.

Chandrasekaran, J., et al., (2016). Development of broad virus resistance in non-transgenic cucumber using CRISPR/Cas9 technology. *Molecular Plant Pathology, 17*(7), 1140–1153.

Chen, W., et al., (2014). Inhibiting replication of *Begomoviruses* using artificial zinc finger nucleases that target viral-conserved nucleotide motif. *Virus Genes, 48*(3), 494–501.

Cheng, X., et al., (2015). Artificial TALE as a convenient protein platform for engineering broad-spectrum resistance to *Begomoviruses*. *Viruses, 7*(8), 4772–4782.

Derbalah, A. S. H., & Elsharkawy, M. M., (2019). A new strategy to control *Cucumber mosaic virus* using fabricated NiO-nanostructures. *Journal of Biotechnology, 306*, 134–141.

Ding, S. W., & Voinnet, O., (2007). Antiviral immunity directed by small RNAs. *Cell, 130*(3), 413–426.

Duan, C. G., Wang, C. H., & Guo, H. S., (2012). Application of RNA silencing to plant disease resistance. *Silence, 3*(1), 1–8.

El Gamal, A. Y., et al., (2022). Antiviral activity of chitosan nanoparticles for controlling plant-infecting viruses. *South African Journal of Science, 118*(1–2), 1–9.

Elazzazy, A. M., Elbeshehy, E. K., & Betiha, M. A., (2017). *In vitro* assessment of activity of graphene silver composite sheets against multidrug-resistant bacteria and *Tomato bushy stunt virus*. *Tropical Journal of Pharmaceutical Research, 16*(11), 2705–2711.

El-Sawy, M. M., et al., (2017). Antiviral activity of 2-nitromethyl phenol, zinc nanoparticles and seaweed extract against *Cucumber mosaic virus* (CMV) in eggplant. *J. Virol. Antivir. Res., 6*(2).

El-Sawy, M., et al., (2018). Inhibition of *Tomato yellow leaf curl virus* by *Zingiber officinale* and *Mentha longifolia* extracts and silica nanoparticles. *Int. J. Antivir. Antiretrovir, 1*, 1–6.

Elsharkawy, M. M., & Derbalah, A., (2019). Antiviral activity of titanium dioxide nanostructures as a control strategy for broad bean strain virus in faba bean. *Pest Management Science, 75*(3), 828–834.

Elsharkawy, M. M., & Mousa, K. M., (2015). Induction of systemic resistance against *Papaya ring spot virus* (PRSV) and its vector *Myzus persicae* by *Penicillium simplicissimum* GP17-2 and silica ($SiO_2$) nanopowder. *International Journal of Pest Management, 61*(4), 353–358.

Eugene, K., & Zholobak, N., (2016). Antiviral activity of cerium dioxide nanoparticles on *Tobacco mosaic virus* model. *Proceedings of the Topical Issues of New Drugs Development*. Kharkiv, Ukraine.

Farooq, T., et al., (2021). Nanotechnology and plant viruses: An emerging disease management approach for resistant pathogens. *ACS Nano, 15*(4), 6030–6037.

Feng, Z., et al., (2021). Development of marker-free rice with stable and high resistance to *Rice black-streaked dwarf virus* disease through RNA interference. *Plant Biotechnology Journal, 19*(2), 212–214.

Finlay, K., (1953). Inheritance of spotted wilt resistance in the tomato II. Five genes controlling spotted wilt resistance in four tomato types. *Australian Journal of Biological Sciences, 6*(2), 153–163.

Foolad, M. R., (2007). Genome mapping and molecular breeding of tomato. *International Journal of Plant Genomics, 2007*.

Fuchs, M., & Gonsalves, D., (1995). Resistance of transgenic hybrid squash ZW-20 expressing the coat protein genes of *Zucchini yellow mosaic virus* and *Watermelon mosaic virus* 2 to mixed infections by both potyviruses. *Bio/technology, 13*(12), 1466–1473.

Gaj, T., & Gersbach, C., (2013). Barbas 3rd CF. ZFN, TALEN, and CRISPR/Cas-based methods for genome engineering. *Trends Biotechnol., 5*(15), 2013.

Gan, D., et al., (2010). Bacterially expressed dsRNA protects maize against SCMV infection. *Plant Cell Reports, 29*(11), 1261–1268.

Gandhi, K., et al., (2021). RNA interference: A novel technology for virus disease management in crop plants. *Madras Agricultural Journal, 108*(1–3), 1.

Gao, L., et al., (2015). Characterization of *Soybean mosaic virus* resistance derived from inverted repeat-SMV-HC-Pro genes in multiple soybean cultivars. *Theoretical and Applied Genetics, 128*(8), 1489–1505.

Gao, L., et al., (2020). Soybean RNA interference lines silenced for eIF4E show broad potyvirus resistance. *Molecular Plant Pathology, 21*(3), 303–317.

Garcia-Ruiz, H., (2018). Susceptibility genes to plant viruses. *Viruses, 10*, 484.

Gardner, R. G., & Panthee, D. R., (2012). Tomato spotted wilt virus-resistant fresh-market tomato breeding lines: NC 58S, NC 123S, NC 127S, and NC 132S. *HortScience, 47*(4), 531–532.

Ghazala, W., & Varrelmann, M., (2007). *Tobacco rattle virus* 29K movement protein is the elicitor of extreme and hypersensitive-like resistance in two cultivars of *Solanum tuberosum*. *Molecular Plant-Microbe Interactions, 20*(11), 1396–1405.

Gill, U., et al., (2019). Ty-6, a major *begomovirus* resistance gene on chromosome 10, is effective against *Tomato yellow leaf curl virus* and *Tomato mottle virus*. *Theoretical and Applied Genetics, 132*(5), 1543–1554.

Gomez, M. A., et al., (2019). Simultaneous CRISPR/Cas9-mediated editing of cassava eIF 4E isoforms nCBP-1 and nCBP-2 reduces cassava brown streak disease symptom severity and incidence. *Plant Biotechnology Journal, 17*(2), 421–434.

Gonsalves, D., (1998). Control of *Papaya ringspot virus* in papaya: A case study. *Annual Review of Phytopathology, 36*(1), 415–437.

Gonsalves, D., (2006). Transgenic papaya: Development, release, impact, and challenges. *Advances in Virus Research, 67*, 317–354.

Gottula, J., & Fuchs, M., (2009). Toward a quarter century of pathogen-derived resistance and practical approaches to plant virus disease control. *Advances in Virus Research, 75*, 161–183.

Grech-Baran, M., et al., (2020). Extreme resistance to *Potato virus Y* in potato carrying the Rysto gene is mediated by a TIR-NLR immune receptor. *Plant Biotechnology Journal, 18*(3), 655–667.

Guo, J., et al., (2015). Transgenic sugarcane resistant to *Sorghum mosaic virus* based on coat protein gene silencing by RNA interference. *BioMed Research International, 2015*.

Gupta, R. M., & Musunuru, K., (2014). Expanding the genetic editing tool kit: ZFNs, TALENs, and CRISPR-Cas9. *The Journal of Clinical Investigation, 124*(10), 4154–4161.

Hameed, A., et al., (2017). RNAi-mediated simultaneous resistance against three RNA viruses in potato. *Molecular Biotechnology, 59*(2), 73–83.

Hančinský, R., et al., (2020). Plant viruses infecting *Solanaceae* family members in the cultivated and wild environments: A review. *Plants, 9*(5), 667.

Hanson, P., Green, S., & Kuo, G., (2006). Ty-2, a gene on chromosome 11 conditioning *Geminivirus* resistance in tomato. *Tomato Genet. Coop. Rep., 56*, 17, 18.

Hao, Y., et al., (2018). Engineered nanomaterials suppress *Turnip mosaic virus* infection in tobacco (*Nicotiana benthamiana*). *Environmental Science: Nano, 5*(7), 1685–1693.

Herrera, M. D. R., et al., (2018). Molecular and genetic characterization of the Ryadg locus on chromosome XI from Andigena potatoes conferring extreme resistance to *Potato virus Y*. *Theoretical and Applied Genetics, 131*(9), 1925–1938.

Hutton, S. F., Ji, Y., & Scott, J. W., (2015). Fla. 8923: A tomato breeding line with *Begomovirus* resistance gene Ty-3 in a 70-kb *Solanum chilense* introgression. *HortScience, 50*(8), 1257–1259.

Ilardi, V., & Tavazza, M., (2015). Biotechnological strategies and tools for *Plum pox virus* resistance: Trans-, intra-, cis-genesis, and beyond. *Frontiers in Plant science, 6*, 379.

Jain, D., & Kothari, S., (2014). Green synthesis of silver nanoparticles and their application in plant virus inhibition. *J. Mycol. Plant Pathol, 44*(1), 21–24.

Ji, X., et al., (2015). Establishing a CRISPR-Cas-like immune system conferring DNA virus resistance in plants. *Nature Plants, 2015.*

Ji, Y., Schuster, D. J., & Scott, J. W., (2007). Ty-3, a *Begomovirus* resistance locus near the *Tomato yellow leaf curl virus* resistance locus Ty-1 on chromosome 6 of tomato. *Molecular Breeding, 20*(3), 271–284.

Jiao, B., et al., (2022). Engineering CRISPR immune systems conferring GLRaV-3 resistance in grapevine. *Horticulture Research, 9.*

Jones, R. A., (2021). Global plant virus disease pandemics and epidemics. *Plants, 10*(2), 233.

Kaniewski, W. K., & Thomas, P. E., (2004). The potato story. *AgBioForum, 7*(1, 2), 41–46.

Kis, A., et al., (2019). Creating highly efficient resistance against *Wheat dwarf virus* in barley by employing CRISPR/Cas9 system. *Plant Biotechnology Journal, 17*(6), 1004.

Kumar, J., Singh, S. P., & Kianian, S., (2017). Engineering resistance to plant viruses: Present status and future prospects. *Current Developments in Biotechnology and Bioengineering*, 75–100.

Kumari, P., et al., (2020). Inhibition of *Potato leafroll virus* multiplication and systemic translocation by siRNA constructs against putative ATPase fold of movement protein. *Scientific Reports, 10*(1), 1–11.

Kung, Y. J., et al., (2012). Multiple artificial microRNAs targeting conserved motifs of the replicase gene confer robust transgenic resistance to negative-sense single-stranded RNA plant virus. *Molecular Plant Pathology, 13*(3), 303–317.

Kuo, Y. W., & Falk, B. W., (2020). RNA interference approaches for plant disease control. *BioTechniques, 69*(6), 469–477.

Lanfermeijer, F. C., et al., (2003). Cloning and characterization of the durable *Tomato mosaic virus* resistance gene Tm-22 from *Lycopersicon esculentum. Plant Molecular Biology, 52*(5), 1039–1051.

Lanfermeijer, F. C., Warmink, J., & Hille, J., (2005). The products of the broken Tm-2 and the durable Tm-2 2 resistance genes from tomato differ in four amino acids. *Journal of Experimental Botany, 56*(421), 2925–2933.

Lapidot, M., et al., (2015). A novel route controlling *Begomovirus* resistance by the messenger RNA surveillance factor pelota. *PLoS Genetics, 11*(10), e1005538.

Liu, H., et al., (2018). CRISPR/Cas9-mediated resistance to cauliflower mosaic virus. *Plant Direct, 2*(3), e00047.

Liu, X. T., et al., (2021). A bean common mosaic virus-resistance gene in the soybean variant V94-5152 was mapped to the Rsv4 locus conferring resistance to *Soybean mosaic virus. Theoretical and Applied Genetics, 134*(8), 2367–2377.

Macovei, A., et al., (2018). Novel alleles of rice eIF4G generated by CRISPR/Cas9-targeted mutagenesis confer resistance to *Rice tungro spherical virus. Plant Biotechnology Journal, 16*(11), 1918–1927.

Mahfouze, H. A., El-Dougdoug, N. K., & Mahfouze, S. A., (2020). Virucidal activity of silver nanoparticles against *Banana bunchy top virus* (BBTV) in banana plants. *Bulletin of the National Research Centre, 44*(1), 1–11.

Makarova, K. S., et al., (2015). An updated evolutionary classification of CRISPR–Cas systems. *Nature Reviews Microbiology, 13*(11), 722–736.

Makarova, K. S., Zhang, F., & Koonin, E. V., (2017). SnapShot: Class 1 CRISPR-Cas systems. *Cell, 168*(5), 946–946. e1.

Malathi, P., et al., (2019). Coat protein 3 of *Rice tungro spherical virus* is the key target gene for development of RNAi mediated tungro disease resistance in rice. *Agri. Gene, 12*, 100084.

Miao, S., et al., (2021). Polycistronic artificial microRNA-mediated resistance to *cucumber green mottle mosaic virus* in cucumber. *International Journal of Molecular Sciences, 22*(22), 12237.

Mitter, N., et al., (2017). Clay nanosheets for topical delivery of RNAi for sustained protection against plant viruses. *Nature Plants, 3*(2), 1–10.

Mubarik, M. S., et al., (2021). Engineering broad-spectrum resistance to cotton leaf curl disease by CRISPR-Cas9 based multiplex editing in plants. *GM Crops & Food, 12*(2), 647–658.

Mumford, R., Macarthur, R., & Boonham, N., (2016). The role and challenges of new diagnostic technology in plant biosecurity. *Food Security, 8*(1), 103–109.

Nahid, N., et al., (2011). RNA interference-based resistance against a legume mastrevirus. *Virology Journal, 8*(1), 1–8.

Newell, C., et al., (1991). Agrobacterium-mediated transformation of *Solanum tuberosum* L. cv. Russet Burbank. *Plant Cell Reports, 10*(1), 30–34.

Noha, K., Bondok, A., & El-Dougdoug, K., (2018). Evaluation of silver nanoparticles as antiviral agent against ToMV and PVY in tomato plants. *Sciences, 8*(01), 100–111.

Ntui, V. O., et al., (2013). Transgenic accumulation of a defective *Cucumber mosaic virus* (CMV) replicase derived double stranded RNA modulates plant defense against CMV strains O and Y in potato. *Transgenic Research, 22*(6), 1191–1205.

Ntui, V. O., et al., (2015). Resistance to *Sri Lankan cassava mosaic virus* (SLCMV) in genetically engineered cassava cv. KU50 through RNA silencing. *PLoS One, 10*(4), e0120551.

Odipio, J., et al., (2014). RNAi-derived field resistance to cassava brown streak disease persists across the vegetative cropping cycle. *GM Crops & Food, 5*(1), 16–19.

Ogwok, E., et al., (2012). Transgenic RNA interference (RNA i)-derived field resistance to cassava brown streak disease. *Molecular Plant Pathology, 13*(9), 1019–1031.

Ohmori, T., Murata, M., & Motoyoshi, F., (1996). Molecular characterization of RAPD and SCAR markers linked to the Tm-1 locus in tomato. *Theoretical and Applied Genetics, 92*(2), 151–156.

Palukaitis, P., (2011). The road to RNA silencing is paved with plant-virus interactions. *The Plant Pathology Journal, 27*(3), 197–206.

Pramanik, D., et al., (2021). CRISPR/Cas9-mediated generation of pathogen-resistant tomato against *Tomato yellow leaf curl virus* and powdery mildew. *International Journal of Molecular Sciences, 22*(4), 1878.

Prins, M., et al., (2008). Strategies for antiviral resistance in transgenic plants. *Molecular Plant Pathology, 9*(1), 73–83.

Pyott, D. E., Sheehan, E., & Molnar, A., (2016). Engineering of CRISPR/Cas9-mediated potyvirus resistance in transgene-free *Arabidopsis* plants. *Molecular Plant Pathology, 17*(8), 1276–1288.

Rai, M., & Ingle, A., (2012). Role of nanotechnology in agriculture with special reference to management of insect pests. *Applied Microbiology and Biotechnology, 94*(2), 287–293.

Ray, P. C., Yu, H., & Fu, P. P., (2009). Toxicity and environmental risks of nanomaterials: Challenges and future needs. *Journal of Environmental Science and Health Part C, 27*(1), 1–35.

Riley, D. G., et al., (2011). Host plant resistance to *Tomato spotted wilt virus* (*Bunyaviridae*: *Tospovirus*) in tomato. *HortScience, 46*(12), 1626–1633.

Roselló, S., et al., (1997). Testing resistance to TSWV introgressed from *Lycopersicon peruvianum* by artificial transmission techniques. *Euphytica, 98*(1), 93–98.

Rubio, L., Galipienso, L., & Ferriol, I., (2020). Detection of plant viruses and disease management: Relevance of genetic diversity and evolution. *Frontiers in Plant Science, 11*, 1092.

Saidi, M., & Warade, S. D., (2008). Tomato breeding for resistance to *Tomato spotted wilt virus* (TSWV): An overview of conventional and molecular approaches. *Czech Journal of Genetics and Plant Breeding, 44*(3), 83–92.

Sanford, J., & Johnston, S., (1985). The concept of parasite-derived resistance—Deriving resistance genes from the parasite's own genome. *Journal of Theoretical Biology, 113*(2), 395–405.

Scorza, R., et al., (1994). Transgenic plums (*Prunus domestica* L.) express the *Plum pox virus* coat protein gene. *Plant Cell Reports, 14*(1), 18–22.

Senthilraja, C., et al., (2018). RNA interference-mediated resistance to *Tobacco streak virus* in transgenic peanut. Australasian *Plant Pathology, 47*(2), 227–230.

Seo, J. K., Lee, S. H., & Kim, K. H., (2009). Strain-specific cylindrical inclusion protein of *Soybean mosaic virus* elicits extreme resistance and a lethal systemic hypersensitive response in two resistant soybean cultivars. *Molecular Plant-Microbe Interactions, 22*(9), 1151–1159.

Sera, T., (2005). Inhibition of virus DNA replication by artificial zinc finger proteins. *Journal of Virology, 79*(4), 2614–2619.

Shafie, R. M., Salama, A. M., & Farroh, K. Y., (2018). Silver nanoparticles activity against *Tomato spotted wilt virus. Middle East J. Appl. Sci., 7*, 1251–1267.

Sharma, A., et al., (2019). Worldwide pesticide usage and its impacts on ecosystem. *SN Applied Sciences, 1*(11), 1–16.

Sharma, S., Kumar, G., & Dasgupta, I., (2018). Simultaneous resistance against the two viruses causing rice tungro disease using RNA interference. *Virus Research, 255*, 157–164.

Sharma, V. K., Basu, S., & Chakraborty, S., (2015). RNAi mediated broad-spectrum transgenic resistance in *Nicotiana benthamiana* to chili-infecting *Begomoviruses*. *Plant Cell Reports, 34*(8), 1389–1399.

Shaw, J., et al., (2014). Coilin, the signature protein of Cajal bodies, differentially modulates the interactions of plants with viruses in widely different taxa. *Nucleus, 5*(1), 85–94.

Shen, W., et al., (2014). Resistance of non-transgenic papaya plants to *Papaya ringspot virus* (PRSV) mediated by intron-containing hairpin dsRNAs expressed in bacteria. *Acta Virologica, 58*(3), 261–266.

Shen, X., et al., (2020). The NLR protein encoded by the resistance gene Ty-2 Is triggered by the replication-associated protein Rep/C1 of *Tomato yellow leaf curl virus*. *Frontiers in Plant Science, 11*, 545306.

Shimizu, T., et al., (2012). Hairpin RNA derived from the gene for Pns9, a viroplasm matrix protein of *Rice gall dwarf virus*, confers strong resistance to virus infection in transgenic rice plants. *Journal of Biotechnology, 157*(3), 421–427.

Shimizu, T., et al., (2013). Strong resistance against *Rice grassy stunt virus* is induced in transgenic rice plants expressing double-stranded RNA of the viral genes for nucleocapsid or movement proteins as targets for RNA interference. *Phytopathology, 103*(5), 513–519.

Soler, N., et al., (2012). Transformation of Mexican lime with an intron-hairpin construct expressing untranslatable versions of the genes coding for the three silencing suppressors of *Citrus tristeza virus* confers complete resistance to the virus. *Plant Biotechnology Journal, 10*(5), 597–608.

Sun, Z. N., et al., (2010). Bacterially expressed double-stranded RNAs against hot-spot sequences of *Tobacco mosaic virus* or *Potato virus Y* genome have different ability to protect tobacco from viral infection. *Applied Biochemistry and Biotechnology, 162*(7), 1901–1914.

Sun, Z. N., et al., (2010). HpRNAs derived from different regions of the NIb gene have different abilities to protect tobacco from infection with *Potato virus Y*. *Journal of Phytopathology, 158*(7, 8), 566–568.

Takahashi, H., et al., (2002). RCY1, an *Arabidopsis thaliana* RPP8/HRT family resistance gene, conferring resistance to *Cucumber mosaic virus* requires salicylic acid, ethylene and a novel signal transduction mechanism. *The Plant Journal, 32*(5), 655–667.

Tashkandi, M., et al., (2018). Engineering resistance against *Tomato yellow leaf curl virus* via the CRISPR/Cas9 system in tomato. *Plant Signal Behav., 13*(10), e1525996.

Tatineni, S., et al., (2020). Transgenic wheat harboring an RNAi element confers dual resistance against synergistically interacting *Wheat streak mosaic virus* and *Triticum mosaic virus*. *Molecular Plant-Microbe Interactions, 33*(1), 108–122.

Tenllado, F., & Dıaz-Ruız, J., (2001). Double-stranded RNA-mediated interference with plant virus infection. *Journal of Virology, 75*(24), 12288–12297.

Tenllado, F., et al., (2003). Crude extracts of bacterially expressed dsRNA can be used to protect plants against virus infections. *BMC Biotechnology, 3*(1), 1–11.

Thomas, P., Kaniewski, W., & Lawson, E., (1997). Reduced field spread of *Potato leafroll virus* in potatoes transformed with the *Potato leafroll virus* coat protein gene. *Plant Disease, 81*(12), 1447–1453.

Tricoll, D. M., et al., (1995). Field evaluation of transgenic squash containing single or multiple virus coat protein gene constructs for resistance to *Cucumber mosaic virus, Watermelon mosaic virus* 2, and zucchini yellow mosaic virus. *Bio/technology, 13*(12), 1458–1465.

Tripathi, J. N., et al., (2019). CRISPR/Cas9 editing of endogenous *Banana streak virus* in the B genome of *Musa* spp. overcomes a major challenge in banana breeding. *Communications Biology, 2*(1), 1–11.

Tudi, M., et al., (2021). Agriculture development, pesticide application and its impact on the environment. *International Journal of Environmental Research and Public Health, 18*(3), 1112.

Usman, M., et al., (2020). Nanotechnology in agriculture: Current status, challenges and future opportunities. *Science of the Total Environment, 721*, 137778.

Verlaan, M. G., et al., (2013). The *Tomato yellow leaf curl virus* resistance genes Ty-1 and Ty-3 are allelic and code for DFDGD-class RNA–dependent RNA polymerases. *PLoS Genetics, 9*(3), e1003399.

Voinnet, O., (2005). Induction and suppression of RNA silencing: Insights from viral infections. *Nature Reviews Genetics, 6*(3), 206–220.

Voorburg, C. M., et al., (2020). Ty-1, a universal resistance gene against *Geminiviruses* that is compromised by co-replication of a betasatellite. *Molecular Plant Pathology, 21*(2), 160–172.

Walsh, H., et al., (2019). RNA silencing of *South African cassava mosaic virus* in transgenic cassava expressing AC1/AC4 hp-RNA induces tolerance. *Biotechnology Reports, 24*, e00383.

Wang, F., et al., (2016). Hairpin RNA targeting multiple viral genes confers strong resistance to *Rice black-streaked dwarf virus*. *International Journal of Molecular Sciences, 17*(5), 705.

Wang, J., et al., (2022). Field application of nanoliposomes delivered quercetin by inhibiting specific hsp70 gene expression against plant virus disease. *Journal of Nanobiotechnology, 20*(1), 1–16.

Wang, X., et al., (2013). Silencing of the host factor eIF (iso) 4E gene confers *Plum pox virus* resistance in plum. *PLoS One, 8*(1), e50627.

Wang, Y., et al., (2016). Preliminary experiments on nano-silver against *Tobacco mosaic virus* and its mechanism. *Tob. Sci. Technol, 49*, 22–30.

Widyaningrum, S., et al., (2021). Induction of resistance to *Sugarcane mosaic virus* by RNA interference targeting coat protein gene silencing in transgenic sugarcane. *Molecular Biology Reports, 48*(3), 3047–3054.

Wilson, T., (1993). Strategies to protect crop plants against viruses: Pathogen-derived resistance blossoms. *Proceedings of the National Academy of Sciences, 90*(8), 3134–3141.

Worrall, E. A., et al., (2018). Nanotechnology for plant disease management. *Agronomy, 8*(12), 285.

Worrall, E. A., et al., (2019). Exogenous application of RNAi-inducing double-stranded RNA inhibits aphid-mediated transmission of a plant virus. *Frontiers in Plant Science, 10*, 265.

Wu, C. H., et al., (2017). NLR network mediates immunity to diverse plant pathogens. *Proceedings of the National Academy of Sciences, 114*(30), 8113–8118.

Wu, X., Kriz, A. J., & Sharp, P. A., (2014). Target specificity of the CRISPR-Cas9 system. *Quantitative Biology, 2*(2), 59–70.

Yamaguchi, H., et al., (2018). An NB-LRR gene, TYNBS1, is responsible for resistance mediated by the Ty-2 *Begomovirus* resistance locus of tomato. *Theoretical and Applied Genetics, 131*(6), 1345–1362.

Yang, R., et al., (1995). Transgenic tomato plants expressing *Cucumber mosaic virus* coat protein and their resistance to CMV. *Jiangsu Journal of Agricultural Sciences/Jiangsu Nongye Xuebao, 11*(1), 40–44.

Yang, X., et al., (2017). Robust RNAi-mediated resistance to infection of seven potyvirids in soybean expressing an intron hairpin NIb RNA. *Transgenic Research, 26*(5), 665–676.

Yang, X., et al., (2018). RNAi-mediated SMV P3 cistron silencing confers significantly enhanced resistance to multiple potyvirus strains and isolates in transgenic soybean. *Plant Cell Reports, 37*(1), 103–114.

Yasmeen, A., et al., (2016). Amplicon-based RNA interference targeting V2 gene of *Cotton leaf curl Kokhran Virus*-Burewala strain can provide resistance in transgenic cotton plants. *Molecular Biotechnology, 58*(12), 807–820.

Yin, G., et al., (2009). Production of double-stranded RNA for interference with TMV infection utilizing a bacterial prokaryotic expression system. *Applied Microbiology and Biotechnology, 84*(2), 323–333.

Yin, K., et al., (2019). Engineer complete resistance to *Cotton leaf curl Multan virus* by the CRISPR/Cas9 system in *Nicotiana benthamiana. Phytopathology Research, 1*(1), 1–9.

Zhan, X., et al., (2019). Generation of virus-resistant potato plants by RNA genome targeting. *Plant Biotechnology Journal, 17*(9), 1814–1822.

Zhang, P., et al., (2020). Multiplex CRISPR/Cas9-mediated metabolic engineering increases soya bean isoflavone content and resistance to soya bean mosaic virus. *Plant Biotechnology Journal, 18*(6), 1384–1395.

Zhang, T., et al., (2018). Establishing RNA virus resistance in plants by harnessing CRISPR immune system. *Plant Biotechnology Journal, 16*(8), 1415–1423.

Zhang, T., et al., (2019). Establishing CRISPR/Cas13a immune system conferring RNA virus resistance in both dicot and monocot plants. *Plant Biotechnology Journal, 17*(7), 1185.

Zhu, Y. X., et al., (1996). Transgenic sweet pepper plants from *Agrobacterium* mediated transformation. *Plant Cell Reports, 16*(1), 71–75.

**CHAPTER 7**

# Detection of Plant Viruses and Their Management Strategies

CHITRA NEHRA,[1] DEEPIKA CHOUDHARY,[2] AVINASH MARWAL,[3] and R. K. GAUR[4]

[1]*Department of Biosciences, Mody University of Science and Technology, Lakshmangarh, Sikar, Rajasthan, India*

[2]*Department of Chemistry, University of Rajasthan, Jaipur, Rajasthan, India*

[3]*Department of Biotechnology, Mohanlal Sukhadia University, Udaipur, Rajasthan, India*

[4]*Department of Biotechnology, Deen Dayal Upadhyaya Gorakhpur University, Gorakhpur, Uttar Pradesh, India*

## ABSTRACT

Viral diseases cause severe crop loss worldwide, posing a significant challenge to food security and crop production. To mitigate the damage caused by viral diseases, it is crucial to employ advanced methods for virus detection and diagnosis. This chapter provides an overview of various serological, molecular, and high-throughput sequencing methods for virus identification and detection. The frequent international trade and the rapid evolution of viruses are the main factors contributing to the emergence of new diseases. Additionally, we will discuss immunization and prophylaxis as major strategies for controlling viral diseases in plants.

Detection and Management of New and Emerging Mystery Plant Virus Sources.
Alireza Golnaraghi, Rajarshi Kumar Gaur, and Rakesh Kumar Verma (Eds.)

## 7.1 INTRODUCTION

In all plant pathogens, viruses, and virus-like pathogens are the causal agents of most emerging plant diseases, presenting a major threat to agricultural production and causing enormous economic losses every year (Maksimov et al., 2019). These viruses infect a wide range of crop plants and affect the production as well as the quality of food crops globally. Worldwide, more than 25 plant virus families have been reported to infect and cause diseases in crop plants, leading to vast agronomic and economic losses every year (He & Krainer, 2020). Plant viruses are spread through vertical and horizontal transmission. In the vertical mode, viral agents are transmitted from parent plants through either vegetative propagation or sexually reproduced infected seeds. In the horizontal mode, viral transmission occurs through vectors or direct contamination (Gallet, Michalakis, & Blanc, 2018).

In crop plants, virus-caused diseases are difficult to predict as they depend on the host plant, virus strain, particular region, and time of infection (Strange, 2005). Upon viral infection, plants develop various symptoms including yellowing, mosaic patterns, chlorosis, reduced growth, and necrosis, which lead to disturbances in growth and other physiological functions of the plant (Seo et al., 2018). Diagnosing viral diseases based on symptoms is more challenging than diagnosing other pathogens, as sometimes plants show very few or no symptoms (Lievens, Grauwet, Cammue, & Thomma, 2005). Additionally, plants can also develop virus-like symptoms in response to abiotic stresses, improper nutrition, and other pathogen infections (Van & Dijkstra, 2006).

Detection and identification are critical steps in disease management and predicting crop loss due to pathogen infection (Aboul-Ata et al., 2011). Currently, there are very few disease control strategies available for managing viral diseases under field conditions. Therefore, accurate and early diagnosis of relevant causative agents is the most important step in preventing viral diseases in plants (Adams et al., 2009). Various accurate and sensitive methods have been continuously developed in viral diagnostics for effective and efficient viral disease management. However, most of these methods are expensive, labor-intensive, time-consuming, and not suitable for detecting emerging viral species. Therefore, updated diagnostic methods are needed that can capture new viruses as well as multiple infections (Rubio, Galipienso, & Ferriol, 2020). For virus detection, various methods such as electron microscopy (EM), immuno-serological methods (ELISA), molecular methods (PCR), and biosensor-based methods are being used to achieve fast and accurate detection (Massart, Olmos, Jijakli, & Candresse, 2014).

The current advancements in advanced molecular biology and genomics tools have allowed us to develop diagnostic methods that are more sensitive, fast, and eliminate the need for culturing viruses. The introduction of next-generation sequencing (NGS) has revolutionized virus diagnosis, enabling us to sequence billions of nucleotides in parallel without any prior knowledge of viral sequences (Maree et al., 2018; Stobbe & Roossinck, 2014). NGS-based sequencing platforms offer reliable and cost-effective diagnostics of plant viruses, as well as predictions about disease emergence and dynamics (Malmstrom, Melcher, & Bosque-Pérez, 2011; Khare, Mishra, & Arora, 2018). In this chapter, we summarize key techniques widely employed for plant viral diagnosis and strategies for viral disease management.

## 7.2 VIRUS DIAGNOSIS METHODS

### 7.2.1 SEROLOGICAL METHODS

Serological detection systems, such as ELISA (enzyme-linked immunosorbent assay), are based on the detection of specific proteins using antibodies. Initially, ELISA was used to detect viruses in plant tissue culture (Atkinson & Matthews, 1970). Clark and Adams were the first to utilize this method for the diagnosis of plant viral diseases (Clark & Adams, 1977). Various ELISA-based methods, including direct tissue blot immunoassay (DTBIA), double antibody sandwich (DAS) ELISA, and tissue-print (TP) ELISA, are widely used for virus identification (Xie, Jiao, Zhou, Liu, Ni, & Wu, 2013; Rani, Donovan, & Mantri, 2019; Abd El-Aziz, 2019). ACP-ELISA and antigen-coated plate ELISA were used for RSMV detection in rice plants (Guo, Wu, Chen, Hong, Zhou, & Wu, 2020). Monoclonal antibodies were developed using purified virions as antigens to identify Rice Dwarf virus. For rapid detection of RDV in rice, a dot enzyme-linked immunosorbent assay (dot-ELISA) and plate-trapped antigen ELISA were developed (Wu, Ni, Liu, Ding, & Zhou, 2014). PTA-ELISA is suitable for virus testing in multiple samples within a short time and under various conditions. In squash, cowpea, and papaya, various plant viruses such as squash mosaic virus (SQMV), Cowpea severe mosaic virus (CPSMV), cucumber mosaic virus (CMV), Cowpea aphid-borne mosaic virus (CABMV), ZYMV, and Papaya lethal yellowing virus (PLYV) have been detected (Nascimento, Lima, & Barbosa, 2017). TAS-ELISA (triple antibody sandwich enzyme-linked immunosorbent assays) was used to detect Tomato yellow leaf curl Thailand virus (TYLCTHV) in tomato, chili, brinjal, okra, and cucurbit plants (Seepiban et al., 2017). Another serological method is immunochromatographic strip-based methods, which are rapid and capable of detecting viruses with sensitivity. Plant viruses, including Potato viruses X and Y, tobacco mosaic virus (TMV), and Tomato zonate spot tospovirus, have been reported to be detected in leaf extract using immunostrips (Byzova et al., 2009 Niu, Wang, Cui, Wang, Pang, & Yu, 2018). Interestingly, this immuntostrip-based assay has been reported to detect Citrus yellow vein clearing virus in the sandwich ELISA technique (Bin, Li, Wu, Wang, Zhou, Li, & Song, 2017). Recently, immunostrips specific to different plant viruses have been developed and are commercially available for detecting viral infections in fields (Mehetre et al., 2021).

### *7.2.2 PCR-BASED METHODS*

Polymerase chain reaction (PCR) is the most commonly used method in pathogen detection and is applied when the genetic information of the pathogen is known (Joo-Jin, Ho-Jong, & Jaejong, 2014). PCR is completed in three basic steps. First, denaturation above 94°C where both strands of DNA are separated. Second, annealing of primers at 50–75°C (temperature depends on the primers). And third, elongation of complementary strands at 72°C (Makkouk & Kumari, 2006). A wide range of adaptations have been developed in PCR-based analysis, and these methods detect viral nucleic acid (DNA or RNA) using molecular primers to amplify a specific segment of the virus genomes (Li, Mock, Huang, Abad, Hartung, & Kinard, 2008; Jeong, Ju, & Noh, 2014). Conventional PCR is the most basic tool in virus diagnostics, although challenges like the requirement of condition optimization, time, and DNA sample-based techniques hinder its popularity in viral diagnosis (Mirmajlessi, Loit, Maend, & Mansouripour, 2015). Reverse transcriptase PCR (RT-PCR) works on the synthesis of cDNA from RNA before the typical PCR steps and is useful for RNA virus diagnosis (Zhang et al., 2020). Single-step RT-PCR is a more efficient, accurate, and cost-effective tool for the detection of a wide range of RNA-genome-based plant viruses. RNA viruses like Iris severe mosaic virus (ISMV) (Nateqi, Koohi, Dizadji, & Parizad, 2015), cherry green ring mottle virus (CGRMV), and cherry necrotic rusty mottle virus (CNRMV) were detected by RT-PCR in virus-infected plant leaf samples (Li & Hartung, 2007). The RT-PCR technique has been employed to detect potato viruses in potato stems, seeds, and aphid vectors (Drygin et al., 2012; Peiman & Xie, 2006; Peter, Gildow, Palukaitis, & Gray, 2009). In addition, for quarantine purposes, this technique was used to detect five previously unreported viruses in Korea (Lee, Cho, Lee, Choi, & Kim, 2011). Pepino mosaic virus (PepMV) has emerged as a virus in tomatoes worldwide over the past few years. RT-PCR has been useful in various studies for the simultaneous detection and identification of different PepMV genotypes in tomato plants (Alfaro-Fernández et al., 2009). RT-PCR has shown great potential in plant viral diagnosis, and its advanced one-step RT digital PCR has been reported as more suitable and reliable for detecting various PepMV genotypes (Mehle, Gregur, Bogožalec, & Dobnik, 2020). Molecular methods have considerably developed in the last few years, and multiplex detection methods have become popular as they offer cost-effective and efficient detection of multiple viruses simultaneously. In multiplex RT-PCR, detection of all members of a specific genus is possible by designing primers for the conserved region. Examples include the identification of *Allexvirus, Begomovirus, Potyvirus* (Majumder & Baranwal, 2014; Van et al., 2014; Zheng, Rodoni, Gibbs, & Gibbs, 2009).

Quantitative real-time multiplex PCR (M-PCR) has been developed to monitor the amplified PCR products in real-time and also allows accurate quantification of the products (Ruiz-Ruiz et al., 2009). Real-time PCR offers multiple advantages over conventional PCR, such as requiring less sample amount and being more accurate and faster as there is no need to run a gel (Lopez et al., 2008). Additionally, real-time PCR can provide accurate results even between closely related organisms (Martinelli et al., 2015). *Citrus tristeza virus* (CTV) in different plant tissues, aphid vectors, and TMV in soil were detected and quantified by real-time PCR (Ruiz-Ruiz, Moreno, Jose, & Ambros, 2007; Yang et al., 2012). However, there are some challenges associated with real-time PCR, such as the amplification stopping when it reaches a certain level or plateau, as well as the requirement of expensive equipment. PCR, RT-PCR, and real-time PCR techniques are widely used in plant viral diagnosis, although there are limitations associated with these detection methods. Therefore, further advancements are needed in these molecular methods for more sensitive and reliable results.

### *7.2.3 NEXT-GENERATION SEQUENCING (NGS)-BASED METHODS*

Next-generation technology-based platforms can generate enormous data within a very short time and at a comparatively lower cost, making them the most popular techniques in various fields of biology research (Mehetre, Paranjpe, Dastager, & Dharne, 2016; Hadidi, 2019). These methods have brought revolutionary changes in plant virology research. NGS-based techniques involve a few basic steps such as nucleic acid extraction (from the environment and plants), sample preparation, sequencing, library preparation, and sequence analysis. The analysis of DNA/RNA from a mixed community of organisms using NGS is called metagenomics (Riesenfield & Schloss, 2004). For the metagenomic analysis of viruses, total nucleic acid (DNA/RNA) is isolated from infected plant samples and sequenced using NGS-based methods, followed by sequence analysis using bioinformatics tools (Blawid, Silva, & Nagata, 2017). Metagenomics has allowed the exploration of plant virus biodiversity in environmental samples, which is referred to as "ecogenomics" (Stobbe & Roossinck, 2014). Viral metagenomics offers advantages over other diagnostic methods as it can identify novel viruses without requiring prior knowledge of viral sequences. However, the unavailability of reference sequences poses a slight challenge as these techniques involve de novo assembly of the genome (Xiao, Li, Al Rwahnih, Dolja, & Meng, 2019). Various studies using NGS-based approaches have been reported to detect both novel and previously identified plant viruses. Plant virus metagenomics allows for the detection of viruses, either as a single agent or in mixed infections, and can reveal the occurrence of novel or unsuspected virus isolates (Roossinck, 2015). Metagenomic-based surveys offer the identification of virus species from both symptomatic and asymptomatic host plants. A recent study using NGS technology to assess viruses associated with Papaya Ring Spot Disease, infecting the important fruit crop papaya in Kenya, showed the presence of Moroccan watermelon mosaic virus (MWMV), a Potyvirus; cowpea mild mottle virus (CpMMV), a Carlavirus, and two novel yet divergent Carlaviruses (namely, papaya mottle-associated virus (PaMV) and papaya mild mottle-associated virus) in symptomatic and asymptomatic papaya leaves collected from Kenyan fields (Mumo et al., 2020). This study has presented the first complete genome sequences of these viruses isolated from papaya in Kenya and primers for their detection. Metagenomic analyses that have evaluated the metagenomes of viruses at the ecosystem scale have disclosed many previously unidentified viral species. More recently, novel georeferenced metagenomic methods have been developed that can specifically link individual sequence reads to both the plant hosts from which they were obtained and the spatial arrangements of these hosts. Besides enlightening the diversity and distribution of plant viruses at the ecosystem scale, the application of these "geo-metagenomics" approaches has allowed the direct testing of hypotheses describing the effects of host spatial variations, host diversity, and environmental conditions on plant virus diversity and occurrence. Metagenomics-based surveys investigating plant virus biodiversity in wild plants, insects, and a few other environments (Roossinck, 2012; Wren et al., 2006). While the metagenomic surveys are a revolutionary start, there are still many challenges ahead.

## 7.3 VIRAL DISEASE MANAGEMENT APPROACHES

Plant viruses are economically important pathogens that are rapidly emerging due to the increasing global trade of agricultural products, plant materials, and intensive farming practices (Gaur, Ali, Cheng, Mäkinen, Agindotan, & Wang, 2021). The complex and dynamic nature of virus epidemics, along with the high evolution rate, makes it difficult to efficiently and effectively control viral diseases in the long term (Acosta-Leal, Duffy, Xiong, Hammond, & Elena, 2011; Elena, Fraile, & Garcia-Arenal, 2014). In agriculture, viral disease management primarily relies on immunization and prophylactic measures. Immunization helps in developing viral resistance in plants, while prophylactic measures control the dispersion of viruses.

## 7.4 IMMUNIZATION METHODS

Introduction of resistance genes into susceptible plants from cultivated or wild plants through backcrossing is the most commonly used method of immunization. There are two types of resistance: active and passive resistance. In active resistance, dominant alleles encode the resistance proteins that recognize viral genes or products and induce a hypersensitive response (HR) in the host, leading to the death of infected cells and restricting virus movement to other cells and systemic infection (De Ronde, Butterbach, & Kormelink, 2014; Peiró et al., 2014). On the other hand, passive resistance is driven by recessive alleles encoding various host factors required for infection, such as eukaryotic translation initiation factors (eIFs) 4E and 4G, and their isoforms (Truniger & Aranda, 2009; Hashimoto, Neriya, Yamaji, & Namba, 2016). Resistant germplasm and cultivars can be analyzed simultaneously on a large-scale using ELISA and molecular hybridization methods (Rubio, Herrero, Sarrio, Moreno, & Guerri, 2003a). Plant breeders generally aim to obtain complete resistance (no systemic infection), but when that is not possible, partial resistance (reduction in viral titer) or tolerance (reduction in symptoms without affecting viral accumulation) are other viable options. Real-time qPCR is the most suitable method for accurately analyzing relative resistance levels (Soler et al., 2015). However, these breeding methods are limited to crops where resistant genes are available in genetically compatible relatives. Genome editing tools, such as the CRISPR-Cas9 system, can be used to induce mutations in plant genes responsible for recessive resistance and prevent viral infection (Chandrasekaran et al., 2016). However, this may disrupt normal plant function and physiology, as the mutated genes could be involved in these biological functions.

Plant genes responsible for dominant resistance generally target viral proteins essential for virus functioning, such as proteins involved in multiplication, movement, and transmission. Therefore, the possibilities of resistance breakdown are comparatively less. Monogenic resistance is more preferred than polygenic resistance as its introgression is simpler in crop plants (Palloix, Ayme, & Moury, 2009).

Plant breeders generally focus on the genetic variability of the host, but virus genetic variability should also be considered to lessen the chances of resistance breakdown. Multiplex real-time qPCR is the most precise tool to analyze the effect of mixed infection in resistance breakdown. Single nucleotide polymorphisms (SNPs) related to resistance failure for beet necrotic yellow vein virus (BNYVV) and tomato spotted wilt virus (TSWV) were detected by real-time qPCR (Acosta-Leal & Rush, 2007; diRienzo, Bubici, Montemurro, & Cillo, 2018). However, these methods cannot be commonly applied to every isolate of each virus species due to epistatic interactions or the presence of other polymorphic sites.

In plants, the RNA silencing mechanism serves as a natural defense mechanism that can also be utilized to develop virus resistance. Resistance can be achieved by introducing a DNA construct that produces viral dsRNA or ssRNA through transformation. Real-time qPCR is a suitable method for evaluating the efficiency of RNA silencing-based resistance. Resistance breakdown can occur due to mixed infection and mutation (De Haan et al., 1992; Syller, 2012). To minimize the possibilities of resistance breakdown, multiple constructs from different virus species or isolates can be introduced (Duan, Wang, & Guo, 2003; Bucher et al., 2006). Another method of immunization involves introducing a transgene that mimics the secondary structure of endogenous miRNA precursors, forming artificial miRNAs (amiRNAs) that target the virus sequences (Qu, Ye, & Fang, 2007; Liu, Zhou, Hu, Wei, & Zhang, 2017). However, amiRNA-based resistance can also be overcome by mutation, selection, and mixed infection with other viruses (Lafforgue et al., 2011; Martinez, Elena, & Daros, 2013). Real-time qPCR is useful for detecting increased virus accumulation, which can indicate resistance breakdown (Peña et al., 2014). The efficiency and durability of resistance can be enhanced by introducing amiRNA constructs that target different regions of a single virus genome (Kis et al., 2016). Another approach for resistance development is cross protection, where mild or attenuated, viral strains are inoculated to provide protection against virulent strains of the same virus species. This method has been successfully used in various crops and viruses, such as CTV in citrus and PepMV in tomatoes (Pechinger, Chooi, MacDiarmid, Harper, & Ziebell, 2019).

## 7.5 PROPHYLACTIC MEASURES

Quarantine and phytosanitary certification are essential measures when trading plant materials to prevent the introduction of new viruses into a specific geographical area. To import and export virus-free plants, precise, broad-spectrum, and rapid diagnosis methods are necessary to avoid delays. NGS-based methods are particularly useful as they can detect both known and unknown viruses. These methods are sensitive, efficient, and cost-effective, making them increasingly utilized in quarantine programs (Villamor, Ho, Al Rwahnih, Martin, & Tzanetakis, 2019). Phytosanitary certificates should only be issued for plant materials free from harmful viruses, as plants may still carry non-virulent viruses (Davino et al., 2013). Phylogeographic analysis, which examines genetic variations in space and time, can help evaluate the sites of introduction and paths of virus dispersal (Maree, Fox, Al Rwahnih, Boonham, & Candresse, 2018; Gómez, Sempere, Aranda, & Elena, 2012). When virus-free plants are not available, sanitation methods such as thermotherapy, chemotherapy, electrotherapy, and tissue culture are employed to remove viruses and obtain virus-free plant materials (Davino et al., 2013). After sanitation, plants must be tested using sensitive techniques like real-time qPCR to confirm their virus-free status. In RNA viruses, the mutation rate can be increased by using nucleoside analogs, leading to lethal mutagenesis. This method has recently been applied to TMV, where the high mutation rate resulted in the loss of virus infectivity (Naik & Buckseth, 2018).

Common agricultural practice for controlling virus dispersal involves removing infected plants from fields. However, this method is only effective when virus incidence is low, and it requires detection techniques such as ELISA and molecular hybridization to identify viruses in field samples (Dıáz-Martıńez et al., 2018). Seed-borne viruses (present on the seed coat) can be eliminated through chemical treatment, while seed-transmitted viruses (present in the seed embryo) can be removed through thermotherapy (Rubio, Janssen, Cuadrado, Moreno, & Guerri, 2003b; Ling, 2003). For simultaneous detection of seed-borne and seed-transmitted viruses in a single crop, M-PCR can be used (Paylan, Erkan, Cetinkaya, Ergun, & Pazarlar, 2014). Mechanically transmissible viruses can be eliminated through hygienic measures such as using gloves, disinfectants, and heat sterilization of tools (Panno et al., 2014). Many plant viruses are transmitted by vectors like whiteflies, aphids, and thrips. The transmission of these viruses can be controlled by limiting vector populations using pesticides, biological control methods, and biotechnology approaches such as RNA silencing, neurotoxins, and protease inhibitors (Broadbent, 1976; Téllez, Simon, Rodriguez, & Janssen, 2017; Nandety, Kuoy, Nouriy, & Falk, 2015; Vogel, Santos, Mingels, Verdonckt, & Broeck, 2019).

## 7.6 CONCLUSION

To provide nutritious food to the growing world population in a sustainable way is the major challenge of agriculture in recent times (Pretty et al., 2010). Plant pathogens, including viruses, cause diseases that damage crop yield in terms of quantity and quality, leading to negative economic impacts in agriculture (Mumford, Macarthur, & Boonham, 2016). The precise and quick identification of viruses is essential to reduce the spread of virus diseases and enable effective management. However, the high evolution rate in viruses poses a major challenge for virus detection and disease management.

In virus detection, ELISA is quite popular for routine assays due to the ease of use and commercial availability of specific antibodies for many plant viruses. However, it is expensive, time-consuming, and cannot manage virus variability. Molecular techniques like PCR-based methods are widely used in virus diagnosis as they are sensitive, cheap, fast, and versatile. However, these common molecular methods also have limitations, such as the requirement of prior knowledge of genome sequences and restricted ability to detect multiple viruses simultaneously.

The development of NGS platforms has enabled the detection of total nucleic acid content in a sample without the need for prior knowledge about genome sequences. Metagenomics has discovered many novel viruses from different sources. Further developments in NGS-based tools could facilitate major advancements in viral disease diagnosis, which will help in managing emerging diseases in crops.

## KEYWORDS

- **immunization**
- **next-generation sequencing**
- **plant viruses**
- **polymerase chain reaction**
- **prophylactic measures**
- **viral disease management**
- **virus diagnosis methods**

## REFERENCES

Abd El-Aziz, M. H., (2019). Three modern serological methods to detect plant viruses. J. Plant. Sci. Phytopathol., 3, 101–106.

Aboul-Ata, A. E., Mazyad, H., El-Attar, A. K., Soliman, A. M., Anfoka, G., Zeidaen, M., Gorovits, R., et al., (2011). Diagnosis and control of cereal viruses in the Middle East. Adv. Virus Res., 81, 33–61.

Acosta-Leal, R., & Rush, C., (2007). Mutations associated with resistance– breaking isolates of Beet necrotic yellow vein virus and their allelic discrimination using TaqMan technology. Phytopathol., 97, 325–330.

Acosta-Leal, R., Duffy, S., Xiong, Z., Hammond, R., & Elena, S. F., (2011). Advances in plant virus evolution: Translating evolutionary insights into better disease management. Phytopathology, 101, 1136–1148.

Adams, I. P., Glover, R. H., Monger, W. A., Mumford, R., Jackeviciene, E., Navalinskiene, M., Samuitiene, M., & Boonham, N., (2009). Next-generation sequencing and metagenomic analysis: A universal diagnostic tool in plant virology. Mol. Plant Pathol., 10, 537–545.

Alfaro-Fernández, A., Sánchez-Navarro, J. Á., Del Carmen, C., M., Del Carmen Córdoba-Sellés, M., Pallás, V., & Jordá, C., (2009). Simultaneous detection and identification of Pepino mosaic virus (PepMV) isolates by multiplex one-step RT-PCR. Europ. J. Plant. Pathol., 125, 143–158.

Atkinson, P. H., & Matthews, R. E., (1970). On the origin of dark green tissue in tobacco leaves infected with tobacco mosaic virus. Virology, 40, 344–356.

Bin, Y., Li, Z., Wu, J., Wang, X., Zhou, Y., Li, T., & Song, Z., (2017). Development of an immunochromatographic strip test for rapid detection of citrus yellow vein clearing virus. Arch. Virol., 163, 349–357.

Blawid, R., Silva, J. M., & Nagata, T., (2017). Discovering and sequencing new plant viral genomes by next generation sequencing: Description of a practical pipeline. Ann. Appl. Biol., 170, 301–314.

Broadbent, L., (1976). Epidemiology and control of tomato mosaic virus. Annu. Rev. Phytopathol., 14, 75–96.

Bucher, E., Lohuis, D., Van, P. P. M. J. A., Geerts-Dimitriadou, C., Goldbach, R., & Prins, M., (2006). Multiple virus resistance at a high frequency using a single transgene construct. J. Gen. Virol., 87, 3697–3701.

Byzova, N. A., Safenkova, I. V., Chirkov, S. N., Zherdev, A. V., Blintsov, A. N., Dzantiev, B. B., & Atabekov, I. G., (2009). Development of immunochromatographic test systems for express detection of plant viruses. Appl. Biochem. Microbiol., 45, 204–209.

Chandrasekaran, J., Brumin, M., Wolf, D., Leibman, D., Klap, C., Pearlsman, M., et al., (2016). Development of broad virus resistance in non-transgenic cucumber using CRISPR/Cas9 technology. Mol. Plant Pathol., 17, 1140–1153.

Clark, M. F., & Adams, A. N., (1977). Characteristics of the microplate method of enzyme-linked immunosorbent assay for the detection of plant viruses. J. Gen. Virol., 34, 475–483.

Davino, S., Willemsen, A., Panno, S., Davino, M., Catara, A., Elena, S. F., et al., (2013). Emergence and phylodynamics of Citrus tristeza virus in Sicily, Italy. PloS One, 8, e66700.

De Haan, P., Gielen, J. J., Prins, M., Wijkamp, I. G., Van, S. A., Peters, D., et al., (1992). Characterization of RNA–mediated resistance to Tomato spotted wilt virus in transgenic tobacco plants. Nat. Biotechnol., 10, 1133–1137.

De Ronde, D., Butterbach, P., & Kormelink, R., (2014). Dominant resistance against plant viruses. Front. Plant Sci., 5, 307.

Dıáz-Martıńez, L., Brichette-Mieg, I., Pineño-Ramos, A., Domıńguez-Huerta, G., & Grande-Pérez, A., (2018). Lethal mutagenesis of an RNA plant virus via lethal defection. Sci. Rep., 8, 1444.

diRienzo, V., Bubici, G., Montemurro, C., & Cillo, F., (2018). Rapid identification of tomato Sw-5 resistance–breaking isolates of Tomato spotted wilt virus using high resolution melting and TaqMan SNP genotyping assays as allelic discrimination techniques. PloS One, 13, e0196738.

Drygin, Y. F., Blintsov, A. N., Grigorenko, V. G., Andreena, I. P., Opsipov, A. P., Varitzev, Y. A., Uskov, A. I., et al., (2012). Highly sensitive eld test lateral ow immunodiagnostics of PVX infection. Appl. Microbiol. Biotechnol., 93, 179–189.

Duan, C., Wang, C., & Guo, H., (2003). Application of RNA silencing to plant disease resistance. Silence, 3, 5.

Elena, S. F., Fraile, A., & Garcia-Arenal, F., (2014). Evolution and emergence of plant viruses. Adv. Virus Res., 88, 161–191.

Gallet, R., Michalakis, Y., & Blanc, S., (2018). Vector-transmission of plant viruses and constraints imposed by virus-vector interactions. Curr. Opin. Virol., 33, 144–150.

Gaur, R. K., Ali, A., Cheng, X., Mäkinen, K., Agindotan, B., & Wang, X., (2021). Editorial: Plant viruses, volume II: Molecular plant virus epidemiology and its management. Front. Microbiol., 12, 756807.

Gómez, P., Sempere, R., Aranda, M., & Elena, S., (2012). Phylodynamics of Pepino mosaic virus in Spain. Eur. J. Plant Pathol., 134, 445–449.

Guo, L., Wu, J., Chen, R., Hong, J., Zhou, X., & Wu, J., (2020). Monoclonal antibody-based serological detection of Rice stripe mosaic virus infection in rice plants or leafhoppers. Virol. Sin., 35, 227–234.

Hadidi, A., (2019). Next generation sequencing and CRISPER/cas13 editing in viroid research and molecular diagnostics. Viruses, 11, 120.

Hancˇinský, R., Mihálik, D., Mrkvová, M., Candresse, T., & Glasa, M., (2020). Plant viruses infecting Solanaceae family members in the cultivated and wild environments: A review. Plants, 9, 667.

Hashimoto, M., Neriya, Y., Yamaji, Y., & Namba, S., (2016). Recessive resistance to plant viruses: Potential resistance genes beyond translation initiation factors. Front. Microbiol., 7, 1695.

He, S., & Krainer, K. M., (2020). Pandemics of people and plants: Which is the greater threat to food security? Mol. Plant, 13, 933–934.

Joo-Jin, J., Ho-Jong, J., & Jaejong, N., (2014). A review of detection methods for the plant viruses. Res. Plant Dis., 20(3), 173–181.

Khare, E., Mishra, J., & Arora, N. K., (2018). Multifaceted interactions between endophytes and plant: Developments and prospects. Front. Microbiol., 9, 2732.

Kis, A., Tholt, G., Ivanics, M., Varallyay, E., Jenes, B., & Havelda, Z., (2016). Polycistronic artificial miRNA-mediated resistance to Wheat dwarf virus in barley is highly efficient at low temperature. Mol. Plant Pathol., 17, 427–437.

Lafforgue, G., Matinez, F., Sardanyes, J., delaIglesia, F., Niu, Q. W., Lin, S. S., et al., (2011). Tempo and mode of plant RNA virus escape from RNA interference- mediated resistance. J. Virol., 85, 9686–9695.

Lee, J. S., Cho, W. K., Lee, S. H., Choi, H. S., & Kim, K. H., (2011). Development of RT-PCR based method for detecting five non-reported quarantine plant viruses infecting the family Cucurbitaceae or Solanaceae. Plant Pathology J., 27, 93–97.

Li, R., & Hartung, J. S., (2007). Reverse transcription-polymerase chain reaction-based detection of plant viruses. Curr. Protoc. Microbiol., 6, 16C-1.

Li, R., Mock, R., Huang, Q., Abad, J., Hartung, J., & Kinard, G., (2008). A reliable and inexpensive method of nucleic acid extraction for the PCR-based detection of diverse plant pathogens. J. Virol. Methods, 154, 48–55.

Lievens, B., Grauwet, T. J. M. A., Cammue, B. P. A., & Thomma, B. P. H. J., (2005). Recent developments in diagnostics of plant pathogens: A review. Recent Res. Dev. Microbiol., 9, 57–79.

Ling, K., (2003). Effectiveness of chemo–and thermotherapeutic treatments on Pepino mosaic virus in tomato seed. Plant Dis., 94, 325–328.

Liu, S., Zhou, J., Hu, C., Wei, C., & Zhang, J., (2017). MicroRNA–mediated gene silencing in plant defense and viral counter–defense. Front. Microbiol., 8, 1801.

Lopez, M. M., Llop, P., Olmos, A., Marco-Noales, E., Cambra, M., & Bertolini, E., (2008). Are molecular tools solving the challenges posed by detection of plant pathogenic bacteria and viruses? Curr. Issues Mol. Biol., 11, 13–45.

Majumder, S., & Baranwal, V. K., (2014). Simultaneous detection of four garlic viruses by multiplex reverse transcription PCR and their distribution in Indian garlic accessions. J. Virol. Methods, 202, 34–38.

Makkouk, K. M., & Kumari, S. G., (2006). Molecular diagnosis of plant viruses. Arab. J. Plant Protect., 24, 135–138.

Maksimov, I. V., Sorokan, A. V., Burkhanova, G. F., Veselova, S. V., Alekseev, V. Y., Shein, M. Y., Avalbaev, A. M., Dhaware, P. D., Mehetre, G. T., Singh, B. P., et al., (2019). Mechanisms of plant tolerance to RNA viruses induced by plant-growth-promoting microorganisms. Plants, 8, 575.

Malmstrom, C. M., Melcher, U., & Bosque-Pérez, N. A., (2011). The expanding field of plant virus ecology: Historical foundations, knowledge gaps, and research directions. Virus Res., 159, 84–94.

Maree, H. J., Fox, A., Al Rwahnih, M., Boonham, N., & Candresse, T., (2018). Application of HTS for routine plant virus diagnostics: State of the art and challenges. Front. Plant. Sci., 9, 1082.

Martinelli, F., Scalenghe, R., Davino, S., Panno, S., Scuderi, G., Ruisi, P., Villa, P., Stroppiana, D., Boschetti, M., Goulart, L. R., et al., (2015). Advanced methods of plant disease detection: A review. Agron. Sustain. Dev., 35, 1–25.

Martinez, F., Elena, S. F., & Daros, J., (2013). Fate of artificial micro RNA-mediated resistance to plant viruses in mixed infections. Phytopathol., 103, 870–876.

Massart, S., Olmos, A., Jijakli, H., & Candresse, T., (2014). Current impact and future directions of high throughput sequencing in plant virus diagnostics. Virus Res., 188, 90–96.

Mehetre, G. T., Leo, V. V., Singh, G., Sorokan, A., Maksimov, I., Yadav, M. K., Upadhyaya, K., Hashem, A., Alsaleh, A. N., Dawoud, T. M., et al., (2021). Current developments and challenges in plant viral diagnostics: A systematic review. Viruses, 13, 412.

Mehetre, G. T., Paranjpe, A. S., Dastager, S. G., & Dharne, M. S., (2016). Complete metagenome sequencing based bacterial diversity and functional insights from basaltic hot spring of Unkeshwar, Maharshtra, India. Genomics Data, 7, 140–143.

Mehle, N., Gregur, L., Bogožalec, K. A., & Dobnik, D., (2020). One-step reverse-transcription digital PCR for reliable quantification of different Pepino mosaic virus genotypes. Plants, 9, 326.

Mirmajlessi, S. M., Loit, E., Maend, M., & Mansouripour, S. M., (2015). Real-time PCR applied to study on plant pathogens: Potential applications in diagnosis: A review. Plant. Prot. Sci., 51, 177–190.

Mumford, R., Macarthur, R., & Boonham, N., (2016). The role and challenges of new diagnostic technology in plant biosecurity. Food Secur., 8, 103–109.

Mumo, N. N., Mamati, G. E., Ateka, E. M., Rimberia, F. K., Asudi, G. O., Boykin, L. M., et al., (2020). Metagenomic analysis of plant viruses with papaya ringspot disease in Carica Papaya L. in Kenya. Front. Microbiol., 11, 205.

Naik, P. S., & Buckseth, T., (2018). "Recent advances in virus elimination and tissue culture for quality potato seed production. In: Gosal, S. S., & Wani, S. H., (eds.), Biotechnologies of Crop Improvement (Vol. 1, pp. 131–158). Cham, Switzerland: Springer.

Nandety, R. S., Kuoy, Y. W., Nouriy, S., & Falk, B. W., (2015). Emerging strategies for RNA interference (RNAi) applications in insects. Bioengineered, 6, 8–19.

Nascimento, A. K. Q., Lima, J. A. A., & Barbosa, G. D. S., (2017). A simple kit of plate-trapped antigen enzyme-linked immunosorbent assay for identification of plant viruses. Rev. Ciênc. Agron., 48, 216–220.

Nateqi, M., Koohi, H. M., Dizadji, A., & Parizad, S., (2015). Detection and molecular characterization of the Iris severe mosaic virus-Ir isolate from Iran. J. Plant. Protect. Res., 55, 235–240.

Niu, Y., Wang, D., Cui, L., Wang, B., Pang, X., & Yu, P., (2018). Monoclonal antibody-based colloid gold immunochromatographic strip for the rapid detection of Tomato zonate spot tospovirus. Virol. J., 15, 1–9.

Palloix, A., Ayme, V., & Moury, B., (2009). Durability of plant major resistance genes to pathogens depends on the genetic background, experimental evidence, and consequences for breeding strategies. New Phytol., 183, 190–199.

Panno, S., Ferriol, I., Rangel, E. A., Olmos, A., Han, C., Martinelli, F., et al., (2014). Detection and identification of Fabavirus species by one-step RT-PCR and multiplex RT-PCR. J. Virol. Methods, 197, 77–82.

Paylan, I. C., Erkan, S., Cetinkaya, N., Ergun, M., & Pazarlar, S., (2014). Effects of different treatments on the inactivation of various seedborne viruses in some vegetables. Ozone: Sci. Eng., 36, 422–426.

Pechinger, K., Chooi, K. M., MacDiarmid, R. M., Harper, S. J., & Ziebell, H., (2019). A new era for mild strain cross- protection. Viruses, 11, 670.

Peiman, M., & Xie, C., (2006). Sensitive detection of potato viruses, PVX, PLRV, and PVS, by RT-PCR in potato leaf and tuber. Australas. Plant Dis. Notes., 1, 41–46.

Peiró, A., Cañizares, M. C., Rubio, L., López, C., Moriones, E., Aramburu, J., et al., (2014). The movement protein (NSm) of Tomato spotted wilt virus is the avirulence determinant in the tomato Sw-5 gene-based resistance. Mol. Plant Pathol., 15, 802–813.

Peña, E., Ferriol, I., Sambade, A., Buschmann, H., Nielh, A., Elena, S. F., et al., (2014). Experimental virus evolution reveals a role of plant microtubule dynamics and TORTIFOLIA1/SPIRAL2 in RNA trafficking. PLoS One, 9, e105364.

Peter, K. A., Gildow, F., Palukaitis, P., & Gray, S. M., (2009). The C terminus of the Polerovirus P5 readthrough domain limits virus infection to the phloem. J. Virol., 83, 5419–5429.

Pretty, J., Sutherland, W. J., Ashby, J., Auburn, J., Baulcombe, D., Bell, M., et al., (2010). The top 100 questions of importance to the future of global agriculture. Int. J. Agr. Sustain., 8, 219–236.

Qu, J., Ye, J., & Fang, R., (2007). Artificial microRNA–mediated virus resistance in plants. J. Virol., 81, 6690–6699.

Rani, A., Donovan, N., & Mantri, N., (2019). The future of plant pathogen diagnostics in a nursery production system. Biosens. Bioelectron., 145, 111631.

Riesenfield, C. S., & Schloss, P. D., (2004). Metagenomics: Genomic analysis of microbial communities. Annu. Rev. Genet., 38, 525–552.

Roossinck, M. J., (2012). Plant virus metagenomics: Biodiversity and ecology. Annu. Rev. Genet., 46, 359–369.

Roossinck, M. J., (2015). A new look at plant viruses and their potential beneficial roles in crops. Mol. Plant Pathol., 16, 231–233.

Rubio, L., Galipienso, L., & Ferriol, I., (2020). Detection of plant viruses and disease management: Relevance of genetic diversity and evolution. Front. Plant. Sci., 11, 1092/

Rubio, L., Herrero, J. R., Sarrio, J., Moreno, P., & Guerri, J., (2003a). A new approach to evaluate relative resistance and tolerance of tomato cultivars to Begomoviruses causing the tomato yellow leaf curl disease in Spain. Plant Pathol., 52, 763–769.

Rubio, L., Janssen, D., Cuadrado, I. M., Moreno, P., & Guerri, J., (2003b). Rapid detection of Cucumber vein yellowing virus by tissue–print hybridization with digoxigenin–labelled cDNA probes. J. Virol. Methods, 114, 105–107.

Ruiz-Ruiz, S., Ambros, S., Del Carmen, V. M., Navarro, L., Moreno, P., & Jose, G., (2009). Detection and quantitation of Citrus leaf blotch virus by TaqMan real-time RT-PCR. J. Virol. Methods., 160, 57–62.

Ruiz-Ruiz, S., Moreno, P., Jose, G., & Ambros, S., (2007). A real-time RT-PCR assay for detection and absolute quantitation of Citrus tristeza virus in different plant tissues. J. Virol. Methods, 145, 96–105.

Seepiban, C., Charoenvilaisiri, S., Warin, N., Bhunchoth, A., Phironrit, N., Phuangrat, B., Chatchawankanphanich, O., et al., (2017). Development and application of triple antibody sandwich enzyme-linked immunosorbent assays for Begomovirus detection using monoclonal antibodies against Tomato yellow leaf curl Thailand virus. Virol. J., 14, 99.

Seo, J. K., Kim, M. K., Kwak, H. R., Choi, H. S., Nam, M., Choe, J., Choi, B., et al., (2018). Molecular dissection of distinct symptoms induced by Tomato chlorosis virus and Tomato yellow leaf curl virus based on comparative transcriptome analysis. Virology, 516, 1–20.

Soler, S., Debreczeni, D., Vidal, E., Aramburu, J., López, C., Galipienso, L., et al., (2015). A new Capsicum baccatum accession shows tolerance to wild–type and resistance–breaking isolates of Tomato spotted wilt virus. Ann. Appl. Biol., 167, 343–353.

Stobbe, A. H., & Roossinck, M. J., (2014). Plant virus metagenomics: What we know and why we need to know more. Front. Plant Sci., 5, 150.

Strange, R. N., (2005). Plant disease: A threat to global food security. Annu. Rev. Phytopathol., 43, 83–116.

Syller, J., (2012). Facilitative and antagonistic interactions between plant viruses in mixed infections. Mol. Plant Pathol., 13, 204–216.

Téllez, M. D. M., Simon, A., Rodriguez, E., & Janssen, D., (2017). Control of Tomato leaf curl New Delhi virus in zucchini using the predatory mite Amblyseius swirskii. Biol. Control, 114, 106–113.

Truniger, V., & Aranda, M. A., (2009). Recessive resistance to plant viruses. Adv. Virus Res., 75, 119–159.

Van, B. S. L., Bergervoet, J. H. W., Pagendam, D. E., De Weerdt, M., Geering, A. D. W., Drenth, A., Van, D. V. R. A., (2014). Development of a multiplexed bead-based suspension array for the detection and discrimination of pospiviroid plant pathogens. PLoS One, 9, e84743.

Van, D. W. J. P. H., & Dijkstra, J., (2006). A history of plant virology. Arch. Virol., 151, 1467–1498.

Villamor, D. E. V., Ho, T., Al Rwahnih, M., Martin, R. R., & Tzanetakis, I. E., (2019). High throughput sequencing for plant virus detection and discovery. Phytopathol., 109, 716–725.

Villamor, D. E., Mekuria, T. A., Pillai, S. S., & Eastwell, K. C., (2016). High-throughput sequencing identifies novel viruses in nectarine: Insights to the etiology of stem-pitting disease. Phytopathology, 106, 519–527.

Vogel, E., Santos, D., Mingels, L., Verdonckt, T. W., & Broeck, J. V., (2019). RNA interference in insects: Protecting beneficials and controlling pests. Front. Physiol., 10, 1912.

Wren, J. D., Roossinck, M. J., Nelson, R. S., et al., (2006). Plant virus biodiversity and ecology. PLoS Biol., 4, E80.

Wu, J., Ni, Y., Liu, H., Ding, M., & Zhou, X., (2014). Monoclonal antibody-based serological assays and immunocapture-RT-PCR for detecting Rice dwarf virus in field rice plants and leafhopper vectors. J. Virol. Methods, 195, 134–140.

Xiao, H., Li, C., Al Rwahnih, M., Dolja, V., & Meng, B., (2019). Metagenomic analysis of Riesling grapevine reveals a complex virome including two new and divergent variants of grapevine leafroll-associated virus 3. Plant Dis., 103, 1275–1285.

Xie, Y., Jiao, X., Zhou, X., Liu, H., Ni, Y., & Wu, J., (2013). Highly sensitive serological methods for detecting Tomato yellow leaf curl virus in tomato plants and whiteflies. Virol. J., 10, 142.

Yang, J. G., Wang, F. L., Chen, D. X., Shen, L. L., Qian, Y. M., Liang, J. Y., Zhou, W. C., & Yan, T. H., (2012). Development of a one-step immunocapture real-time RT-PCR assay for detection of Tobacco mosaic virus in soil. Sensors., 12, 16685–16694.

Zhang, M., Ye, J., He, J. S., Zhang, F., Ping, J., Qian, C., & Wu, J., (2020). Visual detection for nucleic acid-based techniques as potential on-site detection methods. A review. Anal. Chim. Acta, 1099, 1–15.

Zheng, L., Rodoni, B. C., Gibbs, M. J., & Gibbs, A. J., (2009). A novel pair of universal primers for the detection of potyviruses. Plant Pathol., 59, 211–220.

## CHAPTER 8

# Plant Virus Disease Management Strategies: Conventional Versus Modern Techniques

ANIRBAN MANDAL,[1] ARINDAM MUKHERJEE,[2] and RITIKA BANDYOPADHYAY[3]

[1]*Department of Microbiology, Mrinalini Datta Mahavidyapith, Kolkata, West Bengal, India*

[2]*Department of Botany, Mrinalini Datta Mahavidyapith, Kolkata, West Bengal, India*

[3]*Department of Microbiology, Barrackpore Rastraguru Surendranath College, Kolkata, West Bengal, India*

## ABSTRACT

Plant viruses infect a wide range of green vegetation and cause massive agricultural damage worldwide. They pose a serious threat to global food safety. The emergence of different plant viral diseases is mainly due to climate change and rapid viral evolution. Controlling these diseases largely depends on rapid and precise identification of the causative organism. Integrated disease management (IDM) approaches, based on different immunization methods and the use of prophylaxis such as efficient control of insect vectors, removal of infected plants, or other procedures, can help restrain this disastrous problem. The selection of ideal measures requires detailed knowledge about the virus and the mechanism of action of individual control measures. The strategies should require minimal labor demands, insignificant expenses, and less interruption to standard practices. In this chapter, we summarize different conventional and modern techniques used for plant virus isolation, detection, quantification,

Detection and Management of New and Emerging Mystery Plant Virus Sources.
Alireza Golnaraghi, Rajarshi Kumar Gaur, and Rakesh Kumar Verma (Eds.)

and disease management. We also discuss the accuracy of virus diagnosis, viral detection, and disease control principles such as exclusion, protection, eradication, and resistance. It is highlighted that a detailed research study and in-depth knowledge of farming practices are essential before a successful integrated program can be established and promoted as a means of improving crop productivity to meet the increasing demands of the human population.

## 8.1 INTRODUCTION

Plant viral diseases are considered to be the primary threat to sustainable agriculture. Approximately 25 families of plant viruses are known to infect a wide range of crop species (He & Creasey Krainer, 2020). These viruses cause various types of emerging infectious diseases among crop varieties. Plant viruses are responsible for approximately 40% of total crop damage and result in several billion dollars in losses worldwide each year (Maksimov et al., 2019). They not only reduce crop yields in terms of quantity, but also negatively impact the quality of the produce. There is clear evidence of many viruses infecting multiple crop plants, and new viruses continue to emerge through outbreaks. Tobacco mosaic virus (TMV), TSWV, cucumber mosaic virus (CMV), cauliflower mosaic virus (CaMV), etc., are among the most common plant viruses (Figure 8.1) (Hancinsky et al., 2020).

Plant viruses exhibit a wide range of symptoms, including leaf curl, bud necrosis, necrotic spots, abnormal dark green and light green mosaic ring patterns or bumps on plant foliage, and abnormal color and structure of flowers (Mandal, 2022). Vectors play a crucial role in the transmission of these viral diseases. Various factors such as climate change, artificial breeding, poor handling, and other environmental factors can influence the spread of these viral diseases. The epidemic and pandemic of plant viral infections have a significant impact on cultivation, the food market, and the export-import system. They can even lead to food shortages in a country or worldwide. Europe has experienced epidemics like tobacco veinal necrosis, sugar beet yellows, and plum pox diseases, while Ghana, West Africa, has faced cocoa swollen shoot disease (CSSD). Pandemics like sugarcane mosaic disease, cassava brown streak disease (CBSD), and cassava mosaic disease (CMD) have affected different parts of the world (Jones, 2020, 2021). British plant pathologist John Michael Thresh, who studied the cocoa swollen shoot virus in West Africa and the cassava mosaic virus in Uganda during the pandemic period, provided a review on the origins and epidemiology of vital plant virus diseases (Thresh, Fargette, & Otirn-Nape, 1994). Table 8.1 lists some important plant viral diseases, their causative organisms, symptoms, and modes of transmission.

**TABLE 8.1** Some Important Plant Viral Diseases, its Causative Organisms, Symptoms, and Mode of Transmission

| Diseases | Causative Virus | Family | Diseased Crop | Vectors | Symptoms | References |
|---|---|---|---|---|---|---|
| Banana bunchy top | Banana bunchy top virus | Nanoviridae | *Musa textilis*, *Musa paradisiaca* | Aphids (*Pentalonia nigronervosa*) | Dark green streaking and flecking of leaves, stunting of suckers. | Thomas (2008) |
| Bean mosaic | Bean common mosaic virus (BBCMV) | Potyviridae | *Vigna* spp., *Vicia* sp. *Plhaseolu* ssp. | Whitefly (*Bemisia tabaci*) | Mottling, curling, and malformation of leaves and general stunting of the plant. | Kumar, Hallan, & Zaidi (2009) |
| Cacao swollen shoot | Cacao swollen shoot virus | Caulimoviridae | *Theobroma cacao*, *Adansonia digitata* | Mealybug (*Planococcoides njalensis*) | Swelling of trunk and shoot nodes, leaf chlorosis, vein banding, tree dieback. | Gilmour, Crozier, & Kouamé (2012) |
| Cassava brown streak | Cassava brown streak virus. | Potyviridae | *Manihot glaziovii*, *Zanha africana* | Whitefly (*Bemisia tabaci*) | Necrotic roots veinal chlorosis and mottle of leaves, brown stem streaking. | Tomlinson et al. (2018) |
| Cassava mosaic | African cassava mosaic virus | Geminiviridae | *Manihot esculenta*, *Ricinus communis* | Whitefly (*Bemisia tabaci*) | Severe leaf mosaic, deformation, and plant stunting. | Thresh, Fargette, & Otirn-Nape (1994) |
| Cauliflower mosaic | Cauliflower mosaic virus | Caulimoviridae | *Brassica juncea*, *Raphanus sativus* | Aphids (*Myzus persicae*) | Chlorotic mottling, chlorotic spotting, fine vein clearing, stunting, necrotic spotting. | Haas et al. (2002) |
| Citrus tristeza | Citrus tristeza virus | Closteroviridae | Citrus x sinensis, citrus reticulate, citrus x paradisi | Citrus aphid (*Toxoptera citricida*) | Veinal yellowing, leaf size reduction, leaf drop, wilting, twig dieback, and tree death. | Moreno et al. (2008) |
| Cucumber green mottle mosaic | Cucumber green mottle mosaic virus. | Virgaviridae | *Amaranthus graecizans*, *A. muricatus* | No insect vector | Green mottling or mosaic leaves and fruit, internal flesh discoloration and necrosis. | Darzi et al. (2018) |
| Faba bean necrotic yellows | Faba bean necrotic yellows virus (FBNYV) | Nanoviridae | *Phaseolus vulgaris*, *Cicer arietinum* | Aphid (*Acyrthosiphon pisum*) | Rolled older leaves, small cupped young severe plant stunting and plant death. | Kumari & Makkouk (2007) |
| Ground nut rosette | Groundnut rosette virus (GRV) | Tombusviridae | *Vigna unguiculata*, *Vigna radiata* | Aphid (*Aphis craccivora*) | Green leaf mosaic, rosette, and chlorotic yellow leaf mosaic. | Naidu et al. (1999) |

**TABLE 8.1** *(Continued)*

| Diseases | Causative Virus | Family | Diseased Crop | Vectors | Symptoms | References |
|---|---|---|---|---|---|---|
| Leaf curl of chili | Chili leaf curl virus | Geminiviridae | *Capsicum* spp., *Solanum lycopersicum* | Whitefly (*Bemisia tabaci*) | Curling, mottling, and of leaves, flowers, fruits, and affected plants are shorter. | Ahmad et al. (2016) |
| Maize streak | Maize streak virus | Geminiviridae | *Zea mays* | Leafhopper (*Cicadulina mbila*) | Circular spots on leaves. | Shepherd et al. (2010) |
| Maize rough dwarf | Maize rough dwarf virus | Reoviridae | *Zea mays* | Planthopper (*Laodelphax striatellus*) | Rough leaf sheaths developed on the veins. | Clemente-Orta et al. (2021) |
| Peanut stunt | Peanut stunt virus (PSV) | Bromoviridae | *Arachis hypogaea*, *Trifolium repens* | Aphids (*Aphis craccivora*) | Discoloration and distortion of the leaves, stunting growth of rhizomes. | Netsu et al. (2008) |
| Pepino mosaic | Pepino mosaic virus | Alphaflexiviridae | *Solanum muricatum* | Fungi (*Olpidium virulentus*) | Leaf mosaic, chlorotic spotting, distortion, and reduced size. | Hanssen & Thomma (2010) |
| Plum pox | Plum pox virus | Potyviridae | *Prunus domestica*, *Prunus avium* | Mite (*Aculus fockeni*) | Chlorotic spots or rings, leaf vein clearing, chlorosis, and deformation. | Hajizadeh et al. (2019) |
| Potato tuber necrotic ringspot. | Potato virus Y | Potyviridae | *Solanum tuberosum* | Aphid (*Aphis fabae*) | Crinkle and rugose mosaic, shoot necrosis, severe leaf mosaic and deformation. | Fuentes et al. (2019) |
| Rice hoja blanca | Rice hoja blanca virus | Phenuiviridae | *Oryza sativa* | Planthopper (*Sogatodes orizicola*) | Chlorotic leaves and white spots on leaves. | Ramirez et al. (1992) |
| Rice tungro disease | Rice tungro spherical virus. | Secoviridae | *Zizania* spp. | Leafhopper (*Nephotettix virescens*) | Mild stunting, mild tungro. | Bunawan et al. (2016) |
| Rice yellow mottle | Rice yellow mottle virus | Sobemoviridae | *Oryza sativa*, *Oryza glaberrima* | Insect (*Sesselia pusilla*) | Leaf yellowing, stunting, reduced tillering, and poor seed setting. | Hebrard, Fargette, & Konaté (2008) |

**TABLE 8.1** *(Continued)*

| Diseases | Causative Virus | Family | Diseased Crop | Vectors | Symptoms | References |
|---|---|---|---|---|---|---|
| Sugarcane mosaic | Sugarcane mosaic virus | Potyviridae | *Saccharum officinarum*, *Zea mays* | Aphid (*Myzus persicae*) | Irregular yellow inlays, mottles with parallel veins on symptomatic leaves. | Addy et al. (2017) |
| Sweet potato virus disease | Sweet potato virus disease | Closteroviridae | *Ipomoea batatas* | Aphid (*Myzus persicae*) | Malformation and plant stunting, severe leaf mosaic and tuberous roots. | Tairo et al. (2005) |
| Tomato brown rugose fruit | Tomato brown rugose fruit virus. | Virgaviridae | *Solanum lycopersicum* | Silver leaf whitefly (*Bemisia tabaci*) | Uneven ripening, deformation, yellow or brown spotting fruit. | Davino et al. (2020) |
| Tomato yellow leaf curl | Tomato yellow leaf virus | Geminiviridae | *Phaseolus vulgaris*, *Petunia atkinsiana* | Silverleaf whitefly (*Bemisia tabaci*) | Curling leaves, yellowing, and diminished size. | Marchant et al. (2020) |
| Wheat streak mosaic | Wheat streak mosaic virus | Potyviridae | *Triticum* spp. | Wheat curl mite (*Aceria tosichella*) | Yellow or pale green leaf streaking, tufted growth habit, plant stunting. | Singh et al. (2018) |
| Yellow dwarf | Barley yellow dwarf virus | Luteoviridae | *Avena sativa*, *Triticale hexaploide* | Aphid (*Rhopalosiphum padi*) | Leaf yellowing or reddening, stiff leaves, diminished root growth. | D'Arcy & Domier (2000) |
| Yellow vein mosaic | Bhendi yellow vein mosaic virus | Geminiviridae | *Abelmoschus esculentus*, *Ageratum* spp. | Whitefly (*Bemisia tabaci*) | Bleaching of vein and veinlets on leaves. | Shetty, Singh, & Singh (2013) |

There are various reasons to identify or measure diseases in plants. Knowledge of the disease is particularly important for effective management as it is closely related to crop yield. In crop protection, assessing plant resistance is crucial for cost-effective pesticide application. The investigation of disease and symptom intensity is crucial for addressing key issues in plant stress biology. Similar to animal viruses, plant viral DNA, RNA, and functional proteins are the main target materials for disease detection and treatment (Mandal, 2021). Over the past 100 years, farmers have utilized conventional techniques for the treatment and prevention of plant viral diseases. However, there is now a high demand for new techniques that can accurately identify and prevent these causative agents. Different types of detection methods are used depending on the requirements, financial condition, time, and sensitivity of the disease. Diagnostic methods are broadly classified into two parts: (i) conventional techniques; and (ii) molecular diagnostics or DNA-based detection. Traditional techniques such as microbial culture, microscopy, and biochemical profiling are used for plant viral disease identification. In molecular diagnostics or DNA-based detection, methods such as polymerase chain reaction (PCR), genotyping, enzyme-linked immunosorbent assays (ELISA), nucleic acid sequencing, and probe hybridization are examples of plant viral disease detection. However, to treat a disease properly, it is always important to use confirmatory modern techniques in addition to common traditional techniques. In this chapter, we will discuss some important traditional and modern techniques for plant virus detection (Mehetre et al., 2021).

## 8.2 CONVENTIONAL METHODS FOR PLANT VIRUS DISEASES DETECTION

### *8.2.1 DETECTION BY SYMPTOMS*

The most commonly used traditional approach for diagnosing plant viral diseases is the interpretation of visual symptoms (Figure 8.2). For a long time, symptoms of virus-infected plants have been used as a parameter for viral taxonomy (Scholthof et al., 2011). As a result of infection, plant viruses produce various symptoms such as mosaics, yellowing, stunting, chlorosis, and necrosis in host plants. Viral infections lead to a loss of proper growth and function in the vegetative and reproductive parts of the plants (Seo et al., 2018). The external symptoms of viral plant infections must affect specific functions in physiological aspects (Garcia-Ruiz, 2019).

**FIGURE 8.1** Symptoms of plant viral infection in different hosts – (A) Cucumber green mottle mosaic virus in *Cucurbita maxima* L.; (B) chili leaf curl virus in *Capsicum annuum* L.; (C) tomato yellow leaf curl virus in *Amaranthus viridis* L.; (D) bitter gourd yellow mosaic virus in *Momordica charantia* L.; (E) luffa yellow mosaic virus in *Luffa acutangula* (L.) Roxb.; and (F) cucumber chlorotic leaf virus in *Cucumis sativus* L.

The pathogenic potential of the virus determines its ability to infect one or more hosts and the types of symptoms that develop thereafter. For clear and specific symptoms, visual inspection is a very easy method for detection and implementing effective remedies. The development of characteristic symptoms is influenced by various factors such as virus strain, plant variety/cultivar, infection time, and environmental parameters (Webster, Wylie, & Jones, 2004; Mutka & Bart, 2015). Various aspects such as soil mineral/nutrient imbalances, unfavorable weather conditions, and damage caused by mite/insect/nematode pests, infection by non-viral pathogens, air pollution, and pesticides may induce similar virus-like symptoms in plants. Few viruses may show no apparent symptoms or induce symptomless infection in plants. Even different plant viruses or different strains of the same virus induce similar types of symptoms in the host plant (Wu et al., 2020). As symptoms provide crucial information on virus diseases, comprehensive field experience is needed when detecting a viral disease based on symptoms

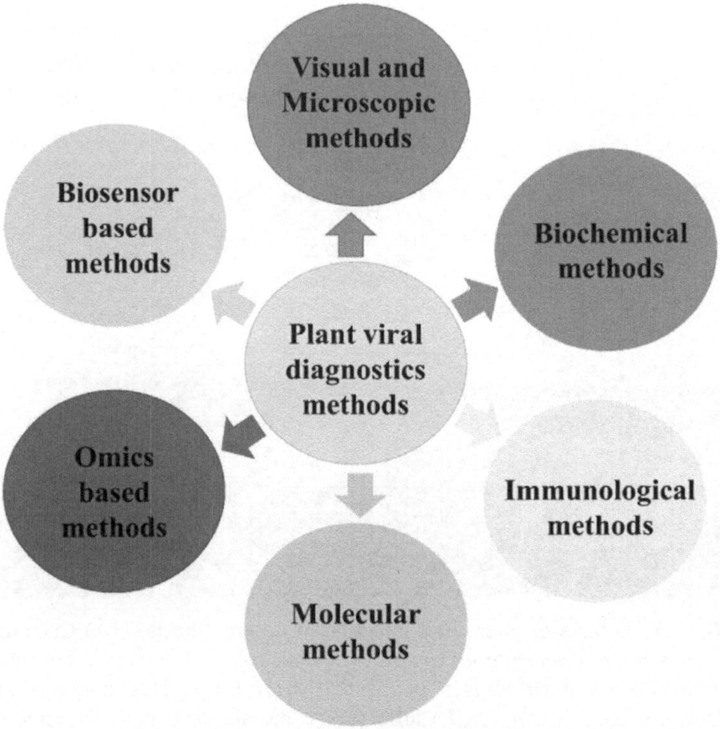


**FIGURE 8.2** Diagnostic techniques used for viral identification and diagnosis of diseases in various plants.

alone (Naidu & Hughes, 2003). Generally, visual detection of symptoms is done along with other modern diagnostics to ensure accurate detection of plant virus infection (Figure 8.3) (Bock, 1982). Nowadays, visual detection can be done using different advanced methods, such as visible light imaging, hyperspectral imaging, thermal imaging, and chlorophyll fluorescence imaging (Balodi et al., 2017).

### *8.2.2 MICROSCOPIC METHODS*

Various groups of plant pathogens, such as algae, fungi, and bacteria, can be easily visualized through light microscopes. However, plant viruses are too small to be seen under a light microscope; only an electron microscope (EM) can detect them (Baker, 1985; Horne, 2018). Viruses produce

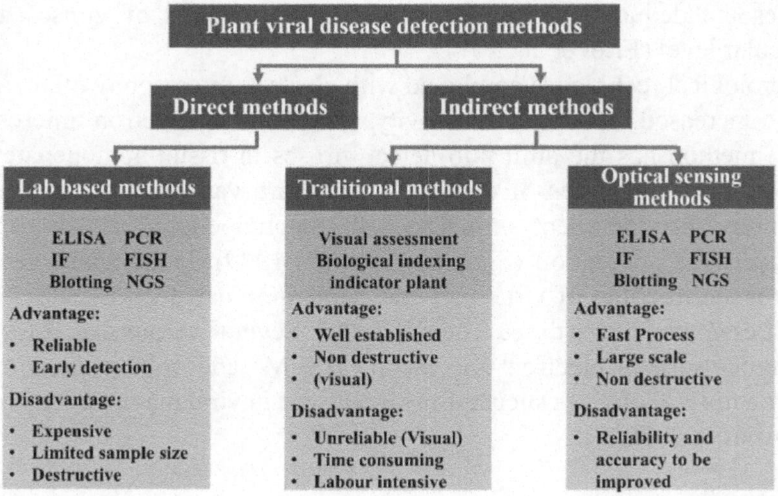


**FIGURE 8.3** Pros and cons of plant viral detection methods.

distinctive intracellular inclusions in host plant cells. EM investigation of these inclusions allows for quick identification and description of numerous plant viruses. The inclusions caused by certain viruses can be used for their identification (Edwardson & Christie, 1978). TMV forms hexagonal and needle-shaped crystals, while wheat mosaic virus (WMV) and potato virus X (PVX) produce spherical amorphous bodies (Baker, 1985). By using negative staining of homogenized infected plant tissue, the transmission electron microscope (TEM) enables direct detection (Richert-Pöggeler et al., 2019). Additionally, several structural changes in viral-infected plant cells and the transport of virions through plasmodesmata via membrane protein-generated tubular structures can be observed (Kozieł, Otulak-Kozieł, & Bujarski, 2018).

Another related microscopic application, such as the confocal microscope (CM), has been utilized for the detection of indicator molecules or compounds (Gorovits & Czosnek, 2017). CM has aided in the identification of co-localized heat shock proteins, such as HSP90 and HSP70, in viral-infected tomato plants (Gorovits & Czosnek, 2017). Furthermore, CM is employed for the detection of viruses labeled with fluorescent proteins in intracellular plant tissues. The most advanced microscopy technique, cryo-electron microscopy (cryo-EM), is currently utilized for the detection of plant-virus interactions (Richert-Pöggeler et al., 2019). Cryo-EM technology enables the deduction of the structural conformations of viruses at the molecular level (Ertel et al., 2017).

Serological techniques combined with electron microscopy offer significantly increased detection sensitivity. The immunoelectron microscopy (IEM) method has the ability to detect viruses in tissue homogenates and requires trace quantities of the virus. There are various ways to perform detections. Pre-treatment of grids with Staphylococcal protein A may enhance virus adsorption (Shukla & Gough, 1979). It is advantageous to monitor the trapping of virus particles on serum-coated grids with decoration (Lesemann, Bozarth, & Koenig, 1980). Several variants of IEM, such as immunosorbent electron microscopy (ISEM) and solid-phase immune electron microscopy, are successfully employed in viral diagnosis (Bhardwaj & Kulshrestha, 2020).

### *8.2.3 PROPHYLACTIC MEASURES*

A crucial initial step in any crop management system is accurate disease diagnosis. Treatment of viral infection in plants may not be effective in controlling the infection. In this case, diseases are most effectively managed if control measures are introduced before the onset of infection. Sowing healthy plant propagules is one of the most powerful approaches adopted by farmers. Effective certification programs to create virus-free propagation material are regarded as essential steps to avoid costly and sensitive diagnostic methods later (Maree et al., 2018). Sanitization is generally done using thermotherapy, electrotherapy, chemotherapy, and tissue culture, either singly or in combination (Ling, 2010). Thermotherapy causes coat protein (CP) rupture or viral particle disruption, viral RNA breakage, viral polymerase inactivation, inhibition of viral movement, translation, and inactivation of viruses. Due to the high cost, antiviral drugs are used only to create disease-free vegetative propagules or for seed production (Panattoni, Luvisi, & Triolo, 2013). Chemotherapeutic effects are mediated by antiviral drugs to disrupt or inhibit specific steps of the virus multiplication cycle, e.g., protease inhibitors interrupt protein processing and nucleoside analogs prevent replication.

### *8.2.4 PLANT VIRAL DISEASES AND FARMER'S PERCEPTIONS*

Small-scale farmers from developing and undeveloped countries lack knowledge or awareness regarding the management of virus-infected crops.

This lack of knowledge leads to a loss of harvestable yield. Therefore, it is important to understand their perceptions of the damage caused by viral infections, their choice of methods for virus control, and their perceived effectiveness (Schreinemachers et al., 2015). Some surveys have shown that farmers generally have awareness of the insect vectors of tomato yellow leaf curl disease and its symptoms. However, only 2% of these farmers are aware that the disease is transmitted by whiteflies, while 86% of them know that it is caused by high temperatures. Almost 90% of farmers are aware of the use of pesticides against tomato yellow leaf curl virus (TYLCV) (Nagaraju et al., 2002). An interview-based survey on farmers regarding sweet potato virus disease (SPDV) revealed that the majority of them (53.6%) believed it was due to unhealthy tubers. In reality, it is caused by the synergistic combination of two viruses: the whitefly-borne sweet potato chlorotic stunt virus (SPCSV) and sweet potato feathery mottle potyvirus (SPFMV) (Gibson et al., 2000).

### *8.2.5 DETECTION WITH DIFFERENTIATED PLANTS*

In some instances, symptoms in a specific host plant are caused by the combined infection of several viruses. PVX can be distinguished from potato virus Y (PVY) by using differentiating plants. Similarly, PVX can be separated from PVY by using *Myzus persicae*, an aphid. PVX can also be isolated from PVY by mechanically inoculating Datura stramonium, which is immune to PVY. This procedure helps in observing the variations in virus virulence caused by the symptoms produced by each isolated virus. Prior to this, the only techniques to identify viruses in plants were using different plants and their variety of hosts, mechanical inoculation, grafting, or the use of vectors. By examining the symptoms in different plants, it is possible to identify viruses such as papaya ringspot virus (PRSV), CMV, and several potato viruses (Kelaniyangoda & Madhubashini, 2010). Comparison charts of symptoms were created to identify all viruses, whether they were spread mechanically, through vectors, or through grafting. Differential plants are still commonly used in various fruit trees, including pome, citrus, and Rosaceae family plants (Zhou, 2018).

## 8.3 MODERN TECHNIQUES FOR PLANT VIRUS DISEASES DETECTION

Symptomatic detection, microscopic methods, prophylactic measures, etc., are not the ultimate methods on which we can base our treatment against plant viruses. To detect the specific virus, we need to examine its inner parts.

Serological, molecular, nucleic acid sequencing methods, etc., are examples of this technique. In this discussion, we will explore some modern techniques.

### *8.3.1 DETECTION BY SEROLOGY*

Serology is the study of any kinds of body fluids in animals and plants. In the laboratory, when we detect the presence of antibodies in a suspected body fluid sample, it is called a serological test. Examples of serological tests include ELISAs, flocculation tests, hemagglutinin-inhibition tests, neutralization tests, and chemiluminescence immunoassays (Slogteren, 1957). These tests help us detect the presence of antibodies produced by the host's body to defend against pathogenic microorganisms such as viruses, bacteria, fungi, protozoa, etc. This method is commonly used to detect viral diseases, particularly in plants where antibodies are also produced to fight pathogenic infections. Serological methods used for plant virus detection include dot blot immunobinding assay (DBIA), direct tissue blot immunoassay (DTBIA), tissue blotting immunoassay (TBIA), western blotting, immunostrip assays, and ELISAs (Abd El-Aziz, 2019; Martinelli et al., 2015). However, before the invention of ELISA, other serological tests had drawbacks such as the need for a large amount of antiserum, limited sample testing, and long processing time. The invention of ELISA opened new doors for serological test methods.

In this process, we can test a huge number of samples within a short time period, making it a highly sensitive process (Boonham et al., 2014). In this assay, extracts from different parts of the plant's body are used as samples. In 1960, scientists used animal antibodies to detect plant viral infections through the ELISA method (Wang et al., 2022). In 1977, Clark et al. were the first to detect and quantify plum pox virus (PPV) (family *Potyviridae*) and Arabis mosaic virus (ArMV) (family *Secoviridae*) in plants using the double-antibody sandwich (DAS) form of ELISA assay (Wang et al., 2022; Clark & Adams, 1977; Boonham et al., 2014). Nowadays, various types of ELISA techniques are used, such as direct antigen-coating, antigen-coated plate, or plate trapped antigen (PTA) ELISA, triple antibody sandwich (TAS) ELISA, protein A-sandwich (PAS) ELISA, etc. (Kumlachew, 2015). Other processes like lateral flow devices (LFDs), spectrophotometry, immuno-chromatographic strip-based assays, etc., are also associated with ELISA to increase the accuracy of the results (Wang et al., 2022; De Boer & López, 2012). PTA ELISA process can be used to detect Potato Y virus, cowpea severe mosaic virus, squash mosaic virus (SQMV), CMV, papaya lethal yellowing virus, and zucchini yellow mosaic virus (ZYMV) (Mehetre et al., 2021; Galvino-Costa et al., 2012). TAS-ELISA process is employed to analyze tomato yellow leaf curl Thailand virus (TYLCTHV) in tomato, pepper, okra, and cucurbit plants (Mehetre et al., 2021; Nascimento, Lima, & Barbosa, 2017; Seepiban et al., 2017). PAS ELISA process is used to detect different types of badnaviruses in yam plants in West Africa (Mehetre et al., 2021).

### *8.3.2 DETECTION BY MOLECULAR METHODS*

Molecular diagnostic methods rely on the detection of pathogen nucleic acid in the host cell. The three key terms – hybridization, amplification, and sequencing – are involuntarily involved or related to molecular diagnostic methods. These methods offer several advantages that make them more preferable compared to traditional methods. They are highly sensitive, specific, and provide rapid results. However, these methods can be costly and require expertise to perform experiments. Before discussing different types of molecular diagnostic techniques, we have listed some significant discoveries. Without these inventions, molecular diagnostic techniques would not have advanced to their current state (Table 8.2).

#### *8.3.2.1 POLYMERASE CHAIN REACTION (PCR)*

Polymerase chain reaction (PCR) is a cyclic DNA amplification technique that depends on primer sequence, Taq polymerase, PCR buffer, and concentration of dNTPs (López et al., 2003). Depending on the working principle, PCR techniques can be divided into several types: reverse transcriptase PCR (RT-PCR), quantitative real-time PCR (qRT-PCR), nested PCR (nPCR), and multiplex PCR (M-PCR). All of these methods are much more advanced and sensitive than normal PCR methods. Both single and double-stranded DNA and RNA can be amplified through different types of PCR methods. The RT-PCR test is widely used for the detection of RNA viruses in both animals and plants. RT plays a main role in converting mRNA into complementary DNA (cDNA). Cucurbit yellow stunting disorder virus, cucumber vein yellowing virus (CVYV), potato aucuba mosaic virus, tomato chlorosis virus, and potato yellow dwarf virus, among others, can be detected through RT-PCR (Jeong, Ju, & Noh, 2014). qRT-PCR is based on the detection of a fluorescent signal that is produced proportionally during the amplification of a PCR product. Generally, fluorescent probes are used in real-time PCR. However, other probes such as TaqMan probes, scorpion probes, hairpin

**TABLE 8.2** Landmark Discoveries in Molecular Biology

| Year | Discovery | Discoverers | References |
|---|---|---|---|
| 1949 | Sickle cell anemia | Linus Pauling and his colleagues | Gabriel & Przybylski (2010) |
| 1953 | DNA double helix | James Watson, Crick | Pray & Education (2008) |
| 1958 | DNA polymerases | Arthur Kornberg and his colleagues | Schekman (2007) |
| 1960 | Hybridization techniques | Joseph Gall and Mary Lou Pardue | Levsky & Singer (2003) |
| 1969 | *In situ* hybridization | Gall and Pardue | Huber, Voith von Voithenberg, & Kaigala (2018) |
| 1970 | Restriction enzymes | Werner Arber | Roberts (2005) |
| 1970 | Reverse transcriptase | David Baltimore and Howard Temin | Coffin & Fan (2016) |
| 1975 | Southern blotting | Edwin Southern | Tofano, Wiechers, & Cook-Deegan (2006) |
| 1977 | DNA sequencing | Fred Sanger and his team | Crossley et al. (2020) |
| 1979 | Purification of dsRNA from plant and fungi | Morris and Dodds | Olmos et al. (2007) |
| 1984 | Polymerase chain reaction | Kary Mullis | Zhu et al. (2020) |
| 1985 | Restriction fragment length polymorphism | Botstein | Kratochwil et al. (2022) |
| 1986 | Fluorescent *in situ* hybridization (FISH) | Gall and Pardue | Huber, Voith von Voithenberg, & Kaigala (2018) |
| 1988 | Thermostable DNA polymerase | Kary Mullis | Ishino & Ishino (2014) |
| 1996 | DNA microarrays | Patrick O. Brown | Bumgarner (2013) |
| 2001 | Human genome sequence | James Watson | Nurk et al. (2022) |

probes, and LUX primers are also utilized in this type of PCR (Zhu et al., 2020). Pepino mosaic virus (PepMV), CTV in citrus plants, and aphid vectors are detected through qRT-PCR (Mehetre et al., 2021; Jeong & Noh, 2014). The portable real-time PCR machine (Cepheid SmartCycler) was the first attempt to bring PCR technology from the laboratory to the field, providing an opportunity to conduct PCR assays outside of the conventional research laboratory. This machine has already been experimentally used for field detection of *Phytophthora ramorum* by PCR (De Boer & López, 2012). nPCR is a modified version of PCR used to detect viral genomic material and 16S rRNA

(Green & Sambrook, 2019). This process is employed for the detection of plant RNA viruses such as PNRSV, PDV, PPV, lettuce mosaic virus (LMV), and CTV (Figure 8.4) (Olmos et al., 1999; Saeed, Musheer, & Jamil, 2019).

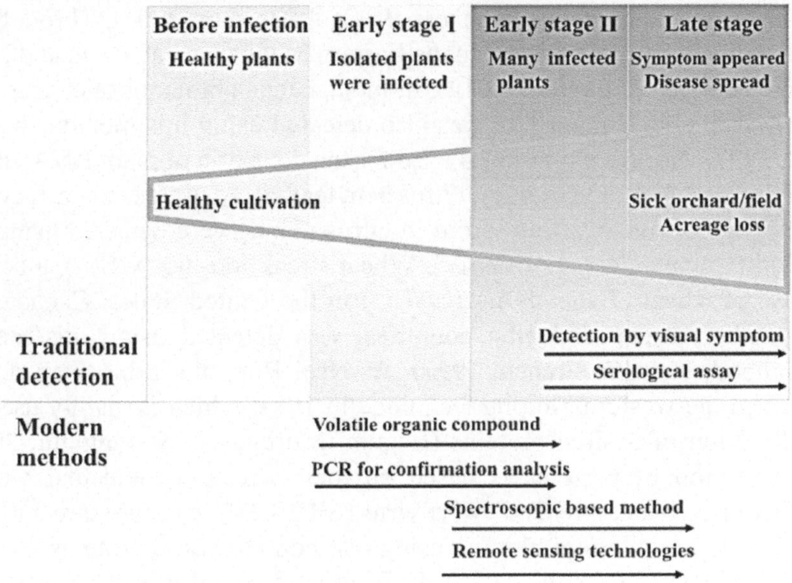


**FIGURE 8.4** Traditional and modern methods and their timing of use during plant disease progression were indicated. Four disease stages were considered.

### *8.3.2.2 HYBRIDIZATION OF NUCLEIC ACIDS*

Another important molecular biological technique is the hybridization of nucleic acids. In this method, a probe that is complementary to the targeted DNA segment is used to detect the presence of a pathogenic genome. The hybridization can occur as either a DNA-DNA hybrid or a DNA-RNA hybrid. The stability of these hybrids depends on both electrostatic and hydrophobic forces, which in turn are influenced by reaction conditions such as temperature, salt concentration, and length of the probe, among other factors (Pallás, Sánchez-Navarro, & James, 2018). In 1981, Robert & Diener first used this method as a diagnostic tool for plant virus detection (Owens & Diener, 1981). This process is highly specific and can work with small amounts of genomic material. However, there are instances where probes may bind to host proteins, leading to false positive signals, as observed in the detection of peach latent mosaic viroid (PLMVd) in peach samples. Occasionally, in the case of PLMVd, probes may bind to host proteins and provide false positive signals (WenXing et al., 2009). Different types of hybridization processes, such as southern, western, northern blotting, fluorescence *in situ* hybridization (FISH), etc., are used for the detection of bacterial and viral genomes in animals and plants (Mandal, 2020).

Southern blotting is used for the detection of specific DNA sequences in DNA samples (Tofano, Wiechers, & Cook-Deegan, 2006). Through this process, TYLCV was detected in the Eastern Mediterranean basin and North Africa (Hanssen & Lapidot, 2012). In India, banana bract mosaic virus from Tamil Nadu and Maharashtra was also detected using this method (Rodoni et al., 1997). Northern blotting is used for the detection of plant RNA viruses (Cohn et al., 1984). In Uruguay, citrus bent leaf viroid, citrus exocortis viroid (CEV), and citrus dwarfing viroid in citrus plants were detected using this method (Umana-Cast et al., 2017). Wheat streak mosaic (WSM), a serious disease of wheat (*Triticum aestivum* L.) in the United States, Canada, and many other wheat-producing countries, was detected using this method (Mahmood, Hein, & French, 1997). W. Neal Burnette, a biochemist, first described the western blotting technique in 1981, which is mainly used for the detection of desired proteins (Begum, Murugesan, & Tangutur, 2022). The detection of proteins is based on their size and molecular weight. Southern rice black-streaked dwarf virus (SRBSDV), a severe threat to rice, was detected by western blotting using two non-structural proteins, P-1 and P-7, as targets (Xie et al., 2021). In Europe, especially in the central and southeastern parts of the continent, different strains of PPV were detected through this technique (Kollerova, Glasa, & Subr, 2008). FISH is another important molecular cytogenetic hybridization technique where fluorescent probes are used to detect the presence of viral genetic material (Chen & Chen, 2013). These probes are labeled with different fluorescence colors like Texas red, FITC green, and acridine orange, and they can be observed under a fluorescent microscope. This process combines microscopy and DNA hybridization, allowing us to locate viruses in plant tissues for a better understanding of viral epidemiology (Wang et al., 2022). This method also provides information on chromosomal deletions, translocations, and gene amplifications. Viruses such as TYLCV, potato leaf roll virus (PLRV), and cucumber green mottle mosaic virus (CGMMV) have been detected using this method (Kliot et al., 2014; Kliot & Ghanim, 2016).

#### *8.3.2.3 SEQUENCING (OMICS STUDY)*

The thought of sequencing began in 1977 with the experiment of Frederick Sanger, an English biochemist, known as Sanger sequencing (Crossley et al., 2020). Since then, numerous sequencing methods have been discovered using various modern technologies. Next-generation sequencing (NGS) has played a significant role in genomic DNA and RNA sequencing. In the context of plant virus detection, we have summarized the application of several sequencing technologies used in the detection of plant viral infections (Table 8.3).

**TABLE 8.3** Detection of Plant Viruses Through Different Types of Sequencing Method

| Name of Plant Virus | Infecting Plants | Sequencing Method | References |
|---|---|---|---|
| Citrus vein enation virus, citrus leprosis virus cytoplasmic type 2 | Citrus plant | Illumina technology | Barba, Czosnek, & Hadidi (2013) |
| Moroccan watermelon mosaic virus, cowpea mild mottle virus | Papaya plant | Illumina MiSeq sequencing | Mumo et al. (2020) |
| Grapevine virus A, grapevine virus B, grapevine leafroll-associated virus 1 | Grapevine | Illumina HiSeq 2500 platform | Xiao et al. (2019) |
| Sweet potato virus | Sweet potato plant | Illumina MiSeq platform, deep sequencing | Nhlapo et al. (2018) |
| Blackeye cowpea mosaic virus, cowpea aphid borne mosaic virus | Cowpea legume plant | 454 (GS FLX) sequencing | Palanga et al. (2016) |
| Little cherry virus 1 (LChV1) | Sweet and sour cherry plant | Ion torrent sequencing | Katsiani et al. (2018) |
| Maize streak virus, plum poxvirus, maize yellow mosaic virus | Maize crops plant, plums | MinION nanopore sequencing | Lu, Giordano, & Ning (2016) |
| Potato virus Y, potato virus X, Arabis mosaic virus (ArMV) | Potato plant | Nanopore sequencing | Bartola, Byrne, & Mullins (2020) |
| Sri Lankan cassava mosaic virus | Cassava plant | Nanopore sequencing | Sun et al. (2022) |
| Tomato severe rugose virus (ToSRV) | *Physalis angulata* | ONT sequencing | Campos (2021) |
| Wheat streak mosaic virus (WSMV) | Wheat plant | Oxford nanopore sequencing technology (ONT) | Fellers et al. (2019) |
| Yam mild mosaic virus (YMMV), yam chlorotic necrosis virus (YCNV) | Yam plant | Oxford nanopore MinION sequencing | Filloux et al. (2018) |
| Cucumber mosaic virus | Cucumber, other cucurbits plants | Deep (pyro) sequencing | Morroni, Jacquemond, & Tepfer (2013) |

#### *8.3.2.4 MICROARRAY (OLIGONUCLEOTIDE ARRAY)*

DNA microarray is a two-dimensional high-density matrix of biomolecules that is printed or synthesized on glass, silicon, plastic, or nylon membranes (Kurkela & Brown, 2020). By using this tool, we can detect both nucleic acids and antibodies that are attached to the oligonucleotide and protein, respectively (Kurkela & Brown, 2020). This approach was first used for the detection and identification of several potato viruses and four species of selected cucurbit-infecting tobamoviruses (Boonham et al., 2003; Bystricka et al., 2005). To improve the efficiency level, scientists used a 40-mer nucleotide microarray chip to analyze 10 important tomato viruses, six pospiviroid species, and multiple artichoke viruses (Tiberini et al., 2010). A 70-mer oligonucleotide microarray chip containing 570 unique probes has been used for the detection of 44 grapevine viral genomes. Another microarray chip with 27 to 75 nucleotides was used for the detection of vitiviruses, nepoviruses, foveavirus, closterovirus ampelovirus, sadwavirus, and maculavirus (Vickers, 2017; Abdullahi, Gryshan, & Rott, 2011). In 2011, Nicolaisen developed a microarray chip containing 150 probes that can detect 52 different viruses from a wide range of genera. DNA microarrays have been used to diagnose plant pathogens in potatoes, tomatoes, and apples. The oligo-chip can detect potato viruses such as potato virus M (PTM), potato virus A, and potato virus S (Bystricka et al., 2005). A large-scale oligonucleotide (LSON) chip was developed for the determination and identification of 538 plant viruses (Zhang et al., 2010).

### *8.3.3 SPECTROSCOPY-BASED METHODS*

Spectroscopy-based methods are widely used in both the medical field and for the detection of plant viruses. Several pieces of evidence support the ability of Raman spectroscopy and nuclear magnetic resonance (NMR) spectroscopy to detect plant viruses (Sanchez et al., 2020; Leiss, 2011). Raman spectroscopy has been employed for plant virus detection for many years and can detect viral presence at very early stages (Mandrile et al., 2019). NMR spectroscopy is utilized for detecting the chemical and physical properties of viruses, as well as changes in plant metabolites caused by virus infection (Choi et al., 2006). Other types of spectroscopies, such as fluorescence spectroscopy, visible-IR spectroscopy, and hyperspectral imaging, can also be used for the detection of plant viruses, assessment of fruit quality, monitoring photosynthetic activities, analyzing tissue structures, identifying disease symptoms, evaluating stress levels, and detecting nutrient deficiencies in plants. Fluorescence spectroscopy, for example, can monitor physiological stress levels in plants by measuring changes in fluorescence at different wavelengths (Muhammed, 2005; Belasque, Gasparoto, & Marcassa, 2008). X-ray imaging, which employs low wavelengths, has been used to detect fungal diseases in wheat (Chaerle et al., 2004). Imaging spectroscopy has been utilized to monitor wheat kernels for head blight disease using machine vision techniques (Okamoto et al., 2007).

### 8.3.4 ANALYSIS OF VOLATILE COMPOUNDS AS BIOMARKERS

Volatile organic compounds, such as cis-3-hexenol, cis-hexenyl acetate, and hexyl acetate, are natural compounds that have a very low molecular weight, high vapor pressure, and low boiling point. They are secreted from the leaf section of many plants (Martinelli et al., 2015; Wang et al., 2022). These compounds exist in the gaseous phase under standard temperature and pressure and are generally present at ultra-low concentrations below the human olfactory threshold (Martinelli et al., 2015). They are naturally secreted by healthy plants, but when a plant is infected by a pathogen, the VOC secretion level gradually increases to attract insect vectors for invading other surrounding plants and to warn them. It has been found that plants infected with CMV produce approximately 50% more total volatile compounds than healthy varieties, which attract the vectors and help the virus invade healthy plants (Mauck, De Moraes, & Mescher, 2010). TMV-infected tobacco plants release methyl salicylate, thereby inducing the defense systems of neighboring plants (Shulaev, Silverman, & Raskin, 1997). Therefore, by testing the VOC, we can detect the physical state of a plant and differentiate between healthy and diseased plants. It is also possible to detect multiple identified and unidentified pathogens and identify the early onset of disease by using VOC. During VOC detection, devices like E-noses of metal oxides or conductive polymer coatings, nano-plasmonic sensor arrays comprising AuNPs, molecularly imprinted sol-gel (AuNPs@MISG), and smartphone-integrated VOC sensing platforms, etc., are essential for the identification of plant viruses (Li et al., 2020). Smartphone-integrated VOC sensing platforms have already been used in the differentiation of tomato late blight from fungal infection (Li et al., 2020). Cultivated squash was used to investigate the mode of action of the CMV, host response, and effects on insect vectors (Martinelli et al., 2015).

### 8.3.5 REMOTE SENSING OF PLANT DISEASE

Remote sensing is a process in which we can detect and monitor the physical characteristics of a targeted area by measuring its reflected and emitted radiation from a distance. Special cameras are used to collect remotely sensed images, which help scientists gather information about the targeted area. It is an indirect assessment technique that also involves the use of spectroradiometers (also known as proximal sensing) to monitor plant physiology (Martinelli et al., 2015). This technique can monitor the absorbed photosynthetic active radiation and stress condition of different plants, allowing us to differentiate between diseased and healthy plants. Radiative transfer modeling of remote sensing is used to monitor various physical properties of plants, such as chlorophyll concentration, water content, and leaf angle distribution. This modeling is an important tool in plant virus detection (Jacquemoud & Baret, 1990). According to Polischuk et al. (1997), spectral reflectance measurements were used for early diagnosis of symptoms in *Nicotiana debneyi* plants at different stages of tomato mosaic tobamovirus infection. The results showed a decrease in leaf reflectance due to a reduction in chlorophyll content a few days after inoculation in a healthy plant (Polischuk et al., 1997).

### 8.3.6 CONTROL OF PLANT VIRAL DISEASES BY CRISPR/CAS9

Clustered regularly interspaced short palindromic repeats (CRISPR) are a type of nucleotide sequence present in the genomes of archaea and bacteria. These sequences are derived from bacteriophage DNA that had previously infected the same microorganism. CRISPR-associated protein 9 (Cas9) is an enzyme that helps the CRISPR sequence identify and cleave sequence-specific cDNA strands (Doudna & Charpentier, 2014). In 2020, Emmanuelle Charpentier, a French professor, and Jennifer Doudna, an American biochemist, were honored with the Nobel Prize for developing the CRISPR-Cas9 genome editing technique. This technology allows us to edit specific pieces of genes using the CRISPR sequence. CRISPR/Cas9 technology is a promising method in plant disease research, particularly in viral diseases (Kumar & Jain, 2015). The accuracy level of this technology is very high. It can help activate plant defense mechanisms against plant viruses by identifying and degrading pathogenic genes. It provides enhanced resistance to plant varieties against different viruses (Hsu, Lander, & Zhang, 2014).

In CRISPR/Cas9 expression, the large pre-CRISPR RNA (pre-crRNA) is effectively transcribed and obtained from the CRISPR locus, which is then converted into several crRNAs and the tracrRNA molecule. The tracrRNA molecule combines with the crRNA repeat through complementary base pairing, which leads to pre-crRNA processing in crRNA. When activated, it can bind with the CRISPR-associated antiviral protection complex and subsequently identify the target site of foreign DNA through base-pairing (Arora & Narula, 2017; Shahriar et al., 2021). The disruption is controlled by the CRISPR-mediated promoter, which can block the expression of genes and the effector-binding site of disease-prone plants (Cao et al., 2020). The CRISPR/Cas9 tool can control and enhance the disease resistance property by altering the susceptible genes that disrupt the proteins in plants (Shahriar et al., 2021).

## 8.4 CURRENT CHALLENGES AND FUTURE PROSPECTS

In this chapter, we have discussed various techniques for the detection of plant viruses. Initially, we focused on common primary steps such as symptom-based detection and microscopic methods. Farmers need to acquire proper knowledge about plant viral infections for this purpose. Among the microscopic methods, advanced electron microscopes (EMs) like TEM, scanning electron microscope (SEM), and fluorescence microscopes are the preferred choices, despite their high cost. Therefore, it is important to explore modern techniques that are already being used for plant virus detection, while also considering their ongoing development to improve detection methods.

Molecular methods are mostly preferred in detection, due to their accuracy, cost-effectiveness, and time management. In the serological technique, we discussed the application of ELISA techniques. However, there are other serological techniques available, such as agglutination, immunodiffusion (ID), lateral flow assay (LFA), counter immuno-electrophoresis (CIE), and radioimmunosorbent assay (RIA), which can also be used for the detection of plant viruses. Additionally, we can focus on different imaging techniques, including magnetic resonance imaging, photoacoustic imaging, tomography, thermography, spectroscopic and imaging technologies, multispectral imaging, hyperspectral imaging, and 3D imaging. These techniques are very expensive and require expert hands to perform. We should also focus on new upcoming methods such as CRISPR-Cas9, nanotechnology, and biosensor-based nanomaterials (NMs). Different nanopore sensors like MinION nanopore and Oxford nanopore can be used for detection purposes. Pyrosequencing, whole-genome shotgun sequencing, NGS, and high-throughput sequencing are a bit costly but highly sensitive and specific. Therefore, these modern techniques should be developed in a cost-effective manner.

## 8.5 CONCLUSION

Due to severe plant viral diseases, a significant economic loss is calculated worldwide each year. The challenge with diagnosing plant viral diseases is that plants are often infected with multiple viruses or different variants of the same virus simultaneously. Therefore, detecting different viruses simultaneously poses a great difficulty in managing plant viral diseases. Additionally, the extensive genetic diversity hinders the use of a specific viral detection method. While symptomatic diagnosis is valuable, it often yields erroneous results due to confusion caused by highly variable symptoms resulting from interactions between hosts, viruses, or abiotic stresses. Consequently, reliable diagnostic platforms are necessary for effective management. Currently, various methods are available, and many more are being developed for disease diagnosis in plants. Since there is no universal detection method that fulfills all requirements, it is crucial to develop effective and appropriate techniques that will aid in controlling viral diseases worldwide while ensuring sustainable agriculture.

## KEYWORDS

- **green vegetation**
- **integrated disease management**
- **nucleic acids**
- **oligonucleotide array**
- **plant viruses**
- **polymerase chain reaction**
- **prophylaxis**

## REFERENCES

Abd El-Aziz, M. H., (2019). Three modern serological methods to detect plant viruses. *Journal of Plant Science and Phytopathology, 3*(3), 101–106. doi: 10.29328/journal.jpsp.1001039.

Abdullahi, I., Gryshan, Y., & Rott, M., (2011). Amplification-free detection of grapevine viruses using an oligonucleotide microarray. *Journal of Virological Methods*, *178*(1, 2), 1–15. Elsevier B.V. doi: 10.1016/j.jviromet.2011.07.009.

Addy, H. S., Nurmalasari, Setyo, W. A. H., Sholeh, A., Anugrah, C., Shandy, I. F. E., Nurmalasari, Wahyudi, A. H. S., Sholeh, A., Anugrah, C., Iriyanto, F. E. S., et al., (2017). Detection and response of sugarcane against the infection of *Sugarcane mosaic virus* (SCMV) in Indonesia. *Agronomy, 7*(50), 1–11. doi: 10.3390/agronomy7030050.

Ahmad, A., Sharma, A., Zehra, S. B., & Kang, S. S., (2016). Potential reservoirs of whitefly transmitted *Begomo-virus* causing leaf curl disease in chilli. *Applied Biological Research, 18*(3), 270–276. doi: 10.5958/0974–4517.2016.00042.2.

Arora, L., & Narula, A., (2017). Gene editing and crop improvement usingCRISPR-Cas9 system. *Frontiers in Plant Science, 8*(1932), 1–21. doi: 10.3389/fpls.2017.01932.

Baker, K. K., Ramsdell, D. C., & Gillett, J. M., (1985). Electron microscopy: Current applications to plant virology. *Plant Disease, 69*(1), 85–90. doi: 10.1094/pd-69–85.

Balodi, R., Bisht, S., Ghatak, A., & Rao, K. H., (2017). Plant disease diagnosis: Technological advancements and challenges. *Indian Phytopathology, 70*(3), 275–281. doi: 10.24838/ip.2017.v70.i3.72487.

Barba, M., Czosnek, H., & Hadidi, A., (2013). Historical perspective, development, and applications of next-generation sequencing in plant virology. *Viruses, 6*(1), 106–136. doi: 10.3390/v6010106.

Bartola, M. D., Byrne, S., & Mullins, E., (2020). Characterization of *Potato virus Y* isolates and assessment of nanopore sequencing to detect and genotype potato viruses. *Viruses, 12*(4), 1–25. doi: 10.3390/v12040478.

Begum, H., Murugesan, P., & Tangutur, A. D., (2022). Western blotting: A powerful staple in scientific and biomedical research. *Biotechniques, 73*(1), 59–70. doi: 10.2144/btn 2022-0003.

Belasque, J., Gasparoto, M. C. G., & Marcassa, L. G., (2008). Detection of mechanical and disease stresses in citrus plants by fluorescence spectroscopy. *Applied Optics, 47*(11), 1922–1926. doi: 10.1364/AO.47.001922.

Bhardwaj, A., & Kulshrestha, S., (2020). Advance methods for the isolation and characterization of plant viruses infecting crops. In: *Applied Plant Virology*, 39–53. doi: 10.1016/b978-0-12-818654-1.00003-7.

Bock, K. R., (1982). The identification and partial characterization of plant viruses in the tropics. *Tropical Pest Management, 28*(4), 399–411. doi: 10.1080/09670878209370747.

Boonham, N., Kreuze, J., Winter, S., Vlugt, R. V. D., Bergervoet, J., Tomlinson, J., et al., (2014). Methods in virus diagnostics: From ELISA to next generation sequencing. *Virus Research, 186*, 20–31. doi: 10.1016/j.virusres.2013.12.007.

Boonham, N., Walsh, K., Smith, P., Madagan, K., Graham, I., & Barker, I., (2003). Detection of potato viruses using microarray technology: Towards a generic method for plant viral disease diagnosis. *Journal of Virological Methods, 108*, 181–187. doi: 10.1016/S0166–0934(02)00284-7.

Bumgarner, R., (2013). Overview of DNA microarrays: Types, applications, and their future. *Current Protocols in Molecular Biology*, 1–11. doi: 10.1002/0471142727.mb2201s101.

Bunawan, S. N., Uda, M. N. A., Salam, F., Rosmi, M. N. M., Saad, M. M., Mat, M., et al., (2016). Set potential determination for *Rice tungro spherical virus* (RTSV) detection using chronoamperometry measurement. *Procedia Chemistry, 20*, 121–123. doi: 10.1016/j.proche.2016.07.022.

Bystricka, D., Lenz, O., Mraz, I., Piherova, L., Kmoch, S., & Sip, M., (2005). Oligonucleotide-based microarray: A new improvement in microarray detection of plant viruses. *Journal of Virological Methods, 128*, 176–182. doi: 10.1016/j.jviromet.2005.04.009.

Campos, M. D., Félix, M. D. R., Patanita, M., Materatski, P., & Varanda, C., (2021). High throughput sequencing unravels tomato-pathogen interactions towards a sustainable plant breeding. *Horticulture Research, 8*(171), 1–12. doi: 10.1038/s41438-021-00607-x.

Cao, Y., Zhou, H., Zhou, X., & Li, F., (2020). Control of plant viruses by CRISPR/Cas system-mediated adaptive immunity. *Frontiers in Microbiology, 11*(593700), 1–9. doi: 10.3389/fmicb.2020.593700.

Chen, A. Y. Y., & Chen, A., (2013). Fluorescence in situ hybridization. *Journal of Investigative Dermatology, 133*(e8), 1–4. doi: 10.1038/jid.2013.120.

Choi, Y. H., Kim, H. K., Linthorst, H. J. M., Hollander, J. G., Lefeber, A. W. M., Erkelens, C., et al., (2006). NMR metabolomics to revisit the *Tobacco mosaic virus* infection in *Nicotiana tabacum* leaves. *Journal of Natural Products, 69*(5), 742–748. doi: 10.1021/np050535b.

Clark, M. F., & Adams, A. N., (1977). Characteristics of the microplate method of enzyme linked immunosorbent assay for the detection of plant viruses. *Journal of General Virology, 34*(3), 475–483. doi: 10.1099/0022-1317-34-3-475.

Clemente-Orta, G., Albajes, R., Batuecas, I., & Achon, M. A., (2021). Planting period is the main factor for controlling maize rough dwarf disease. *Scientific Reports, 11*(977), 1–12. doi: 10.1038/s41598-020-79994-5.

Coffin, J. M., & Fan, H., (2016). The discovery of reverse transcriptase. *Annual Review of Virology, 3*, 29–51. doi: 10.1146/annurev-virology-110615-035556.

Cohn, D. V., Eliting, J. J., Frick, M., & Elde, R., (1984). Selective localization of the parathyroid secretory protein-I/adrenal medulla chromogranin a protein family in a wide variety of endocrine cells of the rat. *Endocrinology, 114*(6), 1963–1974. doi: 10.1210/endo-114-6-1963.

Crossley, B. M., Bai, J., Glaser, A., Maes, R., Porter, E., Killian, M. L., et al., (2020). Guidelines for sanger sequencing and molecular assay monitoring. *Journal of Veterinary Diagnostic Investigation, 32*(6), 767–775. doi: 10.1177/1040638720905833.

D'Arcy, C. J., & Domier, L. L., (2000). Barley yellow dwarf. *The Plant Health Instructor*. doi: 10.1094/phi-i-2000-1103-01.

Darzi, E., Smith, E., Shargil, D., Lachman, O., Ganot, L., & Dombrovsky, A., (2018). The honeybee *Apis mellifera* contributes to *cucumber green mottle mosaic virus* spread via pollination. *Plant Pathology, 67*(1), 244–251. doi: 10.1111/ppa.12702.

Davino, S., Caruso, A. G., Bertacca, S., Barone, S., & Panno, S., (2020). *Tomato brown rugose fruit virus*: Seed transmission rate and efficacy of different seed disinfection treatments. *plants, 9*(1615), 1–13. doi:10.3390/plants9111615.

De Boer, S. H., & López, M. M., (2012). New grower-friendly methods for plant pathogen monitoring. *Annual Review of Phytopathology, 50*, 197–218. doi: 10.1146/annurev-phyto-081211–172942.

Doudna, J. A., & Charpentier, E., (2014). The new frontier of genome engineering with CRISPR-Cas9. *Science, 346*(6213), 1–11. doi: 10.1126/science.1258096.

Edwardson, J. R., & Christie, R. G., (1978). Use of virus-induced inclusions in classification and diagnosis. *Annual Review of Phytopathology, 16*, 31–55. doi: 10.1146/annurev.py.16.090178.000335.

Ertel, K. J., Benefield, D., Castaño-Diez, D., Pennington, J. G., Horswill, M., Boon, J. A. D., et al., (2017). 'Cryo-electron tomography reveals novel features of a viral RNA replication compartment. *eLife, 6*(e25940), 1–24. doi: 10.7554/eLife.25940.

Fang, Y., & Ramasamy, R. P., (2015). Current and prospective methods for plant disease detection. *Biosensors, 5*(3), 537–561. doi: 10.3390/bios5030537.

Fellers, J. P., Webb, C., Fellers, M. C., Rupp, J. S., & Wolf, E. D., (2019). Wheat virus identification within infected tissue using nanopore sequencing technology. *Plant Disease, 103*(9), 2199–2203. doi: 10.1094/PDIS-09-18-1700-RE.

Filloux, D., Fernandez, E., Loire, E., Claude, L., Galzi, S., Candresse, T., et al., (2018). Nanopore-based detection and characterization of yam viruses. *Scientific Reports, 8*(17879), 1–11. doi: 10.1038/s41598-018-36042-7.

Fuentes, S., Jones, R. A. C., Matsuoka, H., Ohshima, K., Kreuze, J., & Gibbs, A. J., (2019). *Potato virus Y*; the Andean connection. *Virus Evolution, 5*(2), 1–15. doi: 10.1093/ve/vez037.

Gabriel, A., & Przybylski, J., (2010). Sickle-cell anemia: A look at global haplotype distribution. *Nature Education, 3*(3), 2–5.

Galvino-Costa, S. B. F., Figueira, A. D. R., Rabelo-Filho, F. D. A. C., Moraes, F. H. R., Nikolaeva, O. V., & Karasev, A. V., (2012). Molecular and serological typing of *Potato virus Y* isolates from Brazil reveals a diverse set of recombinant strains. *Plant Disease, 96*(10), 1451–1458. doi: 10.1094/PDIS-02-12-0163-RE.

Garcia-Ruiz, H., (2019). When viruses infect plants. *Scientia (Bristol), 123*, 40–43.

Gibson, R. W., Jeremiah, S. C., Aritua, V., Msabaha, R. P., Mpembe, I., & Ndunguru, J., (2000). Sweet potato virus disease in sub-Saharan Africa: Evidence that neglect of seedlings in the traditional farming system hinders the development of superior resistant landraces. *Journal of Phytopathology, 148*, 441–447. doi: 10.1046/j.1439-0434.2000.00529.x.

Gilmour, M., Crozier, J., & Kouamé, C., (2012). *Minimizing the Risk of Spreading Cocoa Swollen Shoot Virus Disease,* 1–6.

Gorovits, R., & Czosnek, H., (2017). The involvement of heat shock proteins in the establishment of *Tomato yellow leaf curl virus* infection. *Frontiers in Plant Science, 8*(355), 1–12. doi: 10.3389/fpls.2017.00355.

Green, M. R., & Sambrook, J., (2019). Nested polymerase chain reaction (PCR). *Cold Spring Harbor Protocols*, 175–179. doi: 10.1101/pdb.prot095182.

Haas, M., Bureau, M., Geldreich, A., Yot, P., & Keller, M., (2002). *Cauliflower mosaic virus*: Still in the news. *Molecular Plant Pathology, 3*(6), 419–429. doi: 10.1046/j.1364-3703.2002.00136.x.

Hajizadeh, M., Gibbs, A. J., Amirnia, F., & Glasa, M., (2019). The global phylogeny of *Plum pox virus* is emerging. *Journal of General Virology, 100*(10), 1457–1468. doi: 10.1099/JGV.0.001308.

Hancinsky, R., Mihálik, D., Mrkvová, M., Candresse, T., & Glasa, M., (2020). Plant viruses infecting *Solanaceae* family members in the cultivated and wild environments: A review. *Plants, 9*(5), 1–17. doi: 10.3390/plants9050667.

Hanssen, I. M., & Lapidot, M., (2012). Chapter 2 – major tomato viruses in the mediterranean basin. *Advances in Virus Research, 84*, 31–66. doi: 10.1016/B978-0-12-394314-9.00002-6.

Hanssen, I. M., & Thomma, B. P. H. J., (2010). Pepino mosaic virus: A successful pathogen that rapidly evolved from emerging to endemic in tomato crops. *Molecular Plant Pathology, 11*(2), 179–189. doi: 10.1111/j.1364-3703.2009.00600.x.

He, S., & Creasey, K. K. M., (2020). Pandemics of people and plants: Which is the greater threat to food security? *Molecular Plant, 13*, 933–934. doi: 10.1016/j.molp.2020.06.007.

Hebrard, E., Fargette, D., & Konaté, G., (2008). Rice yellow mottle virus. *Encyclopedia of Virology*, 485–490. doi: 10.1016/B978-012374410-4.00715-9.

Horne, R. W., (2018). The development and application of electron microscopy to the structure of isolated plant viruses. In: *Molecular Plant Virology*. doi: 10.1201/9781351074797-1.

Hsu, P. D., Lander, E. S., & Zhang, F., (2014). Development and applications of CRISPR-Cas9 for genome engineering. *Cell, 157*(6), 1262–1278. doi: 10.1016/j.cell.2014.05.010.

Huber, D., Voith, V. V. L., & Kaigala, G. V., (2018). Fluorescence in situ hybridization (FISH): History, limitations and what to expect from micro-scale FISH? *Micro and Nano Engineering, 1*, 15–24. doi: 10.1016/j.mne.2018.10.006.

Ishino, S., & Ishino, Y., (2014). DNA polymerases as useful reagents for biotechnology – the history of developmental research in the field. *Frontiers in Miceobiology, 5*(465), 1–8. doi: 10.3389/fmicb.2014.00465.

Jacquemoud, S., & Baret, F., (1990). PROSPECT: A model of leaf optical properties spectra. *Remote Sensing of Environment, 34*(2), 75–91. doi: 10.1016/0034-4257(90)90100-Z.

Jones, R. A. C., (2020). Disease pandemics and major epidemics arising from new encounters between indigenous viruses and introduced crops. *Viruses, 12*(12), 1–24. doi: 10.3390/v12121388.

Jones, R. A. C., (2021). Global plant virus disease pandemics and epidemics. *Plants, 10*(2), 1–42. doi: 10.3390/plants10020233.

Katsiani, A., Maliogka, V. L., Katis, N., Svanella-Dumas, L., Olmos, A., Ruiz-García, A. B., et al., (2018). High-throughput sequencing reveals further diversity of little cherry virus 1 with implications for diagnostics. *Viruses, 10*(7), 1–14. doi: 10.3390/v10070385.

Kelaniyangoda, D. B., & Madhubashini, L. W. M., (2010). Indicator plants: Tools for detecting *Papaya ring spot potyvirus* and *Cucumber mosaic cucumo virus*. *Journal of Food and Agriculture*, 64–69. doi: 10.4038/jfa.v1i2.1800.

Kliot, A., & Ghanim, M., (2016). Fluorescent in situ hybridization for the localization of viruses, bacteria and other microorganisms in insect and plant tissues. *Methods, 98*, 74–81. doi: 10.1016/j.ymeth.2015.12.003.

Kliot, A., Kontsedalov, S., Lebedev, G., Brumin, M., Cathrin, P. B., Marubayashi, J. M., et al., (2014). Fluorescence in situ hybridizations (FISH) for the localization of viruses and endosymbiotic bacteria in plant and insect tissues. *Journal of Visualized Experiments*, (84), 1–8. doi: 10.3791/51030.

Kollerova, E., Glasa, M., & Subr, Z. W., (2008). Western blotting analysis of the *Plum pox virus* capsid protein. *Journal of Plant Pathology, 90*(1), 19–22.

Kozieł, E., Otulak-Kozieł, K., & Bujarski, J. J., (2018). Ultrastructural analysis of *Prune dwarf virus* intercellular transport and pathogenesis. *International Journal of Molecular Sciences, 19*(9), 1–22. doi: 10.3390/ijms19092570.

Kratochwil, C. F., Kautt, A. F., Rometsch, S. J., & Meyer, A., (2022). Benefits and limitations of a new genome-based PCR-RFLP genotyping assay (GB-RFLP): A SNP-based detection method for identification of species in extremely young adaptive radiations. *Ecology and Evolution, 12*(3), 1–11. https://doi.org/10.1002/ece3.8751.

Kumar, V., & Jain, M., (2015). The CRISPR-Cas system for plant genome editing: Advances and opportunities. *Journal of Experimental Botany, 66*(1), 47–57. doi: 10.1093/jxb/eru429.

Kumar, Y., Hallan, V., & Zaidi, A. A., (2009). Identification and characterization of *Bean yellow mosaic virus* infecting freesia. *Journal of Plant Biochemistry and Biotechnology, 18*(2), 253–255. doi: 10.1007/BF03263331.

Kumari, S. G., & Makkouk, K. M., (2007). Virus diseases of faba bean (*Vicia faba* L.) in Asia and Africa. *Plant Viruses, 1*(1), 93–105.

Kumlachew, A., (2015). Detection of diseases, identification, and diversity of viruses: A review. *Journal of Biology, Agriculture and Healthcare, 5*(1), 204–214.ISSN 2224-3208 (Paper) ISSN 2225-093X (Online).

Kurkela, S., & Brown, D. W. G., (2009). *Molecular Diagnostic Techniques* (Vol. 37, No. 10, pp. 535–540). Elsevier. doi: 10.1016/j.mpmed.2009.07.012.

Leiss, K. A., Choi, Y. H., Verpoorte, R., & Klinkhamer, P. G. L., (2011). An overview of NMR-based metabolomics to identify secondary plant compounds involved in host plant resistance. *Phytochemistry Reviews, 10*(2), 205–216. doi: 10.1007/s11101-010-9175-z.

Lesemann, D. E., Bozarth, R. F., & Koenig, R., (1980). The trapping of tymovirus particles on electron microscope grids by adsorption and serological binding. *Journal of General Virology, 48*, 257–264. doi: 10.1099/0022-1317-48-2-257.

Levsky, J. M., & Singer, R. H., (2003). Fluorescence in situ hybridization: Past, present, and future. *Journal of Cell Science, 116*(14), 2833–2838. doi: 10.1242/jcs.00633.

Ling, K. S., (2010). Effectiveness of chemo- and thermotherapeutic treatments on *Pepino mosaic virus* in tomato seed. *Plant Disease*. doi: 10.1094/PDIS-94-3-0325.

López, M. M., Bertolini, E., Olmos, A., Caruso, P., Gorris, M. T., Llop, P., et al., (2003). 'Innovative tools for detection of plant pathogenic viruses and bacteria. *International Microbiology, 6*(4), 233–243. doi: 10.1007/s10123-003-0143-y.

Lu, H., Giordano, F., & Ning, Z., (2016). Oxford nanopore MinION sequencing and genome assembly. *Genomics, Proteomics and Bioinformatics, 14*(5), 265–279. doi: 10.1016/j.gpb.2016.05.004.

Mahmood, T., Hein, G. L., & French, R. C., (1997). Development of serological procedures for rapid and reliable detection of *Wheat streak mosaic virus* in a single wheat curl mite. *Plant Disease, 81*(3), 250–253. doi: 10.1094/PDIS.1997.81.3.250.

Maksimov, I. V., et al., (2019). Mechanisms of plant tolerance to RNA viruses induced by plant-growth-promoting microorganisms. *Plants, 8*(575), 1–15. doi: 10.3390/plants8120575.

Mandal, A., Biswas, D., & Hazra, B., (2020). Natural products from plants with prospective anti-HIV activity and relevant mechanisms of action; Chapter 9. *Studies in Natural Products Chemistry, 66*, 225–271. doi: 10.1016/B978-0-12-817907-9.00009-X.

Mandal, A., Jha, A. K., & Hazra, B., (2021). Plant products as inhibitors of coronavirus 3CL protease. *Frontiers in Pharmacology, 12*, 1–16. doi: 10.3389/fphar.2021.583387.

Mandal, A., Mukherjee, A., & Jha, A. K., (2022). Exploring the functional interactions between *Geminivirus* and host during viral replication, assembly, and movement; Chapter 28. *Geminivirus: Detection, Diagnosis and Management*, 488–502. doi: 10.1016/B978-0-323-90587-9.00016-X.

Mandrile, L., Rotunno, S., Miozzi, L., Vaira, A. M., Giovannozzi, A. M., Rossi, A. M., et al., (2019). Nondestructive Raman spectroscopy as a tool for early detection and discrimination of the infection of tomato plants by two economically important viruses. *Analytical Chemistry, 91*(14), 9025–9031. doi: 10.1021/acs.analchem.9b01323.

Marchant, W. G., Gautam, S., Hutton, S. F., & Srinivasan, R., (2020). Tomato yellow leaf curl virus-resistant and -susceptible tomato genotypes similarly impact the virus population genetics. *Frontiers in Plant Science, 11*(599697), 1–17. doi: 10.3389/fpls.2020.599697.

Maree, H. J., Fox, A., Rwahnih, M. A., Boonham, N., & Candresse, T., (2018). Application of HTS for routine plant virus diagnostics: State of the art and challenges. *Frontiers in Plant Science, 9*(1082), 1–4. doi: 10.3389/fpls.2018.01082.

Martinelli, F., Scalenghe, R., Davino, S., Panno, S., Scuderi, G., Ruisi, P., et al., (2015). 'Advanced methods of plant disease detection: A review. *Agronomy for Sustainable Development, 35*(1), 1–25. doi: 10.1007/s13593-014-0246-1.

Mauck, K. E., De Moraes, C. M., & Mescher, M. C., (2010). Deceptive chemical signals induced by a plant virus attract insect vectors to inferior hosts. *Proceedings of the National Academy of Sciences of the United States of America, 107*(8), 3600–3605. doi: 10.1073/pnas.0907191107.

Mehetre, G. T., Leo, V. V., Singh, G., Sorokan, A., Maksimov, I., Yadav, M. K., et al., (2021). Current developments and challenges in plant viral diagnostics: A systematic review. *Viruses, 13*(412), 1–31. doi: 10.3390/v13030412.

Moreno, P., Ambros, S., Albiach-Marti, M. R., Guerri, J., & Pena, L., (2008). *Citrus tristeza virus*: A pathogen that changed the course of the citrus industry. *Molecular Plant Pathology, 9*(2), 251–268. doi: 10.1111/j.1364-3703.2007.00455.x.

Morroni, M., Jacquemond, M., & Tepfer, M., (2013). Deep sequencing of recombinant virus populations in transgenic and nontransgenic plants infected with *Cucumber mosaic virus. Molecular Plant-Microbe Interactions, 26*(7), 801–811. doi: 10.1094/MPMI-02-13-0057-R.

Muhammed, H. H., (2005). Hyperspectral crop reflectance data for characterizing and estimating fungal disease severity in wheat. *Biosystems Engineering, 91*(1), 9–20. doi: 10.1016/j.biosystemseng.2005.02.007.

Mumo, N. N., Mamati, G. E., Ateka, E. M., Rimberia, F. K., Asudi, G. O., Boykin, L. M., et al., (2020). Metagenomic analysis of plant viruses associated with papaya ringspot disease in *Carica papaya* L. in Kenya. *Frontiers in Microbiology, 11*(205), 1–16. doi: 10.3389/fmicb.2020.00205.

Mutka, A. M., & Bart, R. S., (2015). Image-based phenotyping of plant disease symptoms. *Frontiers in Plant Science, 5*(734), 1–9. doi: 10.3389/fpls.2014.00734.

Nagaraju, N., Venkatesh, H. M., Warburton, H., Muniyappa, V., Chancellor, T. C. B., & Colvin, J., (2002). Farmers' perceptions and practices for managing *Tomato leaf curl virus* disease in southern India. *International Journal of Pest Management, 48*(4), 333–338. doi: 10.1080/09670870210153164.

Naidu, R. A., & Hughes, J. D. A., (2003). Methods for the detection of plant virus diseases. *Plant Virology in Sub-Saharan Africa: Proceedings of a Conference Organized by IITA*, 233–253.

Naidu, R. A., Kimmins, F. M., Deom, C. M., Subrahmanyam, P., Chiyembekeza, A. J., & Van Der Merwe, P. J. A., (1999). Groundnut rosette: A virus disease affecting groundnut production in Sub-Saharan Africa. *Plant Disease, 83*(8), 700–709. doi: 10.1094/PDIS.1999.83.8.700.

Nascimento, A. K. Q., Lima, J. A. A., & Barbosa, G. Da. S., (2017). A simple kit of plate-trapped antigen enzyme-linked immunosorbent assay for identification of plant viruses. *Revista Ciencia Agronomica, 48*(1), 216–220. doi: 10.5935/1806-6690.20170025.

Netsu, O., Hiratsuka, K., Kuwata, S., Hibi, T., Ugaki, M., & Suzuki, M., (2008). Peanut stunt virus 2b cistron plays a role in viral local and systemic accumulation and virulence in *Nicotiana benthamiana*. *Archives of Virology, 153*, 1731–1735. doi: 10.1007/s00705-008-0166-y.

Nhlapo, T. F., Rees, D. J. G., Odeny, D. A., Mulabisana, J. M., & Rey, M. E. C., (2018). Viral metagenomics reveals sweet potato virus diversity in the Eastern and Western Cape provinces of South Africa. *South African Journal of Botany, 117*, 256–267. doi: 10.1016/j.sajb.2018.05.024.

Nurk, S., Koren, S., Rhie, A., Rautiainen, M., Bzikadze, A. V., Mikheenko, A., et al., (2022). The complete sequence of a human genome. *Science, 376*(6588), 44–53. doi: 10.1126/science.abj6987.

Okamoto, H., Murata, T., Kataoka, T., & Hata, S. I., (2007). Plant classification for weed detection using hyperspectral imaging with wavelet analysis. *Weed Biology and Management, 7*(1), 31–37. doi: 10.1111/j.1445-6664.2006.00234.x.

Olmos, A., Cambra, M., Esteban, O., Gorris, M. T., & Terrada, E., (1999). New device and method for capture, reverse transcription, and nested PCR in a single closed-tube. *Nucleic Acids Research, 27*(6), 1564–1565. doi: 10.1093/nar/27.6.1564.

Olmos, A., Capote, N., Bertolini, E., & Cambra, M., (2007). Molecular diagnostic methods for plant viruses. *Biotechnology and Plant Disease Management*, 227–249. doi: 10.1079/9781845932886.0227.

Owens, R. A., & Diener, T. O., (1981). Sensitive and rapid diagnosis of potato spindle tuber viroid disease by nucleic acid hybridization. *Science, 213*(4508), 670–672. doi: 10.1126/science.213.4508.670.

Palanga, E., Filloux, D., Martin, D. P., Fernandez, E., Gargani, D., Ferdinand, R., et al., (2016). Metagenomic-based screening and molecular characterization of cowpea-infecting viruses in Burkina Faso. *PLoS One, 11*(10), 1–21. doi: 10.1371/journal.pone.0165188.

Pallás, V., Sánchez-Navarro, J. A., & James, D., (2018). Recent advances on the multiplex molecular detection of plant viruses and viroids. *Frontiers in Microbiology, 9*(2087), 1–11. doi: 10.3389/fmicb.2018.02087.

Panattoni, A., Luvisi, A., & Triolo, E., (2013). Review. elimination of viruses in plants: Twenty years of progress. *Spanish Journal of Agricultural Research, 11*(1), 173–188. doi: 10.5424/sjar/2013111–3201.

Patrinos, G. P., & Ansorge, W., (2010). Molecular diagnostics: Past, present, and future. *Molecular Diagnostics*, 1–11.doi: 10.1016/B978-0-12-374537-8.00001-8.

Polischuk, V. P., Shadchina, T. M., Kompanetz, T. I., Budzanivskaya, I. G., Boyko, A. L., & Sozinov, A. A., (1997). Changes in reflectance spectrum characteristic of nicotiana Debneyi plant under the influence of viral infection. *Archives of Phytopathology and Plant Protection, 31*(1), 115–119. doi: 10.1080/03235409709383221.

Pray, L. A., (2008). Discovery of DNA structure and function: Watson and Crick, *1*(34). *Nature Education, 1*(1), 100, 1–6.

Ramirez, B. C., Macaya, G., Calvert, L. A., & Haenni, A. L., (1992). Rice hoja *Blanca virus* genome characterization and expression *in vitro*. *Journal of General Virology, 73*(6), 1457–1464. doi: 10.1099/0022-1317-73-6-1457.

Richert-Pöggeler, K. R., Franzke, K., Hipp, K., & Kleespies, R., (2019). Electron microscopy methods for virus diagnosis and high-resolution analysis of viruses. *Frontiers in Microbiology, 9*(3255), 1–18. doi: 10.3389/fmicb.2018.03255.

Roberts, R. J., (2005). How restriction enzymes became the workhorses of molecular biology. *PNAS, 102*(17), 5905–5908. doi: 10.1073/pnas.0500923102.

Rodoni, B. C., Ahlawat, Y. S., Varma, A., Dale, J. L., & Harding, R. M., (1997). Identification and characterization of *Banana bract mosaic virus* in India. *Plant Disease, 81*(6), 669–672. doi: 10.1094/PDIS.1997.81.6.669.

Sanchez, L., Ermolenkov, A., Biswas, S., Septiningsih, E. M., & Kurouski, D., (2020). Raman spectroscopy enables non-invasive and confirmatory diagnostics of salinity stresses, nitrogen, phosphorus, and potassium deficiencies in rice. *Frontiers in Plant Science, 11*(573321), 1–8. doi: 10.3389/fpls.2020.573321.

Schekman, R., (2007). Arthur Kornberg 1918–2007. *Cell, 131*(4), 637–639. doi: 10.1016/j.cell.2007.11.008.

Scholthof, K. B. G., Adkins, S., Czosnek, H., Palukaitis, P., Jacquot, E., Hohn, T., et al., (2011). Top 10 plant viruses in molecular plant pathology. *Molecular Plant Pathology, 12*(9), 938–954. doi: 10.1111/j.1364-3703.2011.00752.x.

Schreinemachers, P., Balasubramaniam, S., Boopathi, N. M., Ha, C. V., Kenyon, L., Praneetvatakul, S., et al., (2015). Farmers' perceptions and management of plant viruses in vegetables and legumes in tropical and subtropical Asia. *Crop Protection, 75*, 115–123. doi: 10.1016/j.cropro.2015.05.012.

Seepiban, C., Charoenvilaisiri, S., Warin, N., Bhunchoth, A., Phironrit, N., Phuangrat, B., et al., (2017). Development and application of triple antibody sandwich enzyme-linked immunosorbent assays for *Begomovirus* detection using monoclonal antibodies against *Tomato yellow leaf curl Thailand virus*. *Virology Journal, 14*(99), 1–14. doi: 10.1186/s12985-017-0763-z.

Seo, J. K., Kim, M. K., Kwak, H. R., Choi, H. S., Nam, M., Choe, J., et al., (2018). Molecular dissection of distinct symptoms induced by *Tomato chlorosis virus* and *Tomato yellow leaf curl virus* based on comparative transcriptome analysis. *Virology, 516*, 1–20. doi: 10.1016/j.virol.2018.01.001.

Shahriar, S. A., Islam, M. N., Wai, C. C. N., Rahim, M. A., Paul, N. C., Uddain, J., et al., (2021). Control of plant viral diseases by CRISPR/Cas9: Resistance mechanisms, strategies, and challenges in food crops. *Plants, 10*(7), 1–19. doi: 10.3390/plants10071264.

Shepherd, D. N., Martin, D. P., Der Walt, E. V., Dent, K., Varsani, A., & Rybicki, E. P., (2010). Maize streak virus: An old and complex "emerging" pathogen. *Molecular Plant Pathology, 11*(1), 1–12. doi: 10.1111/j.1364-3703.2009.00568.x.

Shetty, A. A., Singh, J. P., & Singh, D., (2013). Resistance to *Yellow vein mosaic virus* in okra: A review. *Biological Agriculture and Horticulture, 29*(3), 159–164. doi: 10.1080/01448765.2013.793165.

Shukla, D. D., & Gough, K. H., (1979). The use of protein A, from *Staphylococcus aureus*, in immune electron microscopy for detecting plant virus particles. *Journal of General Virology, 45*(2), 533–536. doi: 10.1099/0022–1317–45–2-533.

Shulaev, V., Silverman, P., & Raskin, I., (1997). Airborne signaling by methyl salicylate in plant pathogen resistance. *Letters to Nature, 385*(6618), 718–721. doi:10.1038/385718a0.

Singh, K., Wegulo, S. N., Skoracka, A., & Kundu, J. K., (2018). Wheat streak mosaic virus: A century old virus with rising importance worldwide. *Molecular Plant Pathology, 19*(9), 2193–2206. doi: 10.1111/mpp.12683.

Slogteren, E. V., & Slogteren, D. H. M. V., (1957). Serological identification of plant viruses and serological diagnosis of virus diseases of plants. *Annual Review of Microbiology, 11*, 149–164. doi: 10.1146/annurev.mi.11.100157.001053.

Sun, K., Liu, Y., Zhou, X., Yin, C., Zhang, P., Yang, Q., et al., (2022). Nanopore sequencing technology and its application in plant virus diagnostics. *Frontiers in Microbiology, 13*(939666), 1–12. doi: 10.3389/fmicb.2022.939666.

Tairo, F., Mukasa, S. B., Jones, R. A. C., Kullaya, A., Rubaihayo, P. R., & Valkonen, J. P. T., (2005). Unravelling the genetic diversity of the three main viruses involved in *Sweet Potato Virus Disease* (SPVD), and its practical implications. *Molecular Plant Pathology, 6*(2), 199–211. doi: 10.1111/J.1364–3703.2005.00267.X.

Thomas, J. E., (2008). *Banana bunchy top virus* (*Nanoviridae*). *Encyclopedia of Virology*, 272–279. doi: 10.1016/B978-012374410-4.00636-1.

Thresh, J. M., Fargette, D., & Otirn-Nape, G. W., (1994). Effects of African *Cassava mosaic Geminivirus* on the yield of cassava. *Treopical Science, 34*, 26–42.

Tiberini, A., Tomassoli, L., Barba, M., & Hadidi, A., (2010). Oligonucleotide microarray-based detection and identification of 10 major tomato viruses. *Journal of Virological Methods, 168*(1, 2), 133–140. Elsevier B.V. doi: 10.1016/j.jviromet.2010.05.003.

Tofano, D., Wiechers, I. R., & Cook-Deegan, R., (2006). Edwin southern, DNA blotting, and microarray technology: A case study of the shifting role of patents in academic molecular biology. *Genomics, Society and Policy, 2*(2), 50–61. doi: 10.1186/1746-5354-2-2-50.

Tomlinson, K. R., Bailey, A. M., Alicai, T., Seal, S., & Foster, G. D., (2018). Cassava brown streak disease: Historical timeline, current knowledge and future prospects. *Molecular Plant Pathology, 19*(5), 1282–1294. doi: 10.1111/mpp.12613.

Umana-Castro, R., Pritsch, C., Molina-Bravo, R., & Pagliano, G., (2017). Diagnostic parameters of northern blot hybridization technique for detection of citrus viroids in field-grown plants. *Asian Journal of Plant Pathology, 11*(2), 71–80. doi: 10.3923/ajppaj.2017.71.80.

Vainionpää, R., & Leinikki, P., (2008). Diagnostic techniques: Serological and molecular approaches. *Encyclopedia of Virology*, 29–37. doi: 10.1016/B978-012374410-4.00585-9.

Vickers, N. J., (2017). Animal communication: When I'm calling you, Will you answer too? *Current Biology, 27*(14), R713–R715. doi: 10.1016/j.cub.2017.05.064.

Wang, Y. M., Ostendorf, B., Gautam, D., Habili, N., & Pagay, V., (2022). Plant viral disease detection: From molecular diagnosis to optical sensing technology—A multidisciplinary review. *Remote Sensing, 14*(7), 1–24. doi: 10.3390/rs14071542.

Webster, C. G., Wylie, S. J., & Jones, M. G. K., (2004). Diagnosis of plant vairal pathogens. *Current Science, 86*(12), 1604–1607.

WenXing, X., Ni, H., Ting, J. Q., Farooq, A. B. U., ZeQiong, W., YanSu, S., et al., (2009). Probe binding to host proteins: A cause for false positive signals in viroid detection by tissue hybridization. *Virus Research, 145*(1), 26–30. doi: 10.1016/j.virusres.2009.06.011.

Wu, Q., Habili, N., Constable, F., Rwahnih, M. A., Goszczynski, D. E., & Wang, Y., (2020). Virus pathogens in Australian vineyards with an emphasis on Shiraz disease. *Viruses, 12*(8), 1–21. doi: 10.3390/v12080818.

Xiao, H., Li, C., Rwahnih, M. A., Dolja, V., & Meng, B., (2019). Metagenomic analysis of Riesling grapevine reveals a complex virome including two new and divergent variants of *Grapevine leafroll-associated virus* 3. *Plant Disease, 103*(6), 1275–1285. doi: 10.1094/PDIS-09-18-1503-RE.

Xie, X., Jiang, J., Huang, M., Chen, M., Qu, Z., & Li, X., (2021). Detection of Southern *Rice black-streaked dwarf virus* using western botting with P6. *Frontiers in Sustainable Food Systems, 5*(637382), 1–8. doi: 10.3389/fsufs.2021.637382.

Zhang, Y., Yin, J., Li, G., Li, M., Huang, X., Chen, H., et al., (2010). Oligonucleotide microarray with a minimal number of probes for the detection and identification of thirteen genera of plant viruses. *Journal of Virological Methods*, *167*(1), 53–60. doi: 10.1016/j.jviromet.2010.03.010.

Zhu, H., Zhang, H., Xu, Y., Laššáková, S., Korabečná, M., & Neužil, P., (2020). PCR past, present and future. *BioTechniques, 69*(4), 1–9. doi: 10.2144/btn-2020-0057.

CHAPTER 9

# CRISPR/Cas System-Mediated Adaptive Immunity: A Revolutionary Tool for Controlling Plant Viruses

SHAIKHUL ISLAM,[1] SACHIN ASHOK BHOR,[2] and MD. SHAMIM AKHTER[3]

[1]*Plant Pathology Division, Bangladesh Wheat and Maize Research Institute (BWMRI), Dinajpur 5200, Bangladesh*

[2]*Laboratory of Plant Molecular Biology and Virology, Faculty of Agriculture, Ehime University, Matsuyama, Ehime, Japan*

[3]*Plant Pathology Division, Bangladesh Agricultural Research Institute (BARI), Joydebpur, Bangladesh*

## ABSTRACT

Plant viruses pose a threat to food security and agriculture as they cause devastating diseases in various crop species. The control of viral diseases is a significant challenge due to the lack of chemical products specifically targeting the viruses and the difficulties in obtaining virus-resistant varieties. However, the advent of CRISPR-Cas-based systems has revolutionized the process of inducing resistance against phytopathogenic viruses. CRISPR, which stands for "clustered regularly interspaced short palindromic repeats," and Cas, which stands for "CRISPR-associated proteins," refer to the powerful gene editing technique that has been successfully applied to produce virus-resistant plants. In this chapter, we provide a brief overview of the invention of CRISPR technology, its principles, and its application in molecular plant virology. We also discuss the optimization of this groundbreaking technology for plant-based systems, provide information on how the CRISPR/Cas

Detection and Management of New and Emerging Mystery Plant Virus Sources.
Alireza Golnaraghi, Rajarshi Kumar Gaur, and Rakesh Kumar Verma (Eds.)

technique can be used to create virus-resistant varieties, and explore various methods for introducing these systems into plant cells. Finally, we address the challenges in adopting CRISPR/Cas-based gene editing tools in plant virology and discuss the prospects of this emerging technology.

## 9.1 INTRODUCTION

Plant pathogens, such as bacteria, fungi, nematodes, and viruses, cause significant economic losses in various crops worldwide. Plant viruses alone are responsible for more than half of all plant illnesses, resulting in billions of dollars in annual losses (Nicaise, 2014; Sastry & Zitter, 2014). Currently, there are 1,500 categorized plant virus species belonging to 26 families, based on their viral genome sequences (Cao et al., 2020). Plant viruses, as obligate intracellular parasites, have limited coding capacity but possess the ability to cause pathogenesis. Unfortunately, there are few effective antibiotics or substances that can eliminate contagious plant viruses without harming the host cells. This poses a significant challenge for farmers in controlling plant viral infections.

To address this issue, both conventional and non-traditional methods have been employed to develop virus-resistant plants. In the 1930s, conventional methods were utilized to generate resistance against tobamoviruses in tobacco and pepper plants (Holmes, 1938; Scholthof, 2017). Effector-triggered immunity (ETI), regulated by R-genes, has been identified in several plant species (Jones & Dangl, 2006; de Ronde et al., 2014). The concept of pathogen-derived resistance (PDR) (Sanford & Johnston, 1985) was initially applied to create *tobacco mosaic virus* (TMV) resistant tobacco plants by modifying the viral coat protein (CP) (Powell et al., 1986; Wilson, 1993), and has since been extended to other plant species (Lomonossoff, 1995; Baulcombe, 1996; Beachy, 1997). Recent research has shown that RNA interference (RNAi) is crucial for plant antiviral defenses and can also be used to acquire resistance. Activation of RNAi occurs through the introduction of double stranded RNAs (dsRNAs), which suppress or inhibit gene expression (Hannon, 2002; Voinnet, 2005). Various methods, such as sense/antisense RNA, hairpin RNA (hpRNA), microRNA (miRNA), and artificial miRNA (amiRNA) precursors, have been employed to create transgenic plants resistant to viruses (Duan et al., 2012; Taliansky et al., 2021). RNAi technology has effectively targeted more than 60 economically significant plant viruses (Zhao et al., 2020). However, traditional breeding and transgenic approaches, like PDR and RNAi, have certain limitations. Traditional breeding methods are time-consuming and labor-intensive. Additionally, due to the rapid proliferation of plant viruses through recombination and mutation, traditional viral control strategies may fail to provide adequate protection. To overcome these limitations, advanced biotechnological tools, including genome editing, have significantly enhanced breeding efficiency by manipulating the host's genetic traits to induce immune response, resistance, or tolerance against invading plant viruses (Christou, 2013).

CRISPR (clustered regularly interspaced short palindromic repeats)/ CRISPR-associated systems (Cas), which were adapted from bacterial and archaeal origins, are currently receiving attention as a formidable genome-editing tool for providing vital adaptive protection against invading nucleic acids. This method outperforms artificial miRNAs and RNAi in terms of innovative virus resistance in plants by disrupting key viral or host genes rather than suppressing them at the RNA level. The efficient application of the CRISPR/Cas technology is rapidly increasing due to enhanced know-how in this technique.

A simpler CRISPR system currently consists of only a Cas effector protein and a readily produced guide RNA (gRNA). Multi-targeting gRNAs are a powerful tool for removing mutant viruses (escape mutants) created by CRISPR/Cas9 genome editing (Roy et al., 2019; Rybicki, 2019). There are two classes of CRISPR/Cas-based systems, depending on the various structural and functional features of the Cas proteins (Makarova et al., 2015, 2017a, b). Class 1 CRISPR/Cas systems, which employ a multi-protein effector complex, include Types I, III, and IV (Makarova et al., 2017a). On the other hand, CRISPR/Cas system types II, V, and VI carry out target editing using only one effector protein (Shmakov et al., 2017; Makarova et al., 2017b). As a result, more people are using class 2 CRISPR/Cas technologies to alter nucleic acids (Garcia-Doval & Jinek, 2017; Hsu et al., 2013; Ji et al., 2019). Cas proteins of types II and V are used to edit DNA, whereas those of type VI are employed to modify RNA (Cao et al., 2020).

CRISPR/Cas system-mediated gene-editing technology has been quickly integrated into plant genetic engineering in recent years to build resistance to viruses and other diseases (Ma et al., 2016; Hadidi et al., 2016; Mahas & Mahfouz, 2018; Langner et al., 2018; Li & Wang, 2019; Chen et al., 2019). Plant viruses are unlikely to evolve strategies to combat the CRISPR-Cas system since it is a prokaryotic system (Csorba et al., 2015; Zhao et al., 2020). In general, there are two primary ways of controlling plant viruses using CRISPR/Cas technology – (i) constraining virus multiplication by modifying the viral genome; and (ii) interfering with the viral pathogenesis mechanisms by blocking or eliminating the host susceptibility factors (Khan et al., 2022).

## 9.2 CRISPR-CAS9 TECHNOLOGY: REVOLUTIONIZING GENOME EDITING

### *9.2.1 CRISPR'S DISCOVERY IN PROKARYOTES*

The CRISPR system is a powerful adaptive immunological mechanism found in bacteria and Archaea that protects them against bacteriophages and foreign plasmids (Sternberg et al., 2016). It was named in honor of the Dutch researcher who discovered the Cas genes in the *Escherichia coli* genome in 1987 (Ishino et al., 1987; Jansen et al., 2002). Several investigations have verified that the CRISPR spacers' short sequences are homologous to extra-chromosomal DNA sequences, establishing a link between CRISPR and specific immunity (Mojica et al., 2005; Bolotin et al., 2005; Pourcel et al., 2005). Over a decade later, CRISPR-Cas was successfully established as an efficient tool for editing human, animal, and plant genomes (Cong et al., 2013; Shan et al., 2013; Kaboli & Babazada, 2018).

A typical CRISPR-Cas locus consists of an AT-rich leader sequence, a CRISPR pattern with short repetitive components interspersed with invader DNA that target spacers, and a Cas genes operon producing Cas proteins (Hille & Charpentier, 2016). Based on the involvement of different Cas proteins, CRISPR-Cas systems are categorized into three subtypes: type I, II, and III. Type I and III systems use a large, multi-Cas protein complex to bind and target the nucleic acid (Nam et al., 2012; Rouillon et al., 2013). However, the type II system uses a single Cas protein (Cas9) that recognizes double-stranded DNA under the guidance of RNA (Gasiunas et al., 2012). The simplicity of the usage of type II CRISPR (i.e., the CRISPR-Cas9 system) has allowed for enormous improvements in genome editing (Doudna & Charpentier, 2014).

### *9.2.2 THE PRINCIPLE OF THE CRISPR-CAS9-MEDIATED GENOME EDITING TECHNIQUE*

Generally, the bacterial CRISPR-Cas machinery requires three major steps to halt the invasion of viral DNA:

1. **Acquisition Stage:** The invading nucleic acid is recognized and a spacer sequence, derived from the target nucleic acid, is incorporated into the host's CRISPR array to develop immunological memory.
2. **Expression Stage:** The CRISPR array is transcribed into a precursor RNA transcript, and Cas9 protein is synthesized (pre-crRNA). The

pre-crRNA then forms a complex with the Cas9 protein through a non-coding, trans-activating CRISPR RNA (crRNA), resulting in the production of mature RNA units called crRNAs.

3. **Interference Stage:** The mature crRNA guides the Cas9 protein to the specific DNA target, leading to cleavage and degradation of the invading foreign DNA (Amitai & Sorek, 2016; Wang et al., 2016).

   Double-strand breaks (DSBs) are formed by the Cas9 protein cleaving the DNA, which activates cellular DNA repair mechanisms (Figure 9.1). If the homologous repair template is unavailable, the non-homologous end joining (NHEJ) pathway is activated, resulting in random insertions, deletions, or substitutions at the DSB site and the loss of gene functions (Figure 9.1). Conversely, when a donor DNA template with a matching DSB site is used, the homology-directed repair (HDR) pathway is activated, resulting in alterations

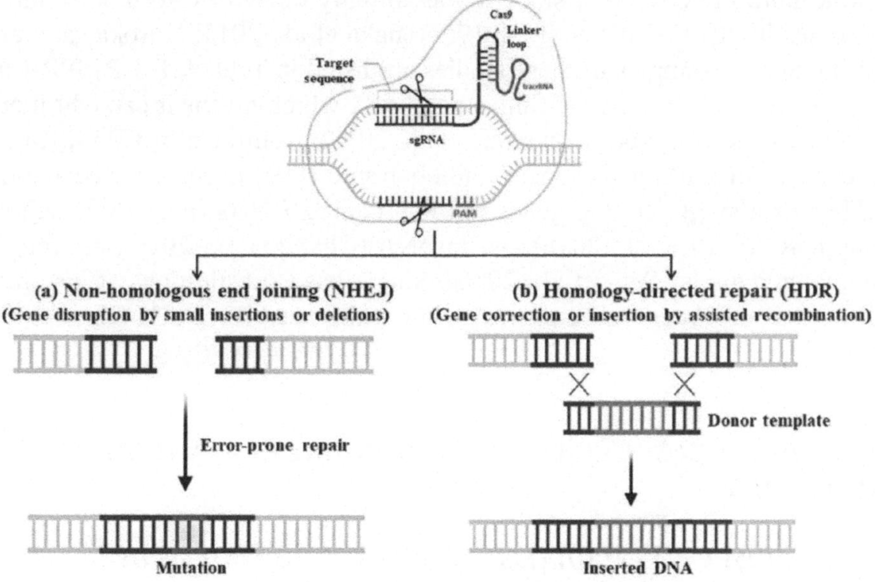


**FIGURE 9.1** The principle of CRISPR-Cas9-mediated plant genetic manipulation. [Using a single guide RNA (sgRNA) to direct a Cas9 protein to cleave DNA at a specific region of the genome adjacent to a PAM sequence. The resulting double-stranded break is then repaired using error-prone non-homologous end joining (NHEJ) and homology-directed repair (HDR) methods].

*Source:* Adapted and modified from: Varanda et al. (2021).

that enable precise gene manipulations such as gene knock-in, deletion, or mutation (Yin et al., 2017). Currently, the most widely used Cas9 protein is derived from *Streptococcus pyogenes* (Sp) (Marraffini et al., 2016). Synthetic single-guide RNAs (sgRNAs) are necessary to generate the CRISPR-Cas9 expression cassettes for using this technology in genome editing purposes. Protospacer adjacent motif (PAM) sequences with the –NGG– pattern is recognized by sgRNAs, which then target DNA sequences through Watson-Crick base pairing and guide Cas9 to specific genomic regions (Liu et al., 2017).

### *9.2.3 APPLICATIONS AND OPTIMIZATION OF THE CRISPR-CAS TECHNOLOGY IN PLANT-BASED SYSTEMS*

Numerous efforts have been undertaken to modify the CRISPR-Cas system into a more effective tool since it was initially developed to modify plant genomes in 2013 (Kaya et al., 2017; Nonaka et al., 2017; Kurokawa et al., 2021). Some examples of these studies are listed in Table 9.1. CRISPR-Cas now possesses multiplex editing capabilities, which means it can edit more than one gene at a time (Donohoue et al., 2018). Additionally, CRISPR-Cas is capable of editing the open reading frame (ORF), untranslated region (UTR) of a single coding gene, and non-coding RNAs (ncRNAs) such as long ncRNA (Li et al., 2018) and miRNA (Chang et al., 2016), as well as promoter regions (Seth et al., 2016). Single-base substitutions at genomic sites have also been accomplished without the need for DSBs (Hess et al., 2017).

## 9.3 CRISPR-CAS9 TECHNOLOGY TO ENGINEER PLANT VIRUS RESISTANCE

### *9.3.1 CRISPR-CAS-MEDIATED RESISTANCE TO VIRUSES BY TARGETING PLANT GENOME*

Plant viruses rely on the host's biology for replication because they interact with several host components necessary for viral replication and transportation within plants, which is needed to complete the infection cycle (Hull, 2001). Recessive resistance, an enduring hereditary trait, can be conferred by CRISPR/Cas through the alteration or deletion of recessive genes that encode crucial host components for viral infection (Hashimoto et al., 2016).

Intensive efforts have been made by scientists to induce disease resistance against invading pathogens. These breakthrough approaches have identified novel genetic resources that have become the major target of plant resistance breeding utilizing transgene technologies (Wittmann et al., 1997; Cillo et al., 2014) (Figure 9.2). Translation initiation factors such as elF4E, elF4G, and elFiso4E, which have been shown to be directly implicated in viral infection, are among a number of significant prospective targets for genome editing. Scientists are using CRISPR to intentionally alter these genes in order to create plant resilience (Sanfaçon, 2019). Therefore, any host genetic factors responsible for regulating the basic compatibility during plant-virus interactions could potentially become a target for CRISPR-mediated genome editing.

Because different endonucleases can target DNA to give plants resistance to RNA viruses by changing host factors/genes associated with viral pathogenesis in the plant, CRISPR-based techniques are extremely intriguing (Chandrasekaran et al., 2016). Additionally, the generation of non-transgenic virus-resistant plants is made possible by CRISPR for plant-mediated resistance, which does not require the conservation of transgenes to produce Cas9 and sgRNAs stably in the plant cell genome (Chandrasekaran et al., 2016).

Multiple investigations have used CRISPR/Cas9 to generate plant-mediated resistance to viruses (Table 9.2). For example, durable resistance to *turnip mosaic virus* (TuMV) in *Arabidopsis thaliana* is achieved through induced mutations caused by the knockout of the elF(iso)4E gene (Pyott et al., 2016). By mutagenesis of the elF4G gene, Macovei et al. (2018) generated rice plants immune to the *rice tungro spherical virus* (RTSV). Similar to this, eliminating the elF4E gene from the cucumber (Cucumis sativus) provided defense against many Potyviridae members, including the *cucumber vein yellowing virus* (CVYV), *zucchini yellow mosaic virus* (ZYMV), and *Papaya ringspot mosaic virus* (PRSV) (Chandrasekaran et al., 2016). Additionally, CRISPR/Cas9 technology was used to target the elF4E1 gene in *A. thaliana* plants to confer resistance to the Clover yellow vein virus (CYVV) (Bastet et al., 2018). In order to increase cassava's resistance to the cassava brown streak virus (CBSV), CRISPR/Cas9 has made it possible to induce double mutations in the new cap-binding protein-1 and -2 (nCBP-1 and nCBP-2) from the elF4E family (Gomez et al., 2019).

However, it is undeniable that changing plant genes poses a risk of interfering with the physiological and biochemical processes associated with

**TABLE 9.1** Modification and Alteration of the CRISPR-Cas Methodologies for Their Efficient Adaptation in Plant-based Systems

| **(i) Optimization of the Different Cas-Variants to Enhance CRISPR Efficiency** | | | | |
|---|---|---|---|---|
| **Cas-Variants** | **Source** | **Types of Mutation** | **Crop Species** | **References** |
| St1Cas9 | *Streptococcus thermophilus* | In-dels | Arabidopsis | Steinert et al. (2015) |
| SaCas9 | *Staphylococcus aureus* | In-dels | Arabidopsis, tobacco | Steinert et al. (2015); Kaya et al. (2016) |
| SpCas9-VQR | *Streptococcus pyogenes* | In-dels | Rice | Hu et al. (2016) |
| SpCas9-VRER | *Streptococcus pyogenes* | In-dels | Rice | Hu et al. (2016) |
| nCas9 | *Streptococcus pyogenes* | In-dels | Arabidopsis, rice, tomato | Fauser et al. (2014); Mikami et al. (2016); Shimatani et al. (2017) |
| dCas9 | *Streptococcus pyogenes* | In-dels | Arabidopsis, maize, rice, tobacco, wheat | Piatek et al. (2015); Shimatani et al. (2017); Zong et al. (2017); Zhang et al. (2017a) |
| FnCas9 | *Francisella novicida* | No cleavage, alanine double point mutations. | Tobacco, Arabidopsis | Zhang et al. (2018) |
| Cas12a (Cpf1) | *Acidaminococcus* sp. (AsCas12a); *Lachnospiraceae bacterium* (LbCas12a); *Acidibacillus sulfuroxidans* (AsCas12f1) | In-dels | Maize, rice, soybean | Endo et al. (2016); Tang et al. (2017); Kim et al. (2017); Lee et al. (2019) |
| LshCas13a | *Leptotrichia shahii* | Cleavage | Rice, tobacco | Abudayyeh et al. (2017); Aman et al. (2018); Zhang et al. (2019); Zhan et al. (2019) |
| LwaCas13a | *Leptotrichia wadei* | For RNAi, replicase gene used as a target. | Tobacco | Mahas et al. (2019) |
| BzCas13b | *Bergeyella zoohelcum* | | | |
| PspCas13b | *Prevotella* sp. P5-125 | | | |
| CasRx (Cas13d) | *Ruminococcus flavefaciens* XPD3002 | | | |

**TABLE 9.1** *(Continued)*

| Cas-Variants | Source | Types of Mutation | Crop Species | References |
|---|---|---|---|---|
| CasRx (Cas13d) | *Ruminococcus flavefaciens* | Cleave target RNA. | Tobacco | Cao et al. (2021) |
| *Cas14/a | Uncharacterized archaea under the super-phylum DPANN (Diapherotrites, Parvarchaeota, Aenigmarchaeota, Nanoarchaeota, and Nanohaloarchaeota) | Selective cleavage of the ssDNA. | – | Harrington et al. (2018); Khan et al. (2019) |
| **(ii) Applying Multitude of Cas-Promoters Increases the Scope of CRISPR Applicability** | | | | |
| Promoters | Activity and Preferential Expression | | Crop Species | References |
| YAO | Tissues with highest cell division activities such as shoot and root apical meristem, embryo, embryo sac, endosperm, and pollen grain. | | Arabidopsis, citrus | Yan et al. (2015); Zhang et al. (2017b) |
| SPL | Pollen or microspore mother cells and microsporocytes. | | Arabidopsis | Mao et al. (2016) |
| EC1.1 and EC1.2 | Zygote cell and egg cells. | | Arabidopsis | Wang et al. (2015); Wolter et al. (2018) |
| ICU2 | Meristematic regions | | Arabidopsis | Hyun et al. (2015) |
| EF1α, hisH4 | Shoot and root meristem cells, oogonia, pollen, etc. | | Arabidopsis | Osakabe et al. (2016) |
| MGE | Meiosis stage | | Arabidopsis | Eid et al. (2016) |
| DMC1 | Meiocytes | | Arabidopsis, maize | Xu et al. (2018); Feng et al. (2018) |
| Arabidopsis RPS5A promoter (At3g11940) | At all developmental stages. | | Arabidopsis | Tsutsui et al. (2017) |
| Cauliflower mosaic virus 35S RNA promoter (CaMV35S), Pea early browning virus (pPEBV) promoter. | Most active to drive crRNAs for RNA interference with the NES signal. | | Tobacco | Mahas et al. (2019) |

**TABLE 9.1** *(Continued)*

| Cas-Variants | Source / Types of Mutation | Crop Species | References |
|---|---|---|---|
| CaMV35S, Arabidopsis RPS5A promoter (At3g11940), maize UBIQUITIN promoter (ZmUBI), Parsley UBIQUITIN promoter (PcUBI). | Apical meristem region of root and shoots and leaf. | Arabidopsis | Kurokawa et al. (2021) |
| PcUBI, AtU6 | Introduction of Cas9 and sgRNA ribonucleoprotein (RNP) into the protoplasts to ensure DNA-free genome editing. | Tobacco | Rahman et al. (2022) |
| TM2-pd35S-dMac (en35S) promoter | Pea hairy roots system. | Pea | Li et al. (2022) |
| **(iii) Optimizing the sgRNA Systems for Precise Genome Editing** | | | |
| Crop Species | Strategies | | References |
| Arabidopsis, maize, populus, rice, tobacco, tomato, pea | Construction of recombinant CRISPR-Cas vector that constitute sgRNA expression cassette. | | Ma et al. (2015); Gao et al. (2015); Fan et al. (2015); Char et al. (2017); Yu et al. (2019); Li et al. (2022) |
| Maize, potato, rice, tomato, wheat | Produce numerous sgRNAs from a single polycistronic gene via the endogenous tRNA-processing system. | | Xie et al. (2015); Qi et al. (2016); Hashimoto et al. (2018); Nakayasu et al. (2018); Li et al. (2018) |
| Tobacco, Arabidopsis, tomato | A sgRNA that targets the most conserved sequences in the intergenic region of most of the Geminiviruses guides Cas9 to simultaneously target different viruses. | | Ali et al. (2018); Tashkandi et al. (2018) |

*Abbreviations:* tRNA: Transfer RNA; crRNA: CRISPR RNA; sgRNA: single-guide RNA; SSBs: single-strand breaks; PFS: protospacer flanking sequence; In-dels: insertion or deletions; and CRISPR-Cas: clustered regularly interspaced short palindromic repeat-CRISPR-associated.

*Cas14a has shown tremendous potential for targeting plant ssDNA viruses of the Geminiviridae and Nanoviridae families. However, application of this Cas-variant in plant-based system needs further studies.

*Source:* Adapted and modified from: Wang et al. (2019).

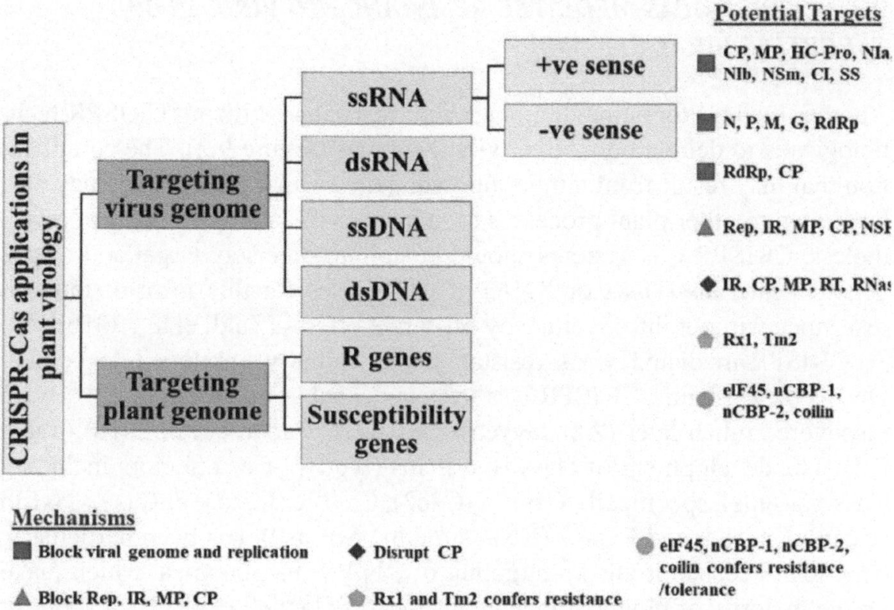


**FIGURE 9.2** Graphical representation of the possible applications of CRISPR-Cas technology in plant virology. [Different DNA and RNA virus genomes (single-stranded RNA: +ssRNA, –ssRNA; double-stranded RNA: dsRNA; single-stranded DNA: ssDNA; and double-stranded DNA: dsDNA) and potential genomic targets for plant genomes and their relevant molecular mechanisms are highlighted. The potential genomic targets of (i) +ve-ssRNA viruses: movement protein (MP), coat protein (CP), helper component-protease (HC-Pro), nuclear inclusion B (NIb), cylindrical inclusion (CI), nuclear inclusion A (NIa), and silencing suppressors (SS); (ii) –ve-ssRNA viruses: putative movement protein (M), viral envelope glycoprotein (G), polymerase-associated phosphoprotein (P), nucleoprotein (N), and RNA-directed RNA polymerase (RdRP) genes; (iii) dsRNA viruses: CP and RdRP; (iv) ssDNA viruses: CP, MP, replication-associated protein (Rep), intergenic region (IR), replication enhancer protein (REn), nuclear shuttle protein (NSP); and (iv) dsDNA viruses: MP, CP, reverse transcriptase (RT), IR, and RNase H. While editing plant genome with the CRISPR-Cas approaches, various R genes (Rx1, Tm2) and susceptibility genes confer resistance or tolerance].

*Source:* Adapted and modified from: Sharma et al. (2021).

the altered genes, which might have a negative impact on the host's fitness. Nevertheless, the above examples clearly illustrate that CRISPR/Cas9 can successfully increase a plant's competitiveness against virus infections by inducing resistance without compromising plant functions.

### *9.3.2 CRISPR-CAS-MEDIATED RESISTANCE TO VIRUSES BY TARGETING VIRAL GENOME*

Another method for achieving plant viral resistance utilizing CRISPR technologies is to deliberately target viral genomes (Figure 9.2). The complication that may result from tampering with genes is that these genes may also be related to other plant processes such as growth and reproduction. Nevertheless, CRISPR/Cas systems should be able to precisely target and cleave genetic materials (DNA or RNA) of the viruses for this form of induced resistance without introducing any off-target effects (Zaidi et al., 2016).

CRISPR-mediated virus resistance was initially employed to combat DNA viruses, while CRISPR/Cas systems capable of cleaving RNA were discovered much later (Abudayyeh et al., 2016; Aman et al., 2018) (Table 9.2). The development of class II systems (type VI Cas effectors including Cas9 variants), specifically Cas13a (C2c2), Cas13c (C2c7), FnCas9, Cas13b (C2c6), Cas13d, and RCas9 (RNA targeting SpCas9), has been particularly beneficial because it allows targeting of RNA virus genomes, which make up the majority of phytopathogenic viruses (O'Connell et al., 2014; Price et al., 2015; Abudayyeh et al., 2016; Shmakov et al., 2017; Cox et al., 2017; Yan et al., 2018).

Numerous studies have demonstrated that CRISPR can confer resistance in plant species by targeting either the DNA or RNA viral genomes, resulting in delayed or reduced virus accumulation and significantly decreased infection symptoms (Mahas & Mahfouz, 2018). Table 9.2 lists some of the experiments that utilize CRISPR/Cas machinery to directly modify DNA and/or RNA viruses in various plants.

Plant DNA viruses can be categorized into two major groups: double-stranded caulimoviruses and single-stranded geminiviruses that replicate as double-stranded DNA inside host tissues (van Regenmortel et al., 2000). According to the current report from the International Committee on Virus Taxonomy (ICTV, 2022), the Geminiviridae is the largest group of plant viruses, consisting of 485 species. Geminiviruses infect several economically important crops, leading to reduced agricultural yields globally (Moffat et al., 1999; Zaidi et al., 2017). As a result, scientists have focused their attention on implementing CRISPR-mediated resistance against DNA viruses, particularly geminiviruses (Table 9.2). In one study, *Nicotiana benthamiana* plants expressing Cas9 were transformed with sgRNA delivered via the *tobacco rattle virus* (TRV) system to modify the coding (rep genes and coat proteins) and non-coding (conserved intergenic region) sequences of

**TABLE 9.2** Recent Developments in the Virus-Resistant Plant Generation Using CRISPR/Cas Technology

| CRISPR-Cas Delivery System | Virus Type/ Target Genome | Virus | Host Plant | Targets for Genome Editing | References |
|---|---|---|---|---|---|
| Agrobacterium; Cas9/sgRNA-transgenic plants | DNA | BSCTV | *Nicotiana benthamiana*, *Arabidopsis thaliana* | CP, Rep genes; IR | Ji et al. (2019) |
| | DNA | BeYDV | *N. benthamiana* | Rep/RepA genes; LIR | Baltes et al. (2015) |
| | DNA | CLCuMuV | *N. benthamiana* | C1 (Rep) gene and IR | Yin et al. (2019) |
| | DNA | WDV | *Hordeum vulgare* (barley) | MP/CP, Rep/RepA genes, and LIR. | Kis et al. (2019) |
| | DNA | TYLCV | *Solanum lycopersicum* (tomato), *N. benthamiana* | CP, Rep genes | Tashkandi et al. (2018) |
| | DNA | ACMV | *Manihot esculenta* (cassava) | AC2 and AC3 genes encoding viral proteins TrAP and Ren, resp. | Mehta et al. (2019) |
| | DNA | CaMV | *A. thaliana* | CP gene | Liu et al. (2018) |
| | DNA | BSV (eBSV) | *Musa* spp. (banana) | Viral ORF1, ORF2, and ORF3 integrated into plant genome. | Tripathi et al. (2019) |
| | Host genome | RTSV | *Oryza sativa* (rice) | eIF4G gene | Macovei et al. (2018) |
| | Host genome | CBSV | *Manihot esculenta* (cassava) | eIF4E isoforms; nCBP-1 and nCBP-2. | Gomez et al. (2019) |
| Agrobacterium; Cas9-transgenic plants; TRV vector-mediated sgRNAs delivery. | DNA | TYLCV, BCTV, MeMV, CLCuKoV | *N. benthamiana* | CP, Rep genes; IR | Ali et al. (2015, 2016) |

**TABLE 9.2** *(Continued)*

| CRISPR-Cas Delivery System | Virus Type/ Target Genome | Virus | Host Plant | Targets for Genome Editing | References |
|---|---|---|---|---|---|
| Agrobacterium; dCas9/ sgRNA-transient expression. | DNA | CLCuV | *N. benthamiana* | IR | Khan et al. (2019) |
| Agrobacterium; sgRNA/Cas9; transient expression | DNA | CLCuV and betasatellite | *N. benthamiana* | Rep, Rep + β<br>C1 | Khan et al. (2020) |
| Agrobacterium; Cas9/sgRNA-transgenic plants with further transgene segregation. | Host genome | CVYV, ZYMV, PRSMV | *Cucumis sativus* (cucumber) | eIF4E gene | Chandrasekaran et al. (2016) |
| | Host genome | TuMV | *A. thaliana* | eIF(iso)4E gene | Pyott et al. (2016) |
| Agrobacterium; FnCas9 or FndCas9/sgRNA-transgenic plants. | RNA | TMV, CMV | *N. benthamiana*, *A. thaliana* | ORF1, ORF2, ORF3, CP gene and 3′UTR. | Zhang et al. (2018) |
| Agrobacterium; Cas13a/ sgRNA-transgenic plants. | RNA | SRBSDV, RSMV | *Oryza sativa* (rice) | Various sequences in RSMV and SRBSDV genomes. | Zhang et al. (2019) |
| | RNA | PVY | *S. tuberosum* (potato) | P3, CI, NIb, CP genes | Zhan et al. (2019) |
| Agrobacterium; nCas9/ sgRNA-transgenic plants with further transgene segregation. | Host genome | ClYVV | *A. thaliana* | eIF4E gene | Bastet et al. (2019) |
| RNP (Cas9/sgRNA); vacuum infiltration or bombardment. | Host genome | PVY | *S. tuberosum* (potato) | Coilin gene | Makhotenko et al. (2019) |
| Agrobacterium; Cas13a transgenic plants; TRV vector-mediated sgRNAs delivery. | RNA | TuMV | *N. benthamiana* | HC-Pro, CP genes; GFP gene (integrated into viral genome). | Aman et al. (2018) |

**TABLE 9.2** *(Continued)*

| CRISPR-Cas Delivery System | Virus Type/ Target Genome | Virus | Host Plant | Targets for Genome Editing | References |
|---|---|---|---|---|---|
| Agrobacterium; Cas13a/ sgRNA-transient expression. | RNA | TMV | *N. benthamiana* | Various sequences in TMV genome. | Zhang et al. (2019) |

*Abbreviations:* BSCTV: Beet severe curly top virus; TYLCV: tomato yellow leaf curl virus; BeYDV: bean yellow dwarf virus; BCTV: beet curly top virus; MeMV: Merremia mosaic virus; CLCuMuV: cotton leaf curl Multan virus; CLCuKoV: cotton leaf curl Kokhran virus; CLCuV: cotton leaf curl virus; WDV: wheat draft virus; ACMV: African cassava mosaic virus; BSV: banana streak virus; eBSV: endogenous BSV; CaMV: cauliflower mosaic virus; TMV: tobacco mosaic virus; CMV: cucumber mosaic virus; TuMV: turnip mosaic virus; SRBSDV: southern rice black-streaked draft virus; RSMV: rice stripe mosaic virus; ClYVV: clover yellow vein virus; PVY: potato virus Y; CVYV: cucumber vein yellowing virus; ZYMV: zucchini yellow mosaic virus; PRSMV: papaya ring spot mosaic virus; CBSV: cassava brown streak virus; CP: coat protein; Rep: replication-associated protein; IR: intergenic region; LIR: long IR; MP: movement protein; TrAP: transcription activator protein; Ren: replication enhancer protein; ORF: open reading frame; UTR: untranslated region; GFP: green fluorescent protein; sgRNA: Single guide RNA; TRV: tobacco rattle virus; and RNP: ribonucleoprotein.

*Source:* Adapted and modified from: Kalinina et al. (2020).

*tomato yellow leaf curl virus* (TYLCV). This resulted in a reduction in viral DNA accumulation and symptomatology in tobacco plants (Ali et al., 2015). Another study targeted viral non-coding, intergenic regions (IRs) using the CRISPR/Cas9 system and a sgRNA that targeted a conserved area in various begomoviruses. This approach decreased the formation of recoverable viral variants that could evade CRISPR-mediated defense (Ali et al., 2016).

## 9.4 CRISPR SYSTEM DELIVERY AND EXPRESSION IN PLANTS

Multiple CRISPR-Cas systems have been designed to modify the genomes of various plant species, including DNA cleaving endonucleases (SaCas9, Cas12a, SpCas9, and Csm1) or RNA cleaving endonucleases (C2c2/Cas13a) (Table 9.2 and Figures 9.3–9.6). Recently developed endonucleases, such as Cas14a, Cas12e, and C2c1 for DNA editing, and Cas13b for RNA editing, have enhanced the potential for gene editing in plant species (Harrington et al., 2018). Cas14a and Cas13b, which can target ssDNA and ssRNA genomes, may be particularly important due to the prevalence of plant viruses with either ssDNA or ssRNA genomes. Endonucleases typically range in size from 400 to 1,368 amino acids. The use of smaller endonucleases like SaCas9, Csm1, and Cas12e may improve the effectiveness of the CRISPR-Cas delivery system to the plant genome. Additionally, these smaller endonucleases can serve as CRISPR effectors for virus-mediated genome editing against plant viruses. Furthermore, Csm1, C2c1, Cas12a, and Cas12e endonucleases are dsDNA specific, creating staggered ends that are beneficial for the HDR pathway.

### *9.4.1 CAS9 NUCLEASE*

Cas9 (formerly known as Cas5, Csn1, or Csx12) is a key component in the defense against bacteriophages and plasmids in certain bacteria. Cas9 is associated with the CRISPR adaptive immune system of bacteria and acts as an RNA-guided DNA endonuclease. Cas9 causes sequence-specific dsDNA breaks and is guided to a target region by a gRNA (Jinek et al., 2012; Maeder et al., 2013; Gilbert et al., 2013). The genome editing cleavage complex of CRISPR also contains tracrRNA, which is synthesized as a sgRNA molecule simultaneously with Cas9 and crRNA (Jinek et al., 2012) (Figure 9.3). By modifying the sequence of its crRNA, CRISPR-Cas9 can be reconfigured to target another location within the genome.

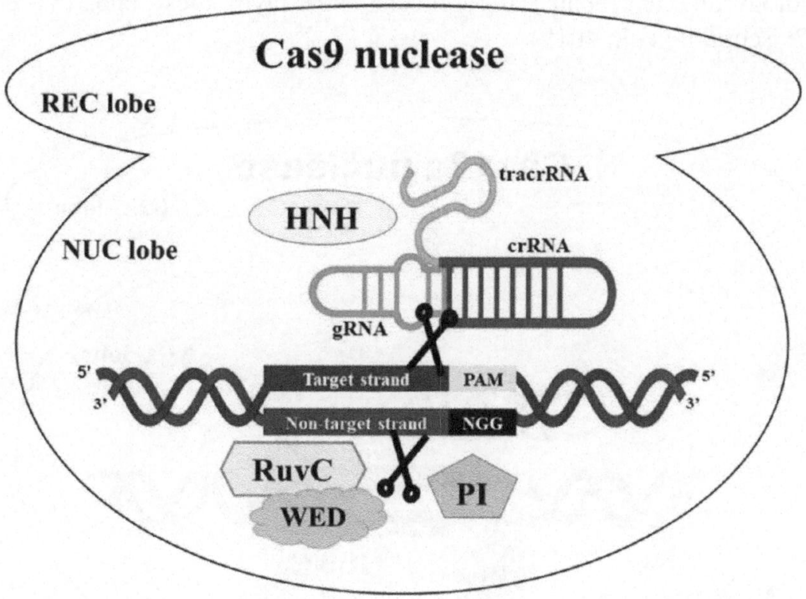


**FIGURE 9.3** Schematics of the principles and mechanisms of action of Cas9 endonucleases. [The traditional CRISPR-Cas9-based genome editing utilizes a specific effector protein, Cas9, a crRNA, and a tracrRNA. The latter two are associated with the Cas9 endonuclease, which hybridizes and binds to the target sequence with the aid of a 20-nt guide RNA (gRNA) sequence upstream of the PAM sequence. The recognition lobe recognizes the crRNA:tracrRNA:target DNA complex (REC). The PAM interacting domain recognizes the PAM sequence (PI). The HNH and RuvC domains of the nuclease (NUC) lobe cleave the target and non-target strands through a blunt-ended double-stranded break (DSB) and non-homologous end joining (NHEJ) upstream of the PAM region, respectively].

*Source:* Adapted and modified from: Shahid et al. (2021).

## 9.4.2 CAS12A NUCLEASE

Cas12a (Cpf1) nuclease belongs to the CRISPR system of class II type V. Unlike the Cas9 nuclease, which requires two separate catalytic domains (HNH and RuvC), Cas12a only requires one (RuvC) to produce double-strand breaks (DSBs) in DNA (Figure 9.4). Cas12a operates correctly with crRNA alone, in contrast to Cas9 which requires a combination of crRNA and tracrRNA. The Cas12a system's effectiveness is further enhanced by its ability to create sticky ends with 50 overhangs during DSBs, as opposed to the blunt ends produced by the Cas9 system (Zetsche et al., 2015; Fonfara et al., 2016). The production of 5′ overhangs is supported by the fact that homology-directed repair (HDR) mechanisms favor sticky ends (Vu et al., 2017; Schmidt et al., 2019).

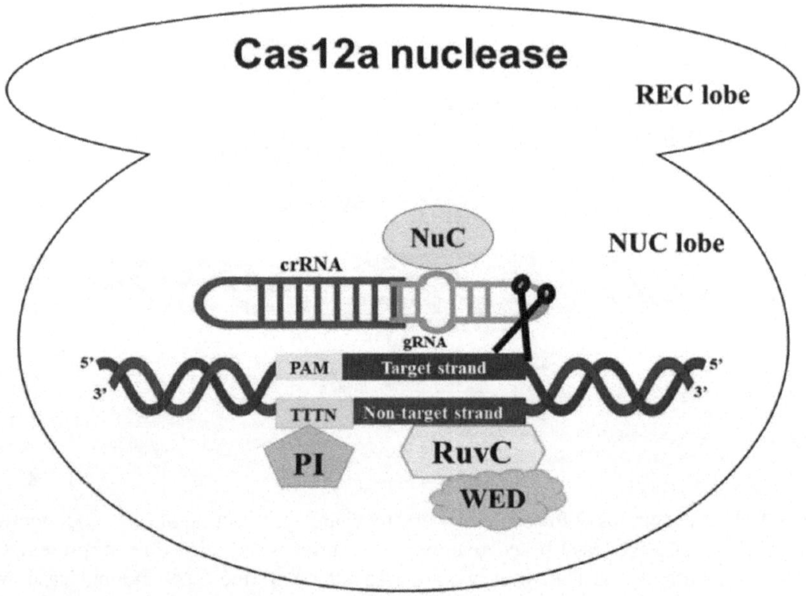


**FIGURE 9.4** Illustrations of the principles and mechanisms of Cas12a endonucleases. [Unlike CRISPR-Cas9, the CRISPR-Cas12a endonuclease is controlled by a single effector protein (Cas12a) that is connected to a single crRNA. The REC lobe recognizes the Cas12a:crRNA complex and attaches to the target strand precisely at the target region downstream of the PAM sequence. The PAM sequence of the 23–25 nucleotide (nt) long gRNA is detected by the PI domain, which aids in targeted DNA binding. The NuC domain is responsible for cleaving the target strand 18 nt downstream of the PAM, while the RuvC domain is responsible for cleaving the non-target strand 23 nt downstream of the PAM].

*Source:* Adapted and modified from: Kalinina et al. (2020); and Shahid et al. (2021).

### 9.4.3 CAS13 NUCLEASE

Recent advances have been made in the application of Cas13 protein by redesigning it to selectively target ssRNA viral genomes with no off-target consequences (Mahas et al., 2019b; Freije et al., 2019). The DNA-nuclease activity of Cas13 orthologs (Cas13a, Cas13b, and Cas13d) has been replaced by two higher eukaryotes and prokaryotes nucleotide-binding (HEPN) domains (Figures 9.5(A) and (B)). Cas13 has been transformed into an RNA-guided RNA endonuclease that specifically cuts ssRNA targets, thanks to the HEPN domains. Cas13 is a dual-effector nuclease that facilitates the conversion of pre-CRISPR RNA into CRISPR RNA (crRNA). Through RNA-RNA hybridization, a single crRNA of Cas13 proteins helps identify target ssRNAs (Wessels et al., 2020). The nuclease activity of Cas13a is mediated by two HEPN domains, which preferentially act on U-rich targets. Experimental evidence has shown that a single point mutation in any HEPN domain can render Cas13a nuclease inactive, resulting in a catalytically inactive or dead Cas13a (dCas13a). However, dCas13a is more susceptible to RNA binding. In contrast, Cas13b's sequence-specific RNA cleavage activity depends on both-sided protospacer flanking sequence (PFS) sites and ssRNA template strands (Hameed et al., 2019). Cas13 nucleases may exhibit some non-specific nuclease activity, but compared to RNAi, their off-target effects are greatly reduced (Abudayyeh et al., 2017). While the restrictions imposed by PFS in the design of Cas13 gRNAs are minimal, the spacer sequences are often intolerant to mismatches between 15 and 21 nt (Wessels et al., 2020). As a result, the potential of catalytically active Cas13a has been tested in plant genomes and ssRNA viruses like potyviruses. Cas13a conferred target-specific resistance to TuMV in *N. benthamiana* and *Arabidopsis* plants, according to these findings (Mahas et al., 2019a; Aman et al., 2018a). Both Cas13a and Cas13b showed effective and highly specific interference against various RNA viruses (Mahas et al., 2019a). However, Cas13d offered the most powerful interference against RNA viruses in both transient and in planta tests (Mahas et al., 2019b). Furthermore, Cas13d has the capability to target two RNA viruses simultaneously, indicating that it might be a potential alternative for developing broad-spectrum resistance in plants against single and multiple viruses.

### *9.4.4 CAS14 NUCLEASE*

Extremophile archaea contain effector proteins called Cas14 nucleases that only cleave ssDNA (Harrington et al., 2018). In comparison to their class 2 counterparts, Cas14 endonucleases (Class 2, type V) are considerably smaller (400–700 amino acids). Contrary to other class 2 Cas effectors, Cas14 nucleases do not always require PAM binding to engage in cleavage activity (Figure 9.6) (Savage, 2019). Nevertheless, the target ssDNAs corresponding to a 20-bp sequence must be present in the gRNA. Cas14 nucleases need sequence specificity in the core 6 bp region of the gRNA to be triggered. Additionally, Cas14 endonucleases lack RNase-III endonuclease activity

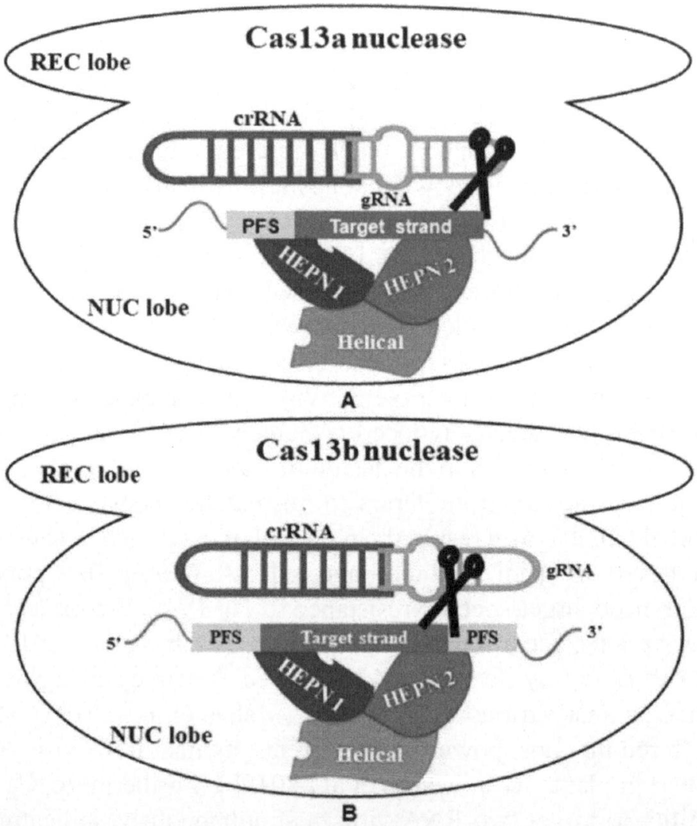


**FIGURE 9.5** Schematic representation of the principles and mechanisms of Cas13 endonucleases. [(A) CRISPR-Cas13-based genome editing is performed using a single effector protein (Cas13) linked to a single crRNA. The REC lobe detects the Cas13 and crRNA complex, which binds to the recognition site through complementary sequence pairing with the ssRNA substrate indicated by the crRNA's gRNA sequence. The sequence-specific cleavage of the ssRNA substrate is carried out by the HEPN1 and HEPN2 domains. Instead of PAM sequences, the Cas13a protein is guided to the ssRNA target by a single protospacer flanking sequence (PFS). And (B) CRISPR-Cas13b is distinguished from Cas13a by the presence of suppressor and enhancer Cas genes regulated by two PFS].

*Source:* Adapted and modified from: Shahid et al. (2021).

(Harrington et al., 2018). Cas14-mediated ssDNA cleavage can confer resistance to viruses with ssDNA genomes or to plasmids, viruses, and transposons (Aquino-Jarquin, 2019). These unique properties make Cas14 nucleases particularly suited for nucleic acid detection. Cas14, in particular, is capable of identifying SNPs in the targeted area with high fidelity.

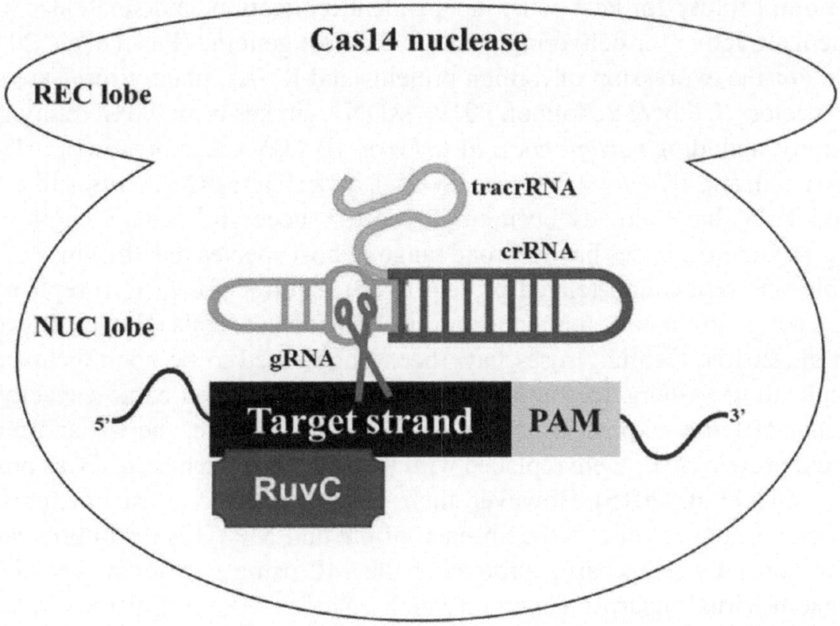


**FIGURE 9.6** Schematic representation of the principles and mechanisms of Cas14 endonucleases. [The single-effector Cas14a, in conjunction with a 130-bp-long tracrRNA and crRNA, mediates CRISPR-Cas14a-based ssDNA editing. The addition of the PAM sequence in the ssDNA substrates is not required for the crRNA:tracrRNA complex. However, ssDNA cleavage by the RuvC domain requires 20 nt of sequence-specific complementarity in the crRNA guide sequence to the ssDNA substrate].

*Source:* Adapted and modified from: Shahid et al. (2021).

## 9.5 THE APPLICATION OF PLANT VIRUSES AS CRISPR DELIVERY VECTORS

CRISPR-Cas targeting efficacy depends on the efficient transport of the CRISPR cassette. Typically, this is achieved through biolistic or Agrobacterium-mediated plant transformation. However, the transformation of a few cells and the addition of extra cargo, such as promoters and degrading DNA fragments, raise concerns about GMOs. To effectively introduce GE chemicals into economically relevant crops, delivery systems must be continually improved. Virus-based vectors with autonomous replication are commonly used to insert GE cassettes into target plants (Zaidi & Mansoor, 2017). The most significant finding is that GE has been successfully manipulated in plant-based systems using DNA or RNA viruses as vectors. Plant viruses' natural affinity for host cells, independent replication, and small size make them attractive for delivering genes to the host genome (Pasin et al., 2019).

For the expression of foreign proteins and RNAs, plant viruses are used as vectors (Cody & Scholthof, 2019). ssDNA viruses from the Geminiviridae family, including *barley yellow dwarf virus* (BYDV), *Cabbage leaf curl virus* (CLCV), and *Wheat dwarf virus* (WDV), as well as ssRNA viruses like TRV and TMV, have already been employed for successful genetic engineering (GE). Geminiviruses have a broad range of host species that they may infect, only one replication-related protein (Rep) required for *in vivo* replication, and a high number of replicons produced inside host cells (Hanley-Bowdoin et al., 2013). Geminiviruses have been engineered to be noninfectious yet replicating systems to compensate for their constrained cargo capacity. To generate hyper-expression of the CRISPR-Cas system, the CP and movement protein (MP) were replaced with the sgRNA sequences and Cas protein (Cermák et al., 2015). However, their use is limited to transient expression in certain tissues due to the absence of CP and MP. This restriction can be overcome by temporarily producing the MP using a different vector or a begomovirus' bipartite genome (which encodes MP on a distinct genomic component). Another approach is to disperse GE components using geminivirus-associated DNA-satellites. These DNA-satellites have been effectively used as cargo vectors for GE chemicals and as vectors for virus-induced gene silencing (VIGS) in a number of crops.

Simple, flexible, and efficient GE tools. TRV-based vectors outperform time-consuming transformation and tissue culture procedures. Cloning, multiplexing, library building, and agroinfiltration are all possible with these vectors. The most important property of TRV-based vectors is the capacity to create transgene-free GE plants (Mahas et al., 2019a). However, the only crops that can use TRV-based CRISPR-Cas systems are transgenic lines that have Cas9 protein generated in a stable manner. Transient gene expression has also been achieved using TMV-based ssRNA vectors (Cody et al., 2017). A modified TMV-based expression vector was generated to express the target area of tobacco microRNA 398 (miR398) (Kaya et al., 2017). After the plants underwent genetic modification, the regenerated plants' miR398 expression eliminated viral RNA throughout shoot development. Additionally, a split-protein technique was used to transiently synthesize the Cas9 protein from two fragments utilizing TMV and Agrobacterium. The virus-free GE plants were produced by reassembling the active Cas9 protein and using it in tissue culture to cause a particular mutation (Kaya et al., 2017). A new era of agricultural biotechnology applications and functional genomics has arrived because of the introduction of viral vectors in GE. Later on, it could be challenging to remove the viral DNA from the GE plants. Meristematic tissue culture would be a good choice for eliminating the virus and any traces of it from GE plants.

## 9.6 CURRENT CHALLENGES FOR ADOPTING CRISPR-CAS-BASED TECHNIQUES IN PLANT VIROLOGY

For the purpose of creating gain-of-function and loss-of-function mutants in plants, particularly for commercial crops, the CRISPR/Cas systems offer a wide range of practical tools and innovative concepts. In the battle against plant viruses, host plants and viruses constantly interact and compete with each other. Several challenges arise when implementing CRISPR/Cas9 technology to inhibit the spread of plant DNA viruses. Mutation or recombination may result in inactivation of endogenous genes, while modifications in the plant virus sequence caused by genome editing technologies would likely accelerate viral development (Figure 9.7). Phytopathogenic RNA viruses are primarily driven by mutation, recombination, and re-assortment, with the first two being characteristic of plant DNA viruses.

Mehta & colleagues (2019) discovered no resistance response in transgenic cassava plants created by targeting the overlapping sequences of AC2 and AC3 genes of *African cassava mosaic virus* (ACMV). A closer look at the full-length geminivirus genomes revealed that 33%–48% of the geminivirus sequences were introduced into the transgenic cassava plants. While these new CRISPR/Cas9 ACMV variants cannot multiply on their own, they require the wild-type ACMV to aggregate in *N. benthamiana* (Mehta et al., 2019). It was also demonstrated that CRISPR/Cas9-mediated alteration of viral ORFs resulted in additional viral variants, which could lead to varying levels of viral escape occurrences, increasing the likelihood of mutant virus creation and transmission from plants even with multiple gRNAs (Liu et al., 2018; Ali et al., 2016). However, compared to a single sgRNA, combinations of two sgRNAs, especially those that are distant, would significantly delay resistance breakdown (Liu et al., 2018). When the CRISPR/Cas9 system was used in *A. thaliana* plants, off-target effects were observed, suggesting that this technology may have unintended consequences (Zhang et al., 2018). Additionally, the folding of dsRNA domains in sgRNAs may result in the formation of small interfering RNAs (siRNAs), which could reduce sgRNA abundance. This concern must be addressed in future CRISPR/Cas9

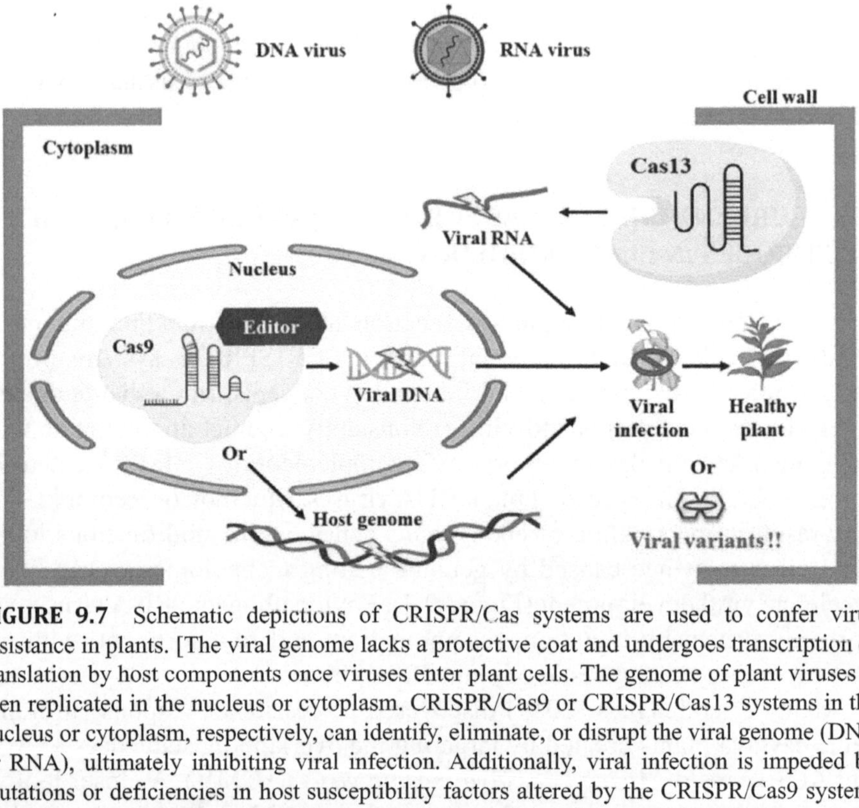


**FIGURE 9.7** Schematic depictions of CRISPR/Cas systems are used to confer virus resistance in plants. [The viral genome lacks a protective coat and undergoes transcription or translation by host components once viruses enter plant cells. The genome of plant viruses is then replicated in the nucleus or cytoplasm. CRISPR/Cas9 or CRISPR/Cas13 systems in the nucleus or cytoplasm, respectively, can identify, eliminate, or disrupt the viral genome (DNA or RNA), ultimately inhibiting viral infection. Additionally, viral infection is impeded by mutations or deficiencies in host susceptibility factors altered by the CRISPR/Cas9 system. However, some novel viral variations may arise as byproducts of CRISPR/Cas systems editing the viral genome, increasing the risk of viral evolution].

*Source:* Adapted and modified from: Cao et al. (2020).

applications. Another key issue is the limited cargo capacity of some viruses (typically <1 kb) for delivering all gene-editing components, including Cas9 (approx. 4.2 kb), as increased cargo can impede systemic movement of the cargo DNA (Shivprasad et al., 1999).

The CRISPR/Cas system should be considered a double-edged sword in antiviral engineering due to the evidence and concerns that viruses may escape from CRISPR/Cas system-mediated antiviral immunity in plants. There is a possibility that new viral variants or species could arise as by-products of genome editing, suggesting that it could accelerate virus evolution or result in transgenic crops losing their specific viral resistance ability, despite its ability to detect, dismantle, or disrupt the viral genome to prevent viral infection.

## 9.7 CONCLUDING REMARKS AND PROSPECTS

The development of CRISPR/Cas technologies would accelerate the synthesis and acquisition of new resistance resources, thereby aiding in the prevention and control of economically significant viral diseases. Molecular breeding has proven to be indispensable in this regard. With the progress and understanding of CRISPR/Cas systems, as well as the expansion and clarification of recessive resistance genes, the use of resistance materials containing mutations or deficiencies in host susceptibility factors generated by CRISPR/Cas systems will become more valuable in preventing viral infections. Instead of targeting, eliminating, or manipulating viral genomes, which could potentially lead to viral evolution, modifying recessive resistance genes could offer a potential approach to designing virus-resistant plants. Furthermore, CRISPR/Cas systems can target host susceptibility genes (S Genes) in insect vectors to inhibit viral transmission and dissemination. It is also worth noting that novel CRISPR/Cas technologies can be employed for antiviral engineering. For instance, the CRISPR Act 2.0-mediated transcriptional activation of several plant-dominant resistance genes enhances plant immunity and provides a wide range of resistance against various diseases (Lowder et al., 2018). Additionally, the use of a primer editor is an appealing option in antiviral engineering (Anzalone et al., 2019). Multiplexed CRISPR techniques, which enable the simultaneous production of multiple gRNAs or Cas proteins, have significantly expanded the scope and effectiveness of genome modification and transcriptional control (McCarty et al., 2020). However, due to practical challenges, such as the limited ability to simultaneously genetically modify multiple Cas proteins into a single plant, as well as the absence of efficient and reliable endogenous targets against various viruses, among other factors, the implementation of these multiplexed CRISPR techniques for disease control in plants is still a distant goal (McCarty et al., 2020).

Finally, numerous efforts have been made in the past decade to develop the CRISPR/Cas-mediated viral resistance system in various crops and model plants. However, this groundbreaking plant virus control technology still needs to be improved in order to provide greater durability and broad-spectrum resistance. The use of multiple combinatorial antiviral techniques may help prevent the formation of CRISPR/Cas escape mutant viruses. By interpreting and elucidating the unique host factors that regulate virus multiplication and targeting these factors using this revolutionary technique, we can propel this technology forward.

## KEYWORDS


- **CRISPR/Cas systems**
- **food security**
- **gene editing**
- **plant genome engineering**
- **plant immune response**
- **plant viruses**
- **plant-virus interactions**
- **viral disease management**
- **virus resistance**


## REFERENCES


Abudayyeh, O. O., Gootenberg, J. S., Essletzbichler, P., Han, S., Joung, J., Belanto, J. J., Verdine, V., et al., (2017). RNA targeting with CRISPR-Cas13. *Nature, 550*, 280–284.

Abudayyeh, O. O., Gootenberg, J. S., Konermann, S., Joung, J., Slaymaker, I. M., Cox, D. B., Shmakov, S., et al., (2016). C2c2 is a single-component programmable RNA-guided RNA-targeting CRISPR effector. *Science, 353*(6299), aaf5573.

Ali, Z., Abulfaraj, A., Idris, A., Ali, S., Tashkandi, M., & Mahfouz, M. M., (2015). CRISPR/Cas9-mediated viral interference in plants. *Genome Biol., 16.*

Ali, Z., Ali, S., Tashkandi, M., Zaidi, S. S. E. A., & Mahfouz, M. M., (2016). CRISPR/Cas9-mediated immunity to *Geminiviruses*: Differential interference and evasion. *Sci. Rep., 6*, 26912.

Ali, Z., Eid, A., Ali, S., & Mahfouz, M. M., (2018). Pea early-browning virus-mediated genome editing via the CRISPR/Cas9 system in *Nicotiana benthamiana* and *Arabidopsis*. *Virus Res., 244*, 333–337.

Aman, R., Ali, Z., & Butt, H., (2018). RNA virus interference via CRISPR/Cas13a system in plants. *Genome Biol., 19*, 1.

Amitai, G., & Sorek, R., (2016). CRISPR–Cas adaptation: Insights into the mechanism of action. *Nat. Rev. Microbiol., 14*, 67–76.

Anzalone, A., Randolph, P., Davis, J., Sousa, A., Koblan, L., Levy, J., Chen, P. J., et al., (2019). Search-and-replace genome editing without double-strand breaks or donor DNA. *Nature, 576*, 149–157.

Aquino-Jarquin, G., (2019). CRISPR-Cas14 is now part of the artillery for gene editing and molecular diagnostic. *Nanomedicine, 18*, 428–431.

Baltes, N. J., Hummel, A. W., Konecna, E., Cegan, R., Bruns, A. N., Bisaro, D. M., & Voytas, D. F., (2015). Conferring resistance to *Geminiviruses* with the CRISPR-Cas prokaryotic immune system. *Nat. Plants., 1.*

Bastet, A., Lederer, B., Giovinazzo, N., Arnoux, X., German-Retana, S., Reinbold, C., Brault, V., et al., (2018). Trans-species synthetic gene design allows resistance pyramiding and broad-spectrum engineering of virus resistance in plants. *Plant Biotechnol. J., 16*, 1569–1581.

Baulcombe, D. C., (1996). Mechanisms of pathogen-derived resistance to viruses in transgenic plants. *Plant Cell, 8*, 1833–1844.

Beachy, R. N., (1997). Mechanisms and applications of pathogen-derived resistance in transgenic plants. *Curr. Opin. Biotechnol., 8*, 215–220.

Bolotin, A., Ouinquis, B., Sorokin, A., & Ehrlich, S. D., (2005). Clustered regularly interspaced short palindrome repeats (CRISPRs) have spacers of extrachromosomal origin. *Microbiology, 151*, 2551–2561.

Cao, Y., Zhou, H., Zhou, X., & Li, F., (2020). Control of plant viruses by CRISPR/Cas system-mediated adaptive immunity. *Front. Microbiol., 11*, 593700.

Cao, Y., Zhou, H., Zhou, X., & Li, F., (2021). Conferring resistance to plant RNA Viruses with the CRISPR/CasRx system. *Virol Sin., 36*(4), 814–817.

Cermák, T., Baltes, N. J., ˇCegan, R., Zhang, Y., & Voytas, D. F., (2015). High-frequency, precise modification of the tomato genome. *Genome Biol., 16*, 232.

Chandrasekaran, J., Brumin, M., Wolf, D., Leibman, D., Klap, C., Pearlsman, M., Sherman, A., et al., (2016). Development of broad virus resistance in non-transgenic cucumber using CRISPR/Cas9 technology. *Mol. Plant Pathol., 17*, 1140–1153.

Chang, H., Yi, B., Ma, R., Zhang, X., Zhao, H., & Xi, Y., (2016). CRISPR/cas9, a novel genomic tool to knock down microRNA *in vitro* and *in vivo*. *Sci. Rep., 6*, 22312.

Char, S. N., Neelakandan, A. K., Nahampun, H., Frame, B., Main, M., Spalding, M. H., Becraft, P. W., et al., (2017). An agrobacterium-delivered CRISPR/Cas9 system for high frequency targeted mutagenesis in maize. *Plant Biotechnol. J., 15*(2), 257–268.

Chen, K., Wang, Y., Zhang, R., Zhang, H., & Gao, C., (2019). CRISPR/Cas genome editing and precision plant breeding in agriculture. *Annu. Rev. Plant Biol., 70*, 667–697.

Christou, P., (2013). Plant genetic engineering and agricultural biotechnology 1983–2013. *Trends Biotechnol., 31*, 125–127.

Cillo, F., & Palukaitis, P., (2014). In: *Transgenic Resistance* (Vol. 90). Academic Press, New York, USA.

Cody, W. B., & Scholthof, H. B., (2019). Plant virus vectors 3.0: Transitioning into synthetic genomics. *Ann. Rev. Phytopathol., 57*, 211–230.

Cody, W. B., Scholthof, H. B., & Mirkov, T. E., (2017). Multiplexed gene editing and protein overexpression using a *Tobacco mosaic virus* viral vector. *Plant Physiol., 175*, 23–35.

Cong, L., Ran, F. A., Cox, D., Lin, S., Barretto, R., Habib, N., Hsu, D. H., et al., (2013). Multiplex genome engineering using CRISPR/Cas systems. *Science, 339*, 819–823.

Cox, D. B. T., Gootenberg, J. S., Abudayyeh, O. O., Franklin, B., Kellner, M. J., Joung, J., & Zhang, F., (2017). RNA editing with CRISPR-Cas13. *Science, 358*, 1019–1027.

Csorba, T., Kontra, L., & Burgyán, J., (2015). Viral silencing suppressors: Tools forged to fine-tune host-pathogen coexistence. *Virology, 479*, 85–103.

De Ronde, D., Butterbach, P., & Kormelink, R., (2014). Dominant resistance against plant viruses. *Front. Plant Sci., 5*, 307.

Donohoue, P. D., Barrangou, R., & May, A. P., (2018). Advances in industrial biotechnology using CRISPR-Cas systems. *Trends Biotechnol., 36*, 134–146.

Doudna, J. A., & Charpentier, E., (2014). The new frontier of genome engineering with CRISPR-Cas9. *Science, 346*, 1258096.

Duan, C. G., Wang, C. H., & Guo, H. S., (2012). Application of RNA silencing to plant disease resistance. *Silence, 3*, 5.

Eid, A., Ali, Z., & Mahfouz, M. M., (2016). High efficiency of targeted mutagenesis in *Arabidopsis* via meiotic promoter-driven expression of Cas9 endonuclease. *Plant Cell Rep., 35*, 1555–1558.

Endo, A., Masafumi, M., Kaya, H., & Toki, S., (2016). Efficient targeted mutagenesis of rice and tobacco genomes using Cpfl from *Francisella novicida*. *Sci. Rep., 6*, 38169.

Fan, D., Liu, T., Li, C., Jiao, B., Li, S., Hou, Y., & Luo, K., (2015). Efficient CRISPR/Cas9-mediated targeted mutagenesis in populus in the first generation. *Sci. Rep., 5*, 12217.

Fauser, F., Schiml, S., & Puchta, H., (2014). Both CRISPR/Cas-based nucleases and nickases can be used efficiently for genome engineering in *Arabidopsis thaliana*. *Plant J., 79*, 348–359.

Feng, C., Su, H., Bai, H., Wang, R., Liu, Y., Guo, X., Liu, C., et al., (2018). High-efficiency genome editing using admc1 promoter controlled CRISPR/Cas9 system in maize. *Plant Biotechnol. J., 16*, 1848–1857.

Fonfara, I., Richter, H., Bratoviˇc, M., Le Rhun, A., & Charpentier, E., (2016). The CRISPR-associated DNA-cleaving enzyme Cpfl also processes precursor CRISPR RNA. *Nature, 532*, 517–521.

Freije, C. A., Myhrvold, C., Boehm, C. K., Lin, A. E., Welch, N. L., Carter, A., Metsky, H. C., et al., (2019). Programmable inhibition and detection of RNA viruses using Cas13. *Mol. Cell., 76*, 826–837.

Gao, J., Wang, G., Ma, S., Xie, X., Wu, X., Zhang, X., Wu, Y., et al., (2015). CRISPR/Cas9-mediated targeted mutagenesis in *Nicotiana tabacum*. *Plant Mol. Biol., 87*, 99–110.

Garcia-Doval, C., & Jinek, M., (2017). Molecular architectures and mechanisms of class 2 CRISPR-associated nucleases. *Curr. Opin. Struct. Biol., 47*, 157–166.

Gilbert, L. A., Larson, M. H., Morsut, L., Liu, Z., Brar, G. A., Torres, S. E., Stern-Ginossar, N., et al., (2013). CRISPR-mediated modular RNA-guided regulation of transcription in eukaryotes. *Cell, 154*, 442–451.

Gomez, M. A., Lin, Z. D., Moll, T., Chauhan, R. D., Hayden, L., Renninger, K., Beyene, G., et al., (2019). Simultaneous CRISPR/Cas9-mediated editing of cassava eIF4E isoforms nCBP-1 and nCBP-2 reduces cassava brown streak disease symptom severity and incidence. *Plant Biotechnol. J., 17*, 421–434.

Hadidi, A., Flores, R., Candresse, T., & Barba, M., (2016). Next-generation sequencing and genome editing in plant virology. *Front. Microbiol., 7*, 1325.

Hameed, A., Shan-E-Ali, Z. S., Sattar, M. N., Iqbal, Z., & Tahir, M. N., (2019). CRISPR technology to combat plant RNA viruses: A theoretical model for *Potato virus Y* (PVY) resistance. *Micr. Pathogen., 133*, 103551.

Hanley-Bowdoin, L., Bejarano, E. R., Robertson, D., & Mansoor, S., (2013). *Geminiviruses*: Masters at redirecting and reprogramming plant processes. *Nat. Rev. Microbiol., 11*, 777–788.

Hannon, G. J., (2002). RNA interference. *Nature, 418*, 244–251.

Harrington, L. B., Burstein, D., Chen, J. S., Paez-Espino, D., Ma, E., Witte, I. P., Cofsky, J. C., et al., (2018). Programmed DNA destruction by miniature CRISPR-Cas14 enzymes. *Science, 362*, 839–842.

Hashimoto, M., Neriya, Y., Yamaji, Y., & Namba, S., (2016). Recessive resistance to plant viruses: Potential resistance genes beyond translation initiation factors. *Front. Microbiol., 7*, 1695.

Hashimoto, R., Ueta, R., Abe, C., Osakabe, Y., & Osakabe, K., (2018). Efficient multiplex genome editing induces precise, and self-ligated type mutations in tomato plants. *Front. Plant Sci., 9*, 00916.

Hess, G. T., Tycko, J., Yao, D., & Bassik, M. C., (2017). Methods and applications of CRISPR-mediated base editing in eukaryotic genomes. *Mol. Cell., 68*, 26–43.

Hille, F., & Charpentier, E., (2016). CRISPR-Cas: Biology, mechanisms, and relevance. *Philos. Trans. R. Soc. Ser. B., 371*, 20150496.

Holmes, F. O., (1938). Inheritance of resistance to tobacco-mosaic disease in tobacco. *Phytopathology, 28*, 553–561.

Hsu, P. D., Scott, D. A., Weinstein, J. A., Ran, F. A., Konermann, S., Agarwala, V., Li, Y., et al., (2013). DNA targeting specificity of RNA-guided Cas9 nucleases. *Nat. Biotechnol., 31*, 827–832.

Hu, X., Wang, C., Fu, Y., Liu, Q., Jiao, X., & Wang, K., (2016). Expanding the range of CRISPR/Cas9 genome editing in rice. *Mol. Plant, 9*, 943–945.

Hull, R., (2001). *Matthews' Plant Virology* (4th edn.). Academic Press: New York, NY, USA.

Hyun, Y., Kim, J., Cho, S. W., Choi, Y., Kim, J. S., & Coupland, G., (2015). Site-directed mutagenesis in *Arabidopsis thaliana* using dividing tissue-targeted RGEN of the CRISPR/Cas system to generate heritable null alleles. *Planta, 241*, 271–284.

ICTV. Available online: https://talk.ictvonline.org/taxonomy/ (accessed on 28 February 2024).

Ishino, Y., Shinagawa, H., Makino, K., Amemura, M., & Nakata, A., (1987). Nucleotide sequence of the IAP gene, responsible for alkaline phosphatase isozyme conversion in *Escherichia coli*, and identification of the gene product. *J. Bacteriol., 169*, 5429–5433.

Jansen, R., Embden, J. D. A. V., Gaastra, W., & Schouls, L. M., (2002). Identification of genes that are associated with DNA repeats in prokaryotes. *Mol. Microbiol., 43*, 1565–1575.

Ji, X., Wang, D., & Gao, C., (2019). CRISPR editing-mediated antiviral immunity: A versatile source of resistance to combat plant virus infections. *Sci. China Life Sci., 62*, 1246–1249.

Ji, X., Zhang, H., Zhang, Y., Wang, Y., & Gao, C., (2015). Establishing a CRISPR–Cas-like immune system conferring DNA virus resistance in plants. *Nat. Plants, 1*, 15144.

Jinek, M., Chylinski, K., Fonfara, I., Hauer, M., Doudna, J. A., & Charpentier, E., (2012). A programmable dual-RNA–guided DNA endonuclease in adaptive bacterial immunity. *Science, 337*, 816–821.

Jones, J. D., & Dangl, J. L., (2006). The plant immune system. *Nature, 444*, 323–329.

Kaboli, S., & Babazada, H., (2018). CRISPR mediated genome engineering and its application in industry. *Curr. Issues Mol. Biol., 26*, 81–92.

Kalinina, N. O., Khromov, A., & Love, A. J., (2020). CRISPR applications in plant virology: Virus resistance and beyond. *Phytopathology, 110*, 18–28.

Kaya, H., Ishibashi, K., & Toki, S., (2017). A split *Staphylococcus aureus* Cas9 as a compact genome-editing tool in plants. *Plant Cell Physiol., 58*, 643–649.

Kaya, H., Mikami, M., Endo, A., Endo, M., & Toki, S., (2016). Highly specific targeted mutagenesis in plants using *Staphylococcus aureus* Cas9. *Sci. Rep., 6*, 26871.

Khan, S., Mahmood, M. S., Rahman, S. U., Rizvi, F., & Ahmad, A., (2020). Evaluation of the CRISPR/Cas9 system for the development of resistance against *Cotton leaf curl virus* in model plants. *Plant Protect. Sci., 56*, 154–162.

Khan, Z. A., Kumar, R., & Dasgupta, I., (2022). CRISPR/Cas-mediated resistance against viruses in plants. *Int. J. Mol. Sci., 23*, 2303.

Kim, H., Kim, S. T., Ryu, J., Kang, B. C., Kim, J. S., & Kim, S. G., (2017). CRISPR/Cpf1-mediated DNA-free plant genome editing. *Nat. Commun., 8*, 14406.

Kurokawa, S., Rahman, H., Yamanaka, N., Ishizaki, C., Islam, S., Aiso, T., Hirata, S., et al., (2021). A simple heat treatment increases SpCas9-mediated mutation efficiency in *Arabidopsis*. *Plant Cell Physiol., 62*(11), 1676–1686.

Langner, T., Kamoun, S., & Belhaj, K., (2018). CRISPR crops: Plant genome editing toward disease resistance. *Annu. Rev. Phytopathol., 56*, 479–512.

Lee, K., Zhang, Y., Kleinstiver, B. P., Guo, J. A., Aryee, M. J., Miller, J., Malzahn, A., et al., (2019). Activities and specificities of CRISPR/Cas9 and Cas12a nucleases for targeted mutagenesis in maize. *Plant Biotechnol. J., 17*, 362–372.

Li, C. Zong, Y., Wang, Y., Jin, S., Zhang, D., Song, Q., Zhang, R., & Gao, C., (2018). Expanded base editing in rice and wheat using a Cas9-adenosine deaminase fusion. *Genome Biol., 19*, 59.

Li, F., & Wang, A., (2019). RNA-targeted antiviral immunity: More than just RNA silencing. *Trends Microbiol., 27*, 792–805.

Li, G., Liu, R., Xu, R., Varshney, R. K., Ding, H., Li, M., Yan, X., et al., (2022). Development of an agrobacterium-mediated CRISPR/Cas9 system in pea (*Pisum sativum* L.), *The Crop Journal*. In-press. https://doi.org/10.1016/j.cj.2022.04.011.

Li, R., Fu, D., Zhu, B., Luo, Y., & Zhu, H., (2018). CRISPR/Cas9-mediated mutagenesis of lncRNA1459 alters tomato fruit ripening. *Plant J., 94*, 513–524.

Liang, W., Wersch, S., Tong, M., & Li, X., (2018). TIR-NB-LRR immune receptor SOC3 pairs with truncated TIR-NB protein CHS1 or TN2 to monitor the homeostasis of E3 ligase SAUL1. *New Phytol., 221*, 2054–2066.

Liu, H., Soyars, C., Li, J., Fei, Q., He, G., Peterson, B. A., Meyers, B. C., et al., (2018). CRISPR/Cas9- mediated resistance to cauliflower mosaic virus. *Plant Direct., 2*, e00047.

Liu, X., Xie, C., Si, H., & Yang, J., (2017). CRISPR/Cas9-mediated genome editing in plants. *Methods, 121, 122*, 94–102.

Lomonossoff, G. P., (1995). Pathogen-derived resistance to plant viruses. *Annu. Rev. Phytopathol., 33*, 323–343.

Lowder, L., Zhou, J., Zhang, Y., Malzahn, A., Zhong, Z., Hsieh, T. F., Voytas, D. F., et al., (2018). Robust transcriptional activation in plants using multiplexed CRISPR-Act2.0 and mTALE-act systems. *Mol. Plant., 11*, 245–256.

Ma, X., Zhang, Q., Zhu, Q., Liu, W., Chen, Y., Qiu, R., Wang, B., et al., (2015). A robust CRISPR/Cas9 system for convenient, high-efficiency multiplex genome editing in monocot and dicot plants. *Mol. Plant., 8*, 1274–1284.

Ma, X., Zhu, Q., Chen, Y., & Liu, Y., (2016). CRISPR/Cas9 platforms for genome editing in plants: Developments and applications. *Mol. Plant, 9*, 961–974.

Macovei, A., Sevilla, N. R., Cantos, C., Jonson, G. B., Slamet-Loedin, I., Cˇermák, T., Voytas, D. F., Choi, I. R., & Chadha-Mohanty, P., (2018). Novel alleles of rice eIF4G generated by CRISPR/Cas9-targeted mutagenesis confer resistance to *Rice tungro spherical virus*. *Plant Biotechnol. J., 16*, 1918–1927.

Maeder, M. L., Linder, S. J., Cascio, V. M., Fu, Y., Ho, Q. H., & Joung, J. K., (2013). CRISPR RNA–guided activation of endogenous human genes. *Nat. Meth., 10*, 977–979.

Mahas, A., & Mahfouz, M., (2018). Engineering virus resistance via CRISPR-Cas systems. *Curr. Opin. Virol., 32*, 1–8.

Mahas, A., Ali, Z., Tashkandi, M., & Mahfouz, M. M., (2019a). Virus-mediated genome editing in plants using the CRISPR/Cas9 system. In: Qi, Y., (ed.), *Plant Genome Editing with CRISPR Systems. Methods in Molecular Biology* (Vol. 1917). Humana Press, New York, USA.

Mahas, A., Aman, R., & Mahfouz, M., (2019b). CRISPR-Cas13d mediates robust RNA virus interference in plants. *Gen. Biol., 20*, 1–16.

Makarova, K. S., Wolf, Y. I., Alkhnbashi, O. S., Costa, F., Shah, S. A., Saunders, S. J., Barrangou, R., et al., (2015). An updated evolutionary classification of CRISPR-Cas systems. *Nat. Rev. Microbiol., 13*, 722–736.

Makarova, K. S., Zhang, F., & Koonin, E. V., (2017a). SnapShot: Class 1 CRISPR/Cas systems. *Cell, 168*, 946–946.e1.

Makarova, K. S., Zhang, F., & Koonin, E. V., (2017b). SnapShot: Class 2 CRISPR/Cas systems. *Cell, 168*, 328–328.e1.

Mao, Y., Yang, X., Zhou, Y., Zhang, Z., Botella, J. R., & Zhu, J. K., (2018). Manipulating plant RNA-silencing pathways to improve the gene-editing efficiency of CRISPR/Cas9 systems. *Genome Biol., 19*, 149.

Mao, Y., Zhang, Z., Feng, Z., Wei, P., Zhang, H., Botella, J. R., & Zhu, J. K., (2016). Development of germ-line-specific CRISPR-Cas9 systems to improve the production of heritable gene modifications in *Arabidopsis*. *Plant Biotechnol. J., 14*, 519–532.

Marraffini, L. A., (2016). Chapter 11. In: Ferretti, J. J., Stevens, D. L., & Fischetti, V. A., (eds.), *Streptococcus Pyogenes: Basic Biology to Clinical Manifestations [Internet]*. University of Oklahoma Health Sciences Center, Oklahoma City.

McCarty, N. S., Graham, A. E., Studená, L., & Ledesma-Amaro, R., (2020). Multiplexed CRISPR technologies for gene editing and transcriptional regulation. *Nat. Commun., 11*, 1281.

Mehta, D., Stürchler, A., Anjanappa, R., Zaidi, S., Hirsch-Hoffmann, M., Gruissem, W., & Vanderschuren, H., (2019). Linking CRISPR-Cas9 interference in cassava to the evolution of editing-resistant *Geminiviruses*. *Genome Biol., 20*, 80.

Mikami, M., Toki, S., & Endo, M., (2016). Precision targeted mutagenesis via Cas9 paired nickases in rice. *Plant Cell Physiol., 57*, 1058–1068.

Moffat, A. S., (1999). *Geminiviruses* emerge as serious crop threat. *Science, 286*, 1835.

Mojica, F. J. M., Díez-Villaseñor, C. S., García-Martínez, J. S., & Soria, E., (2005). Intervening sequences of regularly spaced prokaryotic repeats derive from foreign genetic elements. *J. Mol. Evol., 60*, 174–182.

Nakayasu, M., Akiyama, R., Lee, H. J., Osakabe, K., Osakabe, Y., Watanabe, B., Sugimoto, Y., et al., (2018). Generation of α-solanine-free hairy roots of potato by CRISPR/Cas9 mediated genome editing of the St16DOX gene. *Plant Physiol. Biochem., 131*, 70–77.

Nam, K. H., Haitjema, C., Liu, X., Ding, F., Wang, H., DeLisa, M. P., & Ke, A., (2012). Cas5d protein processes pre-crRNA and assembles into a cascade-like interference complex in subtype I-C/Dvulg CRISPR-Cas system. *Structure, 20*, 1574–1584.

Nicaise, V., (2014). Crop immunity against viruses: Outcomes and future challenges. *Front. Plant Sci., 5*, 660.

Nonaka, S., Arai, C., Takayama, M., Matsukura, C., & Ezura, H., (2017). Efficient increase of γ-aminobutyric acid (GABA) content in tomato fruits by targeted mutagenesis. *Sci. Rep., 7*, 1–14.

O'Connell, M. R., Oakes, B. L., Sternberg, S. H., East-Seletsky, A., Kaplan, M., & Doudna, J. A., (2014). Programmable RNA recognition and cleavage by CRISPR/Cas9. *Nature, 516*, 263–266.

Osakabe, Y., Watanabe, T., Sugano, S. S., Ueta, R., Ishihara, R., Shinozaki, K., & Osakabe, K., (2016). Optimization of CRISPR/Cas9 genome editing to modify abiotic stress responses in plants. *Sci. Rep., 6*, 26685.

Pasin, F., Menzel, W., & Daròs, J. A., (2019). Harnessed viruses in the age of metagenomics and synthetic biology: An update on infectious clone assembly and biotechnologies of plant viruses. *Plant Biotechnol. J., 17*, 1010–1026.

Piatek, A., Ali, Z., Baazim, H., Li, L., Abulfaraj, A., Al-Shareef, S., Aouida, M., & Mahfouz, M. M., (2015). RNA-guided transcriptional regulation in planta via synthetic dCas9-based transcription factors. *Plant Biotechnol. J., 13*, 578–589.

Pourcel, C., Salvignol, G., & Vergnaud, G., (2005). CRISPR elements in *Yersinia pestis* acquire new repeats by preferential uptake of bacteriophage DNA and provide additional tools for evolutionary studies. *Microbiology, 151*, 653–663.

Powell, A. P., Nelson, R. S., De, B., Hoffman, N., Rogers, S. G., Fraley, R. T., & Beachy, R. N., (1986). Delay of disease development in transgenic plants that express the *Tobacco mosaic virus* coat protein gene. *Science, 232*, 738–743.

Price, A. A., Sampson, T. R., Ratner, H. K., Grakoui, A., & Weiss, D. S., (2015). Cas9-mediated targeting of viral RNA in eukaryotic cells. *Proc. Natl. Acad. Sci. USA., 112*, 6164–6169.

Pyott, D. E., Sheehan, E., & Molnar, A., (2016). Engineering of CRISPR/Cas9-mediated potyvirus resistance in transgene-free *Arabidopsis* plants. *Mol. Plant Pathol., 17*, 1276–1288.

Qi, W., Zhu, T., Tian, Z., Li, C., Zhang, W., & Song, R., (2016). High-efficiency CRISPR/Cas9 multiplex gene editing using the glycine tRNA-processing system-based strategy in maize. *BMC Biotechnol., 16*, 58.

Rouillon, C., Zhou, M., Zhang, J., Politis, A., Beilsten-Edmands, V., Cannone, G., Graham, S., et al., (2013). Structure of the CRISPR interference complex CSM reveals key similarities with cascade. *Mol. Cell., 52*(1), 124–134.

Roy, A., Zhai, Y., Ortiz, J., Neff, M., Mandal, B., Mukherjee, S. K., & Pappu, H. R., (2019). Multiplexed editing of a *begomovirus* genome restricts escape mutant formation and disease development. *PLoS ONE, 14*, e0223765.

Rybicki, E. P., (2019). CRISPR-Cas9 strikes out in cassava. *Nat. Biotechnol., 37*, 727–728.

Sanfaçon, H., (2015). Plant translation factors and virus resistance. *Viruses, 7*, 3392–3419.

Sanford, J. C., & Johnston, S. A., (1985). The concept of pathogen derived resistance. *J. Theor. Biol., 113*, 395–405.

Sastry, S. K., & Zitter, T. A., (2014). Management of virus and viroid diseases of crops in the tropics. In: *Plant Virus and Viroid Diseases in the Tropics: Epidemiology and Management* (Vol. 2, pp. 149–480). Dordrecht, Springer.

Savage, D. F., (2019). *Cas14: Big Advances from Small CRISPR Proteins*. ACS Publications, Washington, DC, USA.

Schmidt, C., Pacher, M., & Puchta, H., (2019). DNA break repair in plants and its application for genome engineering. In: Kumar, S., Barone, P., & Smith, M., (eds.), *Transgenic Plants* (pp. 237–266). Humana Press, New York, USA.

Scholthof, K. B., (2017). Spicing up the N gene: F. O. Holmes and *Tobacco mosaic virus* resistance in *Capsicum* and *Nicotiana* plants. *Phytopathology., 107*, 148–157.

Seth, K., & Harish, (2016). Current status of potential applications of repurposed Cas9 for structural and functional genomics of plants. *Biochem. Biophys. Res. Commun., 480*, 499–507.

Shahid, M. S., Sattar, M. N., Iqbal, Z., Raza, A., & Al-Sadi, A. M., (2021). Next-generation sequencing and the CRISPR-Cas nexus: A molecular plant virology perspective. *Front. Microbiol., 11*, 609376.

Shan, Q., Wang, Y., Li, J., Zhang, Y., Chen, K., Liang, Z., Zhang, K., et al., (2013). Targeted genome modification of crop plants using a CRISPR-Cas system. *Nat. Biotechnol., 31*, 686–688.

Sharma, S. K., Gupta, O. P., Pathaw, N., Sharma, D., Maibam, A., Sharma, P., Sanasam, J., et al., (2021). CRISPR-Cas-led revolution in diagnosis and management of emerging plant viruses: New avenues toward food and nutritional security. *Front. Nutr., 8*, 751512.

Shimatani, Z., Kashojiya, S., Takayama, M., Terada, R., Arazoe, T., Ishii, H., Teramura, H., et al., (2017). Targeted base editing in rice and tomato using a CRISPR-Cas9 cytidine deaminase fusion. *Nat. Biotechnol., 35*, 441–443.

Shivprasad, S., Pogue, G. P., Lewandowski, D. J., Hidalgo, J., Donson, J., Grill, L. K., & Dawson, W. O., (1999). Heterologous sequences greatly affect foreign gene expression in tobacco mosaic virus-based vectors. *Virology, 255*, 312–323.

Shmakov, S., Smargon, A., Scott, D., Cox, D., Pyzocha, N., Yan, W., Abudayyeh, O. O., et al., (2017). Diversity and evolution of class 2 CRISPR-Cas systems. *Nat. Rev. Microbiol., 15*, 169–182.

Steinert, J., Schiml, S., Fauser, F., & Puchta, H., (2015). Highly efficient heritable plant genome engineering using Cas9 orthologues from *Streptococcus thermophiles* and *Staphylococcus aureus*. *Plant J., 84*, 1295–1305.

Sternberg, S. H., Richter, H., Charpentier, E., & Qimron, U., (2016). Adaptation in CRISPR-Cas systems. *Mol. Cell, 61*, 797–808.

Taliansky, M., Samarskaya, V., Zavriev, S. K., Fesenko, I., Kalinina, N. O., & Love, A. J., (2021). RNA-based technologies for engineering plant virus resistance. *Plants, 10*, 82.

Tang, X., Lowder, L. G., Zhang, T., Malzahn, A. A., Zheng, X., Voytas, D. F., Zhong, Z., et al., (2017). A CRISPR–Cpf1 system for efficient genome editing and transcriptional repression in plants. *Nat. Plants., 3*, 17018.

Tashkandi, M., Ali, Z., Aljedaani, F., Shami, A., & Mahfouz, M. M., (2018). Engineering resistance against *Tomato yellow leaf curl virus* via the CRISPR/Cas9 system in tomato. *Plant Signal Behav., 13*(10), e1525996.

Tsutsui, H., & Higashiyama, T., (2017). pKAMA-ITACHI vectors for highly efficient CRISPR/Cas9-mediated gene knockout in *Arabidopsis thaliana*. *Plant Cell Physiol., 58*, w191.

Van, R. M. H. V., Fauquet, C. M., Bishop, D. H. L., Carsten, E. B., Estes, M. K., Lemon, S. M., Maniloff, J., et al., (2000). *Virus Taxonomy Seventh Report of the International Committee on Taxonomy of Viruses*. Academic Press, San Diego, CA, USA.

Varanda, C. M., Félix, M. D. R., Campos, M. D., Patanita, M., & Materatski, P., (2021). Plant viruses: From targets to tools for CRISPR. *Viruses, 13*(1), 141.

Voinnet, O., (2005). Induction and suppression of RNA silencing: Insights from viral infections. *Nat. Rev. Genet., 6*, 206–220.

Vu, G. T., Cao, H. X., Fauser, F., Reiss, B., Puchta, H., & Schubert, I., (2017). Endogenous sequence patterns predispose the repair modes of CRISPR-Cas9-induced DNA double-stranded breaks in *Arabidopsis thaliana*. *Plant J., 92*, 57–67.

Wang, H., La Russa, M., & Qi, L. S., (2016). CRISPR/Cas9 in genome editing and beyond. *Annu. Rev. Biochem., 85*, 227–264.

Wang, T., Zhang, H., & Zhu, H., (2019). CRISPR technology is revolutionizing the improvement of tomato and other fruit crops. *Hortic Res., 6*, 77.

Wang, Z., Xing, H., Dong, L., Zhang, H., Han, C., Wang, X., & Chen, Q., (2015). Egg cell-specific promoter-controlled CRISPR/Cas9 efficiently generates homozygous mutants for multiple target genes in *Arabidopsis* in a single generation. *Genome Biol., 16*, 144.

Wessels, H. H., Méndez-Mancilla, A., Guo, X., Legut, M., Daniloski, Z., & Sanjana, N. E., (2020). Massively parallel Cas13 screens reveal principles for guide RNA design. *Nat. Biotechnol., 38*, 722–727.

Wilson, T. M. A., (1993). Strategies to protect crop plants against viruses: Pathogen-derived resistance blossoms. *Proc. Natl. Acad. Sci. USA., 90*, 3134–3141.

Wittmann, S., Chatel, H., Fortin, M. G., & Laliberté, J. F., (1997). Interaction of the viral protein genome linked of *Turnip mosaic potyvirus* with the translational eukaryotic initiation factor (iso) 4E of *Arabidopsis thaliana* using the yeast two-hybrid system. *Virology, 234*, 84–92.

Wolter, F., Klemm, J., & Puchta, H., (2018). Efficient in planta gene targeting in *Arabidopsis* using egg cell-specific expression of the Cas9 nuclease of *Staphylococcus aureus. Plant J., 94*, 735–746.

Xie, K., Minkenberg, B., & Yang, Y., (2015). Boosting CRISPR/Cas9 multiplex editing capability with the endogenous tRNA-processing system. *Proc. Natl. Acad. Sci. USA., 112*, 3570–3575.

Xu, P., Su, H., Chen, W., & Lu, P., (2018). The application of a meiocyte-specific CRISPR/Cas9 (MSC) system and a suicide-MSC system in generating inheritable and stable mutations in *Arabidopsis. Front. Plant Sci., 9*, 01007.

Yan, L., Wei, S., Wu, Y., Hu, R., Li, H., Yang, W., & Xie, Q., (2015). High-efficiency genome editing in *Arabidopsis* using YAO promoter-driven CRISPR/Cas9 system. *Mol. Plant, 8*, 1820–1823.

Yan, W. X., Chong, S., Zhang, H., Makarova, K. S., Koonin, E. V., Cheng, D. R., & Scott, D. A., (2018). Cas13d is a compact RNA-targeting type VI CRISPR effector positively modulated by a WYL-domain-containing accessory protein. *Mol. Cell, 70*, 327–339.e5.

Yin, K., Gao, C., & Qiu, J., (2017). Progress and prospects in plant genome editing. *Nat. Plants, 3*, 17107.

Yu, T., Tzeng, D. T. W., Li, R., Chen, J., Zhong, S., Fu, D., Zhu, B., et al., (2019). Genome-wide identification of long non-coding RNA targets of the tomato MADS box transcription factor RIN and function analysis. *Ann. Bot., 123*, 469–482.

Zaidi, S. S. A., & Mansoor, S., (2017). Viral vectors for plant genome engineering. *Front. Plant Sci., 8*, 539.

Zaidi, S. S. A., Tashkandi, M., Mahfouz, M. M., & Zaidi, S. S. A., (2017). Engineering molecular immunity against plant viruses. *Prog. Mol. Biol. Transl. Sci., 149*, 167–186.

Zaidi, S. S. A., Tashkandi, M., Mansoor, S., & Mahfouz, M. M., (2016). Engineering plant immunity: Using CRISPR/Cas9 to generate virus resistance. *Front. Plant Sci., 7*, 1673.

Zetsche, B., Gootenberg, J. S., Abudayyeh, O. O., Slaymaker, I. M., Makarova, K. S., Essletzbichler, P., Volz, S. E., et al., (2015). Cpf1 is a single RNA-guided endonuclease of a class 2 CRISPR-Cas system. *Cell, 163*, 759–771.

Zhan, X., Zhang, F., Zhong, Z., Chen, R., Wang, Y., Chang, L., Bock, R., et al., (2019). Generation of virus-resistant potato plants by RNA genome targeting. *Plant Biotechnol J., 17*, 1814–1822.

Zhang, F., LeBlanc, C., Irish, V. F., & Jacob, Y., (2017b). Rapid and efficient CRISPR/Cas9 gene editing in citrus using the YAO promoter. *Plant Cell Rep., 36*, 1883–1887.

Zhang, F., Wang, L., Qi, B., Zhao, B., Ko, E. E., Riggan, N. D., Chin, K., & Qiao, H., (2017a). EIN2 mediates direct regulation of histone acetylation in the ethylene response. *Proc. Natl. Acad. Sci. USA., 114*, 10274–10279.

Zhang, Q., Xing, H. L., Wang, Z. P., Zhang, H. Y., Yang, F., Wang, X. C., & Chen, Q. J., (2018). Potential high-frequency off-target mutagenesis induced by CRISPR/Cas9 in *Arabidopsis* and its prevention. *Plant Mol. Biol., 96*, 444–456.

Zhang, T., Zhao, Y., Ye, J., Cao, X., Xu, C., Chen, B., An, H., et al., (2019). Establishing CRISPR/Cas13a immune system conferring RNA virus resistance in both dicot and monocot plants. *Plant Biotechnol J., 17*, 1185–1187.

Zhang, T., Zheng, Q., Yi, X., An, H., Zhao, Y., Ma, S., & Zhou, G., (2018). Establishing RNA virus resistance in plants by harnessing CRISPR immune system. *Plant Biotechnol J., 16*, 1415–1423.

Zhao, Y., Yang, X., Zhou, G., & Zhang, T., (2020). Engineering plant virus resistance: From RNA silencing to genome editing strategies. *Plant Biotechnol. J., 18*, 328–336.

Zong, Y., Wang, Y., Li, C., Zhang, R., Chen, K., Ran, Y., Qiu, J. L., et al., (2017). Precise base editing in rice, wheat, and maize with a Cas9- cytidine deaminase fusion. *Nat. Biotechnol., 35*, 438–440.

## CHAPTER 10

# Biological and Anthropogenic Risk Factors Favoring Persistence and Spread of Harmful Plant Viruses

OLEKSIY SHEVCHENKO, HALYNA SNIHUR, TETIANA SHEVCHENKO, and IRENA BUDZANIVSKA

*Virology Department, ESC 'Institute of Biology and Medicine,' Taras Shevchenko National University of Kyiv, Kyiv, Ukraine*

## ABSTRACT

This chapter reviews different risk factors that are or may be important for the emergence and establishment of plant virus infections in a new environment or host(s), as well as their subsequent spread and reoccurrence, with a focus on Ukraine. In this context, we cover virus properties, transmission routes including vector- or seed-borne spread, host range, availability of wild-growing reservoir plants, and the effect of some anthropogenic factors (i.e., heavy metals and radioactivity).

To showcase this, we analyzed particular plant pathogens that are of specific importance to Ukraine and suggested major issues that may contribute to the emergence or further successful persistence of viral diseases in important crops. Additionally, we also demonstrate that abiotic stressors (heavy metals and radionuclides) may be essential drivers of virus abundance in ecosystems, as well as higher incidence rates of plant infection. The main risk factors favoring the appearance and spread of viruses in Ukraine are summarized.

## 10.1 INTRODUCTION

Plant viruses pose a significant economic factor and cause a notable reduction in crop yields (15–100%, 20–40% on average) (Polischuk et al., 2001).

---

Detection and Management of New and Emerging Mystery Plant Virus Sources.
Alireza Golnaraghi, Rajarshi Kumar Gaur, and Rakesh Kumar Verma (Eds.)

Ukraine is one of the leading producers of agricultural production. According to indirect estimates, the total loss of Ukraine from viral infections of plants could reach 10 billion dollars per year. Thus, the control of viral infections is of strategic importance for sustainable plant production, human nutrition, and the biosafety of the state.

The biosafety of the state is determined by its ability to withstand biological threats. The results of recent research indicate the discovery of new and potentially widespread modified viruses of strategically important crops (cereals, legumes, vegetables, etc.) in Ukraine. Plant viral diseases are incurable, and viruses are the main cause of plant production loss. Viruses are a major bio-barrier that is currently uncontrolled and requires the development of new effective approaches to manage existing risks, especially taking into account the impact of abiotic and anthropogenic factors, as well as the overall decline in crop yields.

Managing the risks of viral infections has always attracted attention, as they are the cause of significant crop losses and the only example of incurable diseases. The logical conclusion is the need to prevent the emergence or development of plant diseases. There are two main approaches to controlling viral plant diseases. The first is the systematic monitoring of plant viral infections for viruses, which involves regular surveys of large areas, selection of large numbers of plant samples, and their analysis for diagnostics of various viral diseases (Polischuk et al., 2001). Currently, there are only a few scientific groups in the world that engage in comprehensive monitoring of crop viruses due to the high cost of such research. One of the few current examples is the group led by Cécile Desbiez (INRA, France) (Desbiez et al., 2020), whose efforts focus on studying particularly dangerous viruses in the Mediterranean because of the region's economic importance for growing vegetables. The main disadvantage of this approach is the large amount of work that requires financial support and a network of specialized agencies. The situation is complicated by the fact that the occurrence of many unknown plant viruses has been confirmed recently, for which no diagnostic tools have been developed yet (MacDiarmid et al., 2013; Dutta et al., 2014; Melcher et al., 2014). The second approach is to mathematically model the development of viral infection. Recently, foreign scientific groups have published a small number of papers predicting epidemics of some viral infections, including those transmitted by vectors and/or seeds (Thackray et al., 2004; Jones, 2006; Maling et al., 2008; Makkouk & Kumari, 2009; Wosula et al., 2018; Singh et al., 2018). Common to such works is the involvement of various parameters and statistical methods in an attempt to model the occurrence and patterns of development of viral plant diseases (Congdon et al., 2017; Gamarra et al., 2020; Van den Bosch & Jeger, 2017). All these sparse models require prior knowledge of the range of viruses relevant to specific ecosystems, are difficult to create, and are suitable only for a specific virus, taking into account the relevant environmental conditions. In our opinion, in Ukraine and most other countries, the existing approaches are not viable in modern realities due to the high cost of total monitoring and lack of proper infrastructure, lack of information on the current profile of viral pathogens, and narrow specialization of models.

During the past few years, we have conducted studies on several viruses affecting commercially important plants. These studies have led to the identification and characterization of new and dangerous pathogens for Ukraine. Some of these viruses include wheat dwarf virus (Tóbiás et al., 2011), High Plains wheat mosaic virus (HPWMoV), turnip mosaic virus (TuMV), zucchini yellow mosaic virus (ZYMV), watermelon mosaic virus (WMV), onion yellow dwarf virus (OYDV), leek yellow stripe virus (LYSV), raspberry leaf blotch virus (RLBV), and sugarcane mosaic virus (SCMV) (Shevchenko et al., 2018; Sherevera et al., 2019; Pozhylov et al., 2018; Snihur et al., 2019, 2020, 2021). Additionally, quarantine viruses such as tomato spotted wilt virus (TSWV) (Dikova et al., 2016) and plum pox virus (PPV) have been detected (Pavlova et al., 2017; Kutsenko et al., 2019).

Our research has laid the foundation for studying the epidemic role of environmental factors, confirming that anthropogenic activities can increase the aggressiveness of viruses (Shevchenko et al., 2004, 2007; Iutinska et al., 2012). The results highlight the growing threat to profitable cultivation of existing and promising crops, the ineffectiveness of current control measures, and the lack of information regarding the current list and actual spread of dangerous crop viruses. These crops are strategically important for the security of our state and play a significant role in its budget. Therefore, it is crucial to urgently conduct sample screening studies to identify biological and epidemiological risk factors, as well as to determine the patterns and mechanisms of emergence and spread of dangerous viruses in Ukrainian agroecosystems in accordance with the International Plant Protection Convention (IPCC).

Compiling a list of evolutionarily 'plastic' harmful viruses relevant to Ukraine's agroecosystems and studying their biological properties will offer specialized methods of counteraction based on the analysis of risk factors inherent in specific viral infections. This will allow for the formulation of principles of biological and environmental risk management as part of Ukraine's biosafety.

The purpose of this work is to identify risk factors and justify approaches to the management of dangerous viral diseases of plants, taking into account their biological characteristics and epidemic potential in ecosystems of Ukraine under different conditions.

The following is based on the data of plant virus monitoring or screening in Ukraine under varying climatic and ecological conditions. However, in our view, this knowledge can be applied to many other countries or regions as well, especially those lacking proper plant health surveillance systems.

## 10.2 EXISTING AND (RE)EMERGING VIRAL THREATS IN UKRAINE

Due to the number of reasons discussed above and below, our current knowledge of the list of harmful viruses endangering the production of important crops is rather limited. We have vast retrospective data dating back to the USSR times, which poorly correlates with modern diagnostic techniques and does not reflect the state-of-the-art situation.

This has also led to the recent discovery of many novel viruses over the last 5–10 years, including pathogens of cereals, vegetable and fruit crops, as well as weeds. Additionally, some widespread plant species have been identified as new hosts for previously known viruses.

We will focus on several viruses of major interest that either persist in Ukrainian ecosystems or can be considered emerging pathogens in subsections.

### *10.2.1 WHEAT STREAK MOSAIC VIRUS (WSMV) IN CEREALS*

The economic significance of viral diseases in cereals is hard to overestimate, as yield losses can reach 80–90% during epidemics. Virus infections affect and redirect physiological and biochemical processes in grain plants. Infected cereals demonstrate decreased stooling productivity, reduced content of chlorophyll and carotenoids, abnormal seed formation, and impairment of seed quality (smaller number of seeds in spikes, reduced seed weight, sharp decrease in gluten content).

In Ukraine, viral diseases of cereals have been repeatedly described by many authors since the 1960s. Early studies focused on the biological properties of infectious agents, their transmission and spread, as well as the visual appearance of the disease on different crops and their virulence. Today, research is mostly centered around virus diagnostics, the effect of viral diseases on plant physiology, and their prevention (Mishchenko et al., 2021a).

According to available data, at least a dozen different viruses (including soil-borne pathogens) have circulated or were circulating in various regions of Ukraine. These include barley yellow dwarf viruses (BYDVs), wheat streak mosaic virus (WSMV), barley stripe mosaic virus (BSMV), brome mosaic virus (BMV), as well as the potentially harmful barley yellow mosaic virus (BYMV) transmitted via the fungi-like organism *Polymyxa graminis*. Additionally, there have been reports of the occurrence of winter wheat (Russian) mosaic virus and wheat dwarf virus (WDV) transmitted by the leafhopper *Psammotettix alienus*. In light of this, we conducted a 15-year monitoring of major cereal viruses in different regions of Ukraine, and the results obtained suggest that BYDV-PAV, WSMV, and WDV have been the most harmful and commercially important viral pathogens endangering cereal cultivation in Ukraine for years (Snihur et al., 2018).

Interestingly, the BYDV virus complex (mostly BYDV-PAV) has long been considered a major virus issue with cereals in Ukraine, claiming to infect approximately 50–70% of symptomatic plants. However, since the beginning of the 21st century, WSMV has started to efficiently 'replace' this established dominator and has now reached a threshold of over half of the mosaic disease cases in cereals (Figure 10.1).

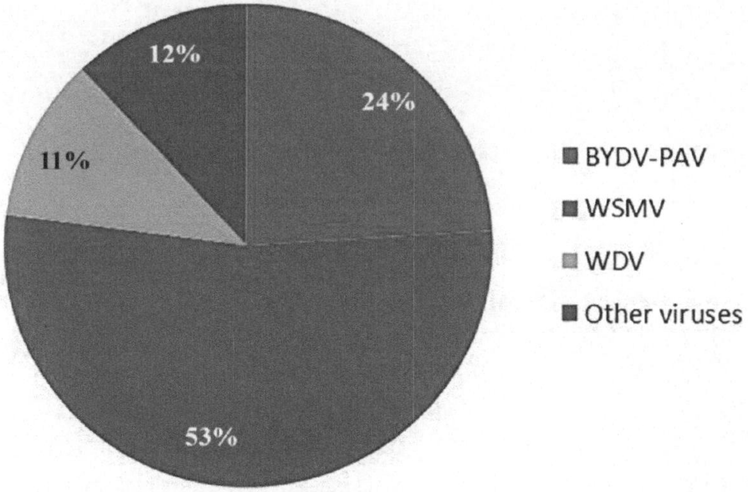


**FIGURE 10.1** Averaged data on the current occurrence of different cereal viruses in Ukraine shows the dominance of wheat streak mosaic virus (2002–2016). [BYDV-PAV: Barley yellow dwarf virus-PAV; WSMV: wheat streak mosaic virus; and WDV: wheat dwarf virus].

WSMV belongs to the *Tritimovirus* genus of the *Potyviridae* family and has filamentous particles approximately 700 nm in length and 13 nm in diameter (King et al., 2012). The WSMV genome consists of a single-stranded positive RNA molecule of approximately 9,300–9,400 nucleotides, encoding a polypolyprotein that is subsequently cleaved by virus-encoded proteases into functional proteins (Choi et al., 2002).

WSMV can naturally infect many species of plants in the *Poaceae* family, which are commonly cultivated in Ukraine, including almost all wheat cultivars. Additionally, most WSMV isolates can infect other cultivated cereals such as oats, barley, rye, maize, and millet, as well as various grasses, ensuring the survival of the virus (Ellis et al., 2004; Singh et al., 2018; Bennypaul et al., 2019).

On grain cereals, WSMV typically causes mosaic symptoms on the leaves and sometimes leads to plant stunting (Figure 10.2).

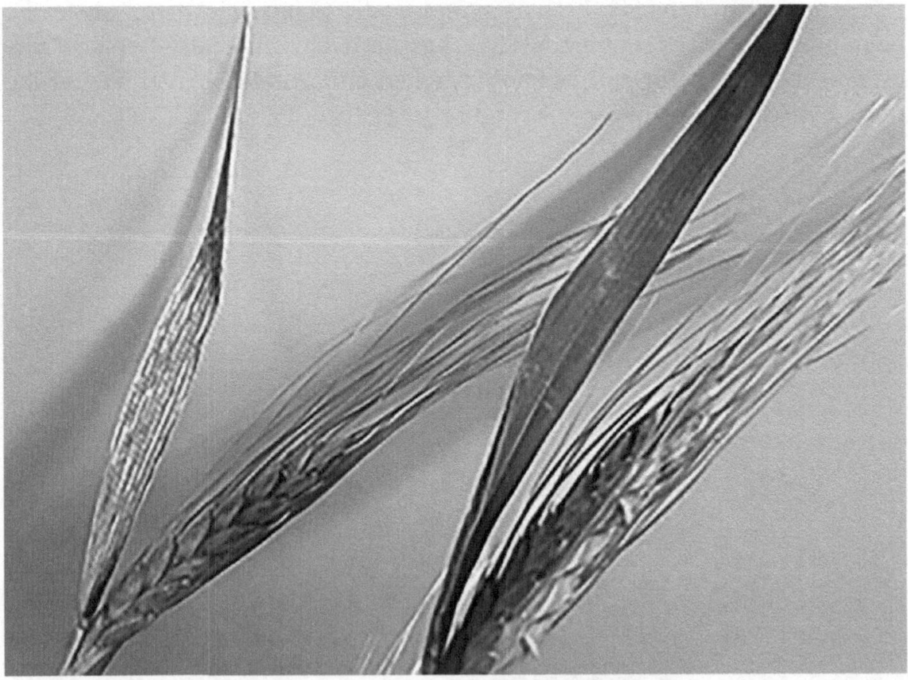

**FIGURE 10.2** The symptoms of leaf mosaic caused by wheat streak mosaic virus on a winter wheat plant (left), compared to a healthy plant (right).

The severity of wheat streak disease depends on various factors such as weather conditions, crop cultivar, sowing period, mite vector population density, and agronomic techniques (Miller et al., 2014; McMechan, 2016; Wosula et al., 2017).

In Ukraine, WSMV has been known for many years, with several outbreaks of the disease recorded in wheat crops (Mischenko & Bojko, 2004; Snihur et al., 2009). Historically, the eastern regions of the country were the most affected by the disease. However, nowadays, the virus is also found in the northern, central, and southern regions. The western regions of Ukraine have not been surveyed yet. However, considering the extensive cultivation of cereals in Ukraine, domestic transportation of seed material, international seed trade, and the mechanical and seed transmission of the virus, it is highly likely that WSMV is widespread throughout the country. Recently, we have confirmed that over 90% of popular winter wheat cultivars currently produced commercially in Ukraine are susceptible to WSMV infection in field conditions (Pozhylov et al., 2022).

The significant spread of WSMV in Ukraine, primarily dependent on the wheat curl mite vector, highlights the crucial need for vector monitoring and control to effectively manage this viral disease. Over the years, *Aceria tosichella* has been predominantly detected in the eastern part of the country, where favorable climatic conditions likely contribute to its spread and overwintering, similar to other regions worldwide (Overmyer, 2020; Karpicka-Ignatowska et al., 2021).

With a few exceptions (Mishchenko et al., 2019), WSMV isolates circulating in Ukraine exhibit relatively low genetic diversity, as observed through partial sequencing. These isolates show high similarity to previously published isolates from Poland, Slovakia, Serbia, Russia, Czech Republic, Italy, Germany, France, Lithuania, Hungary, and Turkey, primarily recovered from wheat but also from barley and millet (Pozhylov et al., 2022). In other words, WSMV populations in Europe demonstrate a high level of overall homogeneity, and Ukraine is no exception.

In summary, several important factors contribute to the spread and long-term persistence of WSMV in Ukraine (and likely in many other countries):

- Several means of transmission (mechanical, by vector and by seed);
- Vector circulation in the fields where cereals are grown (favoring climatic conditions);
- Susceptibility of used cereal cultivars to WSMV;
- A common issue is the absence of crop rotation, where the same fields are often repeatedly used for wheat cultivation;
- The presence of remaining volunteer plants in the fields can favor virus circulation in the absence of cultivated cereals.

Some of these issues can and should be addressed. Primarily, this is about management: crop rotation, vector control (although not very efficient), and eradication of volunteer plants and neighboring grasses on and around the field. Seed transmission is a problem here because Ukraine is a major producer and exporter of wheat, but also imports seed material of foreign selection. Seed contamination by WSMV is considered to be the easiest way for the virus to invade new territories.

Considering the current state of WSMV spread in Ukraine and Europe in general, there is not much hope to eradicate the disease. However, these measures will certainly help control its further spread and minimize its impact.

### *10.2.2 CUCUMBER MOSAIC VIRUS (CMV) IN VEGETABLES AND OTHER HOSTS*

In Ukraine, vegetable crops are considered strategically important as they account for a significant amount of gross plant production and value, reaching approximately 20%. The areas used for the production of these crops and production volumes, including cultivation in glasshouse conditions, increase annually. Additionally, many vegetables such as tomatoes, melons, squashes, and cucumbers traditionally play a crucial role in international trade.

Several highly damaging vegetable viruses have been described in Ukraine, including ZYMV and WMV affecting zucchini, squash, and pumpkin plants. Pepper mild mottle virus (PMMoV) and pepino mosaic virus (PeMV) have been found on tomatoes and sweet peppers. TuMV has been observed on various wild-growing and domesticated brassicas (Tsvigun et al., 2016; Shevchenko et al., 2018).

*Cucumber mosaic virus* (CMV), however, is yet another example of a virus that successfully persists in Ukraine for a very long period of time, primarily due to its extremely wide host range, which includes both mono- and dicotyledonous plants.

CMV poses a threat to the production of a vast majority of vegetables worldwide, including tomato, spinach, and cucurbits. It is known to cause epidemics (Arafati et al., 2013). CMV is efficiently transmitted by aphids in a non-persistent manner, as well as through mechanical inoculation and seed transmission – a situation somewhat similar to WSMV discussed above but amplified by its extremely wide host range.

CMV is classified as a type species of the genus *Cucumovirus* (family *Bromoviridae*) and is known to infect approximately 1,300 species from more than 100 plant families. The genome of CMV consists of a single-stranded, positive-sense RNA. It contains three RNA segments with five open reading frames (ORF), which are packaged in separate icosahedral particles (King et al., 2012). The successful spread of CMV also suggests its ability to rapidly adapt to new hosts and environments (Roossinck, 2002).

Many CMV isolates originating from different hosts have been previously described and classified into two subgroups, designated I and II, with a further subdivision of subgroup I into subgroups IA and IB (Palukaitis & Garcia-Arenal, 2003). Subgroup IA and group II have a worldwide distribution, whereas the subgroup IB isolates were initially supposed to be restricted to Asia only. In Ukraine, isolates of subgroups IA and IB have been reported, and IB isolates of CMV prevail, at least in cucurbits (Shevchenko et al., 2015, 2018).

In Ukraine, CMV is typically found in cucurbits (cucumber, zucchini, squash, melons, etc.), tomato, etc., because these crops are widely cultivated both in greenhouses and open field conditions, so the symptoms can be easily detected. Plants affected by CMV develop mosaic, chlorosis, mottling, necroses on leaves and fruits, and a decrease in the size of the leaf blade is also typically notable (Figure 10.3). Disease severity can vary from no visible symptoms to the death of the host (Delmiglio & Pearson, 2006; Massumi et al., 2007).

In our recent work, we utilized an experimental plot for continuous cultivation of cucurbit crops in central Ukraine. This involved regular virus screening during the vegetative period, which typically occurs in the summer months under open field conditions. The purpose was to investigate the seasonal dynamics of virus populations. From the very beginning of the experiment, CMV was detected, and the number of CMV-positive plants gradually increased over time. Additionally, in cucurbits, CMV is often found in mixed infections with ZYMV and/or WMV, as well as with other viruses in tomato or other solanaceous plants. CMV, however, appears to outcompete these pathogens in terms of spread and transmission efficiency. The results suggest that CMV persists in the ecosystem, where it may circulate off-season by infecting other available hosts, thanks to its opportunistic nature in host selection (Shevchenko et al., 2021).

To provide a few examples specific to Ukraine, CMV has recently been discovered in the medicinal plant *Echinacea purpurea* (Mishchenko et al.,

**FIGURE 10.3** Mosaic symptoms induced by cucumber mosaic virus on pumpkin (A); and cucumber (B).

2021b), the decorative gladiolus plant (Sovinska et al., 2022), and the brassica weed garlic mustard *Alliaria petiolata* (Kyrychenko et al., 2022).

In light of the aforementioned, it should be noted that controlling CMV in open field conditions is purportedly nearly impossible. Various modes of virus transmission (mechanical, via aphid vectors, and through seeds) leave little room for avoiding infection, particularly considering the wide range of aphid species that can transmit this virus. CMV can be transmitted non-persistently by at least 75 aphid species (Mauck et al., 2014). Another aspect of this issue is CMV's ability to infect weed plants, making detection and eradication challenging. Crop rotation is not a reliable option either due to the virus's broad host range. Greenhouses offer a promising approach to managing these risks, albeit at the expense of losing the natural characteristics of seasonal produce.

### *10.2.3 TURNIP MOSAIC VIRUS (TUMV) IN BRASSICAS AND OTHER HOSTS*

Turnip mosaic virus (TuMV) was first described in Ukraine in 2018 (Shevchenko et al., 2018) and should have been recognized as an emerging pathogen. However, the significant number of positive samples and the wide range of infected species indicate that TuMV has likely persisted here for a considerable period of time, which is not surprising after all.

TuMV belongs to the *Potyvirus* genus, which is part of the largest *Potyviridae* family of plant viruses. It has flexible filamentous particles approximately 700–750 nm long, containing a single-stranded positive sense genomic RNA of about 10,000 nt (King et al., 2012).

Like many potyviruses, TuMV has an extremely wide host range, but it primarily infects plant species from the *Brassicaceae* family and induces persistent symptoms such as mosaics, mottling, chlorotic lesions, and more. Among domesticated *Brassica* plants, TuMV is considered one of the most damaging and economically significant viruses (Walsh & Jenner, 2002). TuMV is mainly transmitted non-persistently by various aphid species, as well as mechanically from plant to plant.

TuMV occurs worldwide and has been detected in Africa, Asia, Europe, Oceania, and North and South America. In Europe, TuMV has been reported in nearly every country (Kawakubo et al., 2021).

Ukraine enjoys a strategic position between the eastern EU states and the Black Sea/Middle East region, where TuMV has also been recently detected in Turkey (Korkmaz et al., 2008) and Iran (Farzadfar et al., 2009). Despite

Ukraine's geographical location and long history of cultivating various *Brassica* crops, it is only recently that the authors have registered TuMV in our country (Shevchenko et al., 2018).

When screening just one region of Ukraine for TuMV back in 2017, we collected over 50 symptomatic plants (most belonging to the *Brassicaceae* family) from commercial fields, private gardens, and city streets/outskirts. Surprisingly, the incidence rate of TuMV infection reached 50%. Among the most typical symptoms on cabbage plants were vein banding and leaf deformation (Figure 10.4). Notably, 22% of TuMV-infected plants were also CMV-positive, suggesting a high percentage of mixed infection.

**FIGURE 10.4** Cabbage plant infected by turnip mosaic virus with symptoms of vein banding and leaf deformation.

Several sampling sites in Kyiv, the capital of Ukraine (i.e., where there is no agricultural activity), were characterized by an even higher incidence of TuMV infection, with an average of 55% for symptomatic plants. Such a significant spread of TuMV outside sites for commercial growing of brassicas required further analysis to establish plausible ecological and epidemiological links between the pathogen, host plants, and coexistence conditions. The results of this approach to the analysis of experimental data are shown in Table 10.1.

The results in Table 10.1 show virtually the same frequency of TuMV infection in symptomatic plants, regardless of whether it is an agricultural area or a city street. Thus, it can be argued that despite the effectiveness of classical measures to control viral diseases (Shevchenko et al., 2017), they are not the only and decisive factor in preventing the spread of TuMV. Evidently, in urban areas, associated crops and wild plants play an important role as reservoirs for the conservation and transmission of virus populations. The obtained results testify to the significant spread of TuMV in both agricultural and urban ecosystems and also confirm that this virus has remained unknown in Ukraine for a long time.

**TABLE 10.1** Summary of Detection of Turnip Mosaic Virus in Cultivated, Concomitant, and Wild Plants

| Samples | Overall Quantity | ELISA-Positive Samples | TuMV Incidence Rate (%) |
|---|---|---|---|
| Cultivated plants sampled from cultivation sites (*Brassicaceae*). | 30 | 13 | 43% |
| Concomitant cultivated and wild plants sampled outside cultivation sites (*Brassicaceae*, *Primulaceae*, *Papaveraceae*/*Ranunculaceae*, *Asteraceae*, *Malvaceae*, *Orchidaceae*). | 34 | 14 | 41% |
| Total | 64 | 27 | 42% |

Phylogenetically sequenced Ukrainian TuMV isolate UKR9 showed a high level of homology to the recently described Turkish isolate TUR9 (Korkmaz et al., 2008) and was a recombinant sharing genetic traits of phyletic groups World-B and Asian-BR (Shevchenko et al., 2018). This suggests a possible recent introduction of the virus via international trade of plant material.

To summarize TuMV data, it is worth mentioning that, similarly to CMV, the situation with this pathogen is uncontrollable. Several factors contribute to this: significant spread in cultivated and wild plants, wide host range, multiple means of transmission, and extreme adaptability of the virus.

### *10.2.4 HIGH PLAINS WHEAT MOSAIC VIRUS (HPWMOV) – EMERGING PATHOGEN IN CEREALS*

High plains wheat mosaic virus (HPWMoV) belongs to the *Emaravirus* genus, *Fimoviridae* family. Its genome is (–)ssRNA. This virus is the only known wheat-infecting emaravirus originally reported exclusively from the USA, Argentina, Australia, and New Zealand (Coutts et al., 2014; Alemandri et al., 2017).

In Ukraine, HPWMoV was reported as an emerging wheat and maize pathogen in 2020 after analyzing samples collected back in 2018 (Snihur et al., 2020). This was also the first communication of this pathogen in Europe. Since its first description in Ukraine, HPWMoV has been detected in wheat and maize only in four eastern regions of the country, with an incidence rate of approximately 40% in symptomatic samples (Pozhylov et al., 2022).

Phylogenetic analysis confirmed that Ukrainian isolates of HPWMoV were most similar to the wheat isolates originating from the USA.

In 2018–2019, we studied the patterns of infection of two wheat viruses, WSMV and HPWMoV, in symptomatic wheat plants. The results showed that the average incidence of WSMV monoinfection was 60%, HPWMoV monoinfection accounted for nearly 9%, whereas mixed infection (WSMV+HPWMoV) was found in over 31% of the samples (Figure 10.5). A similar situation was also observed in the USA, as these two viruses are transmitted by the same unique vector, *A. tosichella* or wheat curl mite (Burrows et al., 2009). Mixed infection of winter wheat plants by WSMV and HPWMoV typically correlated with more severe visual mosaic and stunting symptoms.

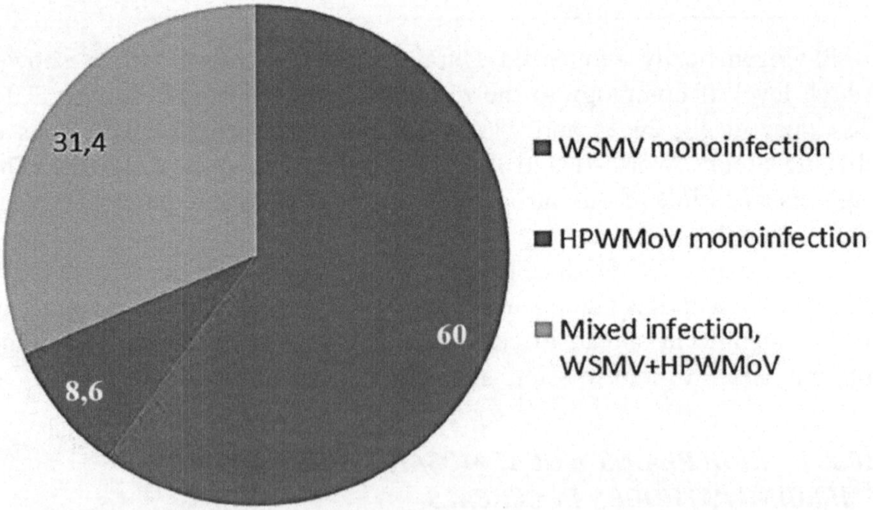


**FIGURE 10.5** The overall patterns of infections induced by wheat streak mosaic virus (WSMV) and high plains wheat mosaic virus (HPWMoV) in symptomatic wheat plants.

A high percentage of mixed infected wheat plants may indicate that this pattern of infection could facilitate more efficient spread of HPWMoV and/or transmission by the vector. If this is the case, we may anticipate a wider spread of HPWMoV in Ukraine in the coming years, particularly in regions where wheat curl mites are present and susceptible cereals are commonly cultivated. Another aspect to consider is that HPWMoV may also be transmitted through maize seeds (Tatineni et al., 2021). The fact that HPWMoV has been detected on both wheat and maize in Ukraine suggests that the virus may have been introduced through contaminated seed material, with subsequent initial circulation on maize, or that multiple introductions of the virus into Ukraine have occurred. Additionally, the possibility of HPWMoV-contaminated mites and/or infected plant material being introduced cannot be ruled out.

### *10.2.5 ONION YELLOW DWARF VIRUS (OYDV) AND LEEK YELLOW STRIPE VIRUS (LYSV) – EMERGING PATHOGENS IN ONIONS AND GARLIC*

Historically, onions and garlic have remained an important part of agriculture and national cuisine worldwide. The increase in production volumes and geographical spread over the last few decades has led to the expected spread of diverse viruses, which in turn has hindered profitable cultivation of these crops. Furthermore, these pathogens are often efficiently transmitted through plant material commonly used for propagation, such as bulbs.

Onion crops can be infected by a variety of viruses, resulting in yield losses (Taglienti et al., 2020). Due to the frequent accumulation of viruses in bulbs and the environment, viral infections of onions may have a mixed etiology and are often caused by a so-called 'garlic viral complex' (Van Dijk, 1994), which may include taxonomically diverse viruses and result in more severe symptoms and heavier crop losses (Majumder et al., 2008).

Until recently, onion and garlic viruses had not been registered or studied in Ukraine. However, since 2015, large areas in different regions of Ukraine have been extensively used for increased cultivation of onions (including onion, leek, Welsh onion, etc.) and especially garlic. Alongside domestic varieties, many new cultivars (seeds and bulbs) have been imported from abroad. This has resulted in numerous cases of disease(s). Research into their cause(s) has rapidly identified two novel viruses that have not been previously reported in Ukraine: OYDV and leek yellow stripe virus (LYSV) (Snihur et al., 2019; Sherevera et al., 2019).

Serological screening conducted in 2018–2019 in four regions showed that both OYDV and LYSV are widespread in Ukraine and are often found in mixed infections. The average incidence rates were 64% and 28% for OYDV and LYSV, respectively. These findings of OYDV and LYSV have already resulted in additional export restrictions on Ukrainian onion/garlic production to the EU.

Both of these viruses belong to the *Potyvirus* genus in the *Potyviridae* family (King et al., 2012) and therefore share many common properties with WSMV and TuMV discussed above. For both OYDV and LYSV, vegetative propagation of onion crops is suggested as the major means of transmission. However, these viruses can also be efficiently transmitted mechanically from plant to plant and non-persistently by aphids, with many widespread aphid species serving as vectors (Katis et al., 2012). In onions and garlic, these two viruses may induce quite similar symptoms. Typically, however, OYDV induces leaf mosaic and deformation of shoots/plant dwarfing, whereas LYSV tends to show up as yellow striping and/or mosaic (Figures 10.6 and 10.7).

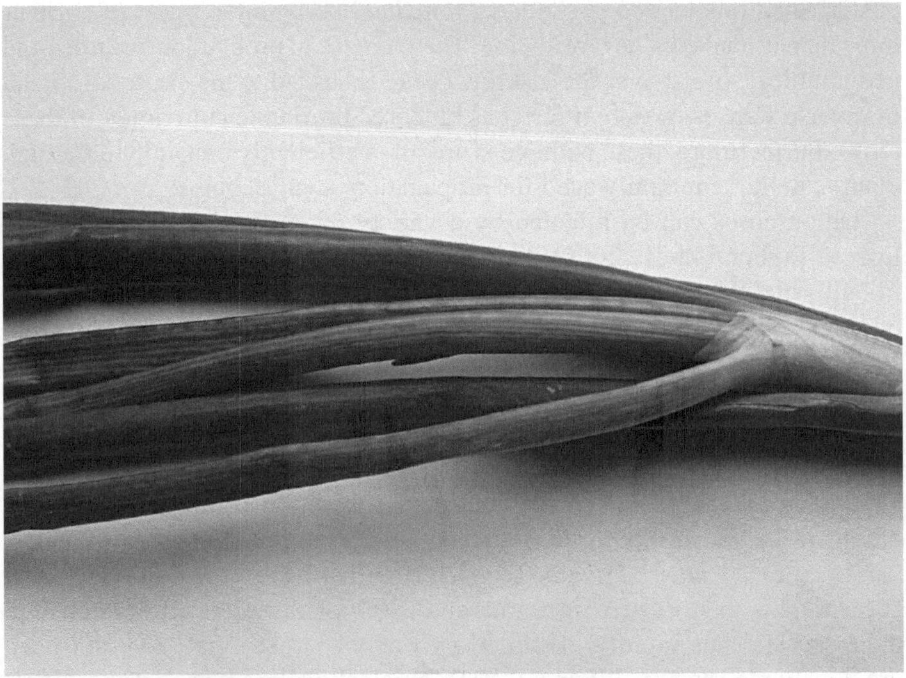

**FIGURE 10.6** Symptoms induced by onion yellow dwarf virus in onion.

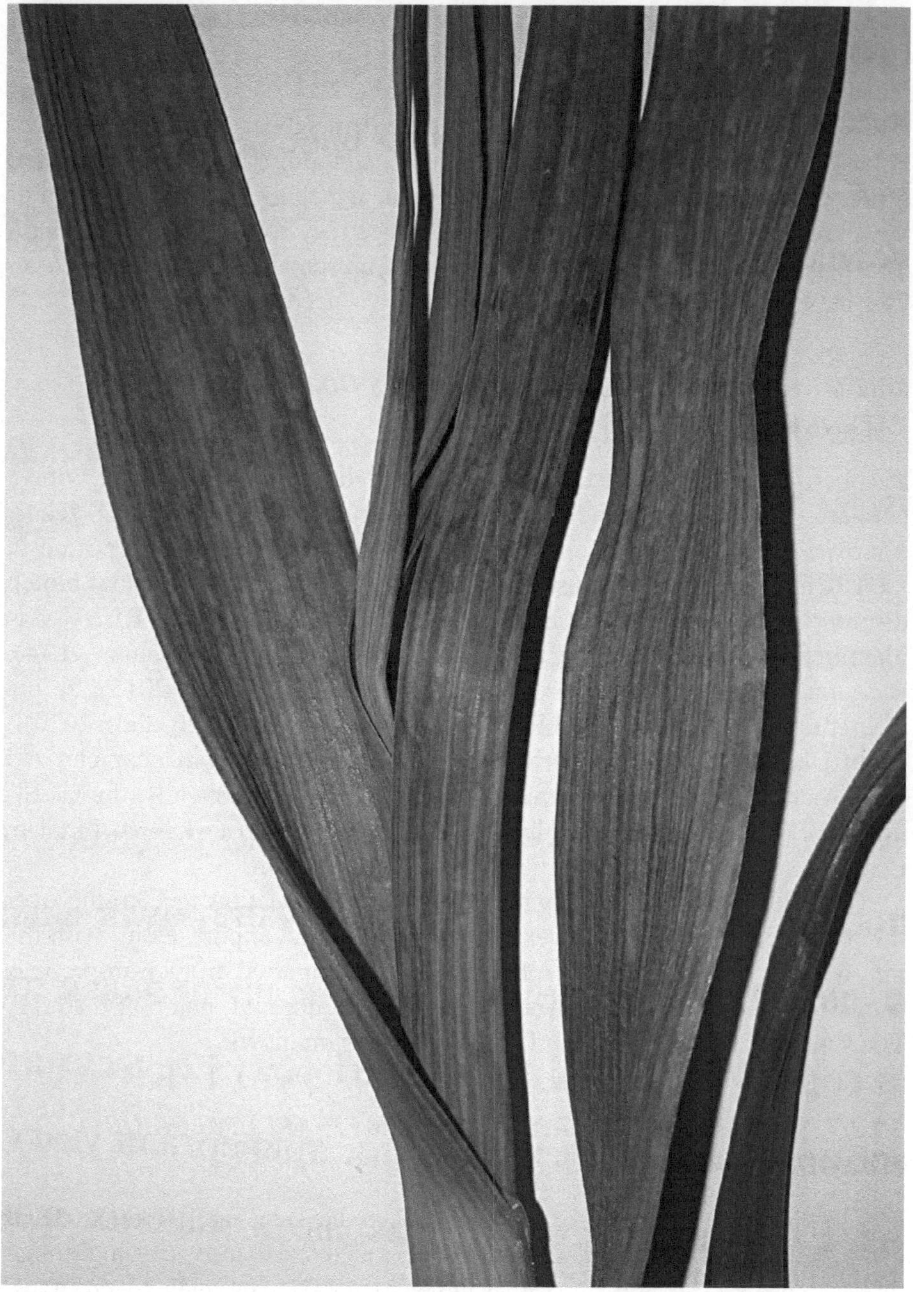

**FIGURE 10.7** Symptoms induced by mixed infection of onion yellow dwarf virus (OYDV) and leek yellow stripe virus (LYSV) in garlic.

Phylogenetic analysis of the Ukrainian isolate of OYDV showed that it was most homologous to published Chinese isolates, suggesting a possible recent introduction with propagative material.

Contrary to OYDV, LYSV was not registered in neighboring countries (except Poland), suggesting that LYSV-infected seed garlic might have been the initial source of this virus in Ukraine. Similarly to the hypothesized Chinese origin of the Ukrainian OYDV isolate, LYSV occurrence may follow the same pattern, as a significant part of (seed) garlic was (and presumably still is) imported to Ukraine from China. However, further research is needed to confirm this.

### *10.2.6 RASPBERRY LEAF BLOTCH VIRUS (RLBV) – EMERGING PATHOGEN IN RASPBERRY*

Raspberry leaf blotch virus (RLBV) is a relative of HPWMoV, with a similar (–)ssRNA genome, and it also belongs to the *Emaravirus* genus, *Fimoviridae* family (King et al., 2012). Initially, this virus was reported in the UK in 2012, where it was found to be the causative agent of leaf blotch disorder in raspberries (McGavin et al., 2012). Subsequently, RLBV was identified in Finland (Bi et al., 2012), Poland (Cieslinska & Tartanus, 2014), Bulgaria (Mavrič et al., 2014), Montenegro (Zindović et al., 2015), Serbia (Jevremovic et al., 2019), and Bosnia and Herzegovina (Delić et al., 2020).

In Ukraine, RLBV was initially discovered in 2018 but characterized three years later (Pozhylov et al., 2018, 2021). Phylogenetically, the Ukrainian RLBV isolate showed a high similarity (93–99%) to known Finnish, British, and other isolates, suggesting a common origin.

Considering the rapid growth of raspberry production in Ukraine over the past decade, both for domestic consumption and export, along with the introduction of many new raspberry varieties imported from Europe, it is suggested that RLBV was introduced to Ukraine through planting material. The spread of this pathogen in Ukraine remains unknown.

### *10.2.7 SUGARCANE MOSAIC VIRUS (SCMV) – (RE)EMERGING PATHOGEN IN MAIZE*

Mosaic disease of maize has been known in Ukraine since the 1970s, and it was concluded at that time that the pathogens were identical to those causing similar diseases in the USA and Europe (Naumenko, 1973). The agents were transmitted mechanically and by aphids in a non-persistent manner. Understandably, proper identification of the viruses was not possible at that time.

However, it is now accepted that there are several viruses that induce mosaic disease in maize: maize dwarf mosaic virus (MDMV), sugarcane mosaic virus (SCMV), sorghum mosaic virus (SrMV), and Johnsongrass mosaic virus (JGMV), which belong to the potyvirus group and infect maize and some other cereals. Among these viruses, only MDMV and SCMV are widespread, while JGMV and SrMV have not been reported in Europe. Despite both MDMV and SCMV circulating in the European area, MDMV is considered to be more damaging to maize, while SCMV is more common in neighboring countries (Pokorný & Porubová, 2000; Trzmiel, 2009).

Like most other potyviruses, both MDMV and SCMV are transmitted mechanically and non-persistently by many aphid species. Additionally, MDMV and SCMV can be transmitted through seeds with an efficiency of up to 0.5% and 0.4%, respectively (Shepherd & Holdeman, 1965; Mikel et al., 1984). Similar to other viruses discussed here, this combination of transmission routes ensures not only the successful introduction of these agents into new ecosystems but also their subsequent persistence.

In our recent work, we conducted screenings of commercial maize fields in 2018 and 2020. During this study, we identified plants exhibiting characteristic mosaic symptoms on their leaves. Subsequent research confirmed that the disease was caused by SCMV, while other potyviruses known to infect maize (MDMV, SrMV, and JGMV) were ruled out. The ELISA data were further validated through transmission electron microscopy and bioassay, which involved mechanical inoculation of different varieties of maize and sweet corn. All tested plants displayed original mosaic symptoms and tested positive in the ELISA (Snihur et al., 2021).

Furthermore, we observed that the presence of mosaic symptoms on maize in the field correlated with a high infestation of *Rhopalosiphum padi* aphids. This suggests that this particular aphid species may be responsible for transmitting SCMV in Ukraine, which aligns with existing data from other countries (Hasan et al., 2003).

The sporadic distribution of symptomatic maize plants, particularly at the early stage with only four to five leaves, indicates that virus-contaminated seeds could potentially be the initial source of the disease in the field.

## 10.3 HEAVY METAL AND RADIOACTIVE POLLUTION AS FACTORS INCREASING RISKS OF VIRUS INFECTIONS

Plants growing in a natural or transformed environment are subjected to a variety of biotic and abiotic stressors. Viruses, being obligate intracellular parasites, are completely dependent on the host plant and therefore one of the most typical biotic stressors. On the other hand, heavy metals and radionuclides are common soil contaminants and powerful abiotic stressors. Their chronic action is likely to affect plant development.

Studies have shown that regions of Ukraine with different levels of soil contaminants exhibit varying percentages of plant viral infections and overall viral diversity. Specifically, screening contaminated ecosystems generally reveals a higher number of virus-infected plants with a higher virus load as well (Polischuk et al., 2006; Dmitriev et al., 2009).

Model experiments using tobacco mosaic virus (TMV) or potato virus X (PVX) with relevant indicator plants (tobacco, tomato, potato) have shown that heavy metals (lead, copper, zinc) in significantly elevated concentrations (5–10 maximum permissible concentration (MPC)) can potentiate viral infections. In particular, virus-infected plants that were additionally exposed to heavy metals exhibited more severe systemic symptoms of viral infection, although their onset was somewhat delayed (Shevchenko et al., 2004, 2007). In addition, the virus content in such plants was 1.3–2.5 times higher than in infected plants that were not exposed to abiotic stressors – heavy metals (Figure 10.8).

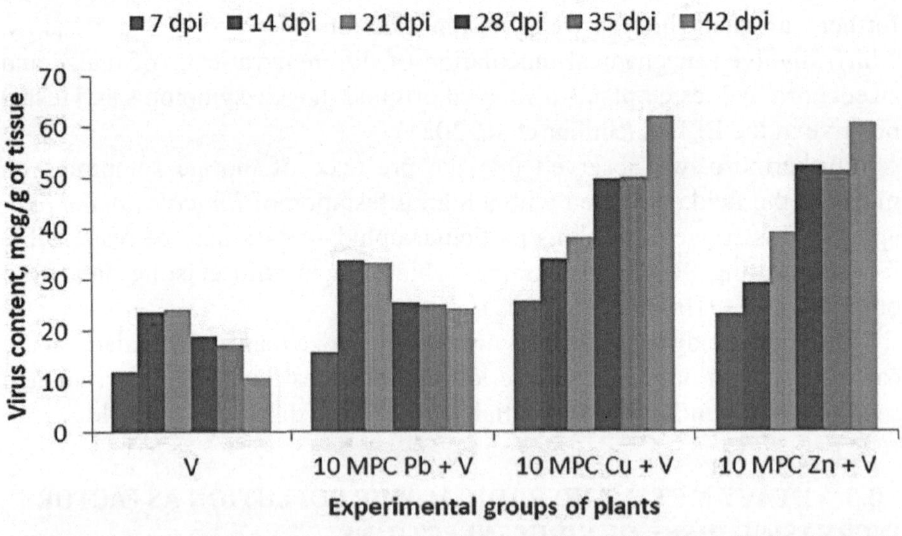


**FIGURE 10.8** Dynamics of virus (V) content in plants chronically affected by heavy metals lead (Pb), copper (Cu), and zinc (Zn) applied in soil at 5 maximum permissible concentrations (MPC) for 7–42 days post inoculation (dpi).

Similar data were obtained both in the laboratory and in open field conditions. Additionally, the effect of heavy metals on the formation of virus-specific inclusions in infected plants has also been demonstrated (Dmitriev et al., 2009).

A consistently high virus content in the tissues of a systemically infected host plant may further facilitate more efficient transmission of the virus to other plants in the ecosystem (i.e., may contribute to a more efficient reservoir of viral inoculum), which is likely to have further consequences for virus ecology and their hosts at the population level.

In general, we believe that there is synergy in the interaction of plant viral diseases and abiotic environmental stressors.

## 10.4 FUTURE PROSPECTS

The identification of numerous emerging viruses and novel hosts for known pathogens (Table 10.2) in such a short period of time indicates that Ukrainian scientists may not have a comprehensive understanding of the current virological situation in crop production.

**TABLE 10.2** Emerging Viruses and Novel Virus Hosts Recently Described in Ukraine

| | **Emerging or Re-Emerging Viruses** | |
|---|---|---|
| **Virus** | **Host** | **References** |
| Turnip mosaic virus | *Brassica* sp., *Raphanus* sp., many other | Shevchenko et al. (2018) |
| Raspberry leaf blotch virus | *Rubus idaeus* | Pozhylov et al. (2018) |
| Bean common mosaic virus | *Phaseolus vulgaris* | Kyrychenko et al. (2019) |
| Onion yellow dwarf virus | *Allium sativum* | Snihur et al. (2019) |
| Leek yellow stripe virus | *Allium sativum* | Sherevera et al. (2019) |
| High plains wheat mosaic virus | *Triticum aestivum, Zea mays* | Snihur et al. (2020) |
| Sugarcane mosaic virus | *Zea mays* | Snihur et al. (2021) |
| | **New Host Plants of Known Viruses** | |
| Tomato spotted wilt virus | *Lophanthus anisatus* | Dikova et al. (2016) |
| Bean yellow mosaic virus | *Gladiolus* sp. | Mishchenko et al. (2020) |
| Cucumber mosaic virus | *Echinacea purpurea* | Mishchenko et al. (2021b) |
| Cherry leaf roll virus | *Sambucus nigra* | Mishchenko et al. (2021c) |
| Potato viruses X, Y, M, and S | *Lamium purpureum* | Kyrychenko et al. (2021) |
| Cucumber mosaic virus | *Alliaria petiolata* | Kyrychenko et al. (2022) |

This situation is also due to the lack of a proper regulatory framework, particularly the absence of ISO-like standards that regulate the necessity and procedure for virological examination of Ukrainian plant products. Currently, crop producers are not required to test their production to confirm that it is free from viruses, except for products intended for export, which is governed by the legislation of the importing countries.

In Ukraine, there is a current situation where monocultures, mainly of foreign selection including transgenic crops, have been cultivated for the past 10–15 years, often without proper phytosanitary control and with an unknown genetic background. With the land reform and land market liberalization starting in 2021, this poses a significant threat to national security.

The range of topical viruses affecting the most important crops in Ukraine is not well-studied. It is likely that there are many more viruses yet to be discovered, and the coming years will confirm this. The current state can be described as a pyramid of knowledge gaps (Figure 10.9). In the near future, this may result in a significant decrease in crop yields, which are strategically important for our country's security and a crucial export commodity that contributes to the national budget, especially wheat, maize, soybean, and rapeseed. Additionally, it will lead to the accumulation of viruses in the agroecosystems of Ukraine, including the soil and reservoir plants, which may render them unusable within 5–10 years.

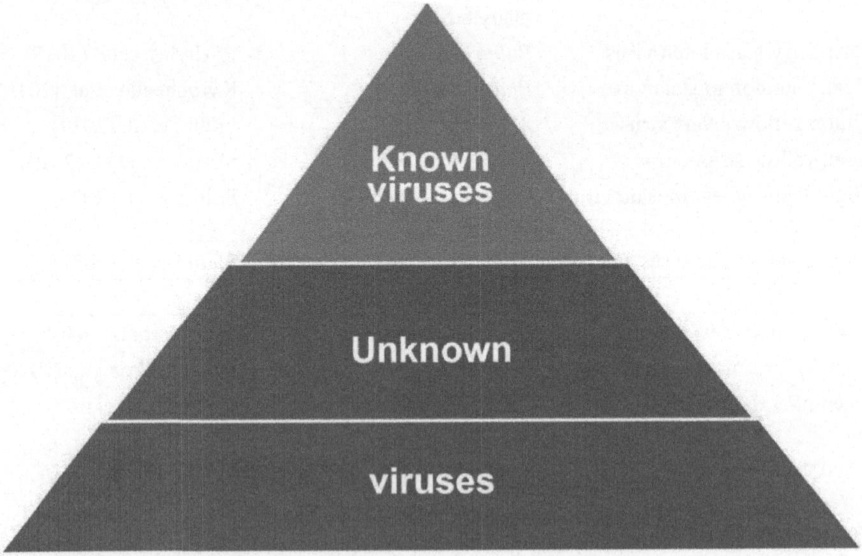


**FIGURE 10.9** The symbolic pyramid of ignorance in plant virology.

In our opinion, given the current conditions for many countries and viruses, modeling their spread is not a promising and cost-effective solution, and is suitable only for individual plant viruses, which indicates the need to develop a different algorithm. With this in mind, it is proposed to focus on a number of key risk factors, taking into account the epidemic chain of plant viruses and ecosystem characteristics, known and new viral pathogens in Ukraine or similar states, ways of their transmission and preservation in the ecosystem, susceptible wild and cultivated plants, and other factors. Based on the obtained data, it is proposed to summarize such information, including retrospective data, which will form the basis for the development of a universal and easy-to-use decision support system (i.e., risk assessment system). Such a system will provide the user with basic information on the probable risk for growing certain crops on a given area and will allow determining further actions to decide on the profitability of the intended use of a particular site.

## 10.5 CONCLUSIONS

Ukraine's showcase data indicates that ignorance is the major risk factor for the emergence and persistence of viral infections. Below, we provide our view on basic issues in plant virology that favor the appearance and spread of viral infections. The list is not exhaustive, but we believe it will also apply to many other parts of the world:

- The absence of regular screening/monitoring of crops for viral diseases (also due to the lack of legislative requirements);
- Uncontrolled (=unscreened) import of seed/plant material;
- Several means of virus transmission (mechanical, by vector and by seed);
- The extreme adaptability of the virus, such as potyviruses capable of recombination;
- Vector circulation in the fields where cereals are grown (favoring climatic conditions, absence of countermeasures);
- Wide susceptibility of used cultivars to certain viruses or absence of resistant cultivars;
- Common non-observance of agricultural practices such as crop rotation and eradication of volunteer plants, which can favor virus circulation in the absence of cultivated crops;
- Anthropogenic contamination of agroecosystems favors virus accumulation and, supposedly, spread.

Many of these issues can be easily addressed through proper training and the implementation of administrative measures. However, considering the high adaptability and diversity of plant viruses, it is unlikely that this battle will be won in the near future. The goal is to minimize losses rather than to control the viruses.

## KEYWORDS


- **cucumber mosaic virus**
- **disease management**
- **emerging disease**
- **pathogens**
- **plant virus**
- **risk factor**
- **turnip mosaic virus**
- **virus monitoring**


## REFERENCES


Alemandri, V., Mattio, M. F., Rodriguez, S. M., & Truol, G., (2017). Geographical distribution and first molecular detection of an *Emaravirus*, high plains *Wheat mosaic virus*, in Argentina. *Eur. J. Plant Pathol., 149*, 743–750.

Arafati, N., Farzadfar, S., & Pourrahim, R., (2013). Characterization of coat protein gene of *Cucumber mosaic virus* isolates in Iran. *Iran J. Biotech., 11*(2), 109–114.

Bennypaul, H., Abdullahi, I., Harding, M., & Neeser, Ch., (2019). First detection of *Wheat streak mosaic virus* in two perennial weed species, *Agropyron cristatum* (L.) Gaertn. and *Hordeum jubatum* subsp. *intermedium* Bowden, in Canada. *Plant Dis., 103*(6), 1441.

Bi, Y., Artola, K., Kurokura, T., Hytönen, T., & Valkonen, J. P. T., (2012). First report of *Raspberry leaf blotch virus* in raspberries in Finland. *Plant Dis., 96*(8), 231.

Burrows, M., Franc, G., Rush, C., Blunt, T., Ito, D., Kinzer, K., Olson, J., et al., (2009). Occurrence of viruses in wheat in the great plains region, 2008. *Plant Health Prog., 10*(1), 14.

Choi, I. R., Horken, K. M., Stenger, D. C., & French, R., (2002). Mapping of the P1 proteinase cleavage site in the polyprotein of *Wheat streak mosaic virus* (genus *Tritimovirus*). *J. Gen Virol., 83*, 443–450.

Cieslinska, M., & Tartanus, M., (2014). *Molecular Diversity of Raspberry Leaf Blotch Virus – A New Pathogen of Rubus sp. Plants in Poland.* Book of Abstracts, The 11th Conference of the European Foundation for Plant Pathology, Krakow, Poland.

Congdon, B. S., Coutts, B. A., Jones, R. A. C., & Renton, M., (2017). Forecasting model for *Pea seed-borne mosaic virus* epidemics in field pea crops in a Mediterranean-type environment. *Virus Res., 241*, 163–171.

Coutts, B. A., Cox, B. A., Thomas, G. J., & Jones, R. A. C., (2014). First report of *Wheat mosaic virus* infecting wheat in Western Australia. *Plant Dis., 98*(2), 285.

Delić, D., Radulović, M., Vakić, M., Sunulahpašić, A., Villamor, D. E. V., & Tzanetakis, I. E., (2020). *Raspberry leaf blotch emaravirus* in Bosnia and Herzegovina: Population structure and systemic movement. *Mol. Biol. Rep., 47*, 4891–4896.

Delmiglio, C., & Pearson, M. N., (2006). Effects and incidence of *Cucumber mosaic virus, Watermelon mosaic virus* and *Zucchini yellow mosaic virus* in New Zealand's only native cucurbit, *Sicyos australis*. *Australasian Plant Pathol., 35*, 29–35.

Desbiez, C., Wipf-Scheibel, C., Millot, P., Berthier, K., Girardot, G., Gognalons, P., Hirsch, J., et al., (2020). Distribution and evolution of the major viruses infecting cucurbitaceous and solanaceous crops in the French Mediterranean area. *Virus Res., 286*, 198042.

Dikova, B., Mishchenko, L., Dunich, A., & Dashchenko, A., (2016). *Tomato spotted wilt virus* on giant hyssop and common valerian in Ukraine and Bulgaria. *Bulg. J. Agricult. Sci., 22*(1), 108–113.

Dmitriev, A., Shevchenko, O., Polischuk, V., & Guscha, N., (2009). Effects of low dose chronic radiation and heavy metals on plants and their fungal and virus infections. *Data Sci J., 8*, PBR49–BR66.

Dutta, M., Sokhandan, B. N., Palmer, M. W., et al., (2014). Genomic characterization of *Ambrosia asymptomatic virus* 1 and evidence of other *Tymovirales* members in the Oklahoma tallgrass prairie revealed by sequence analysis. *Arch. Virol., 159*, 1755–1764.

Ellis, M. H., Rebetzke, G. J., Kelman, W. M., Moore, C. S., & Hyles, J. E., (2004). Detection of *Wheat streak mosaic virus* in four pasture grass species in Australia. *Plant Pathol., 53*(2), 239.

Farzadfar, S., Tomitaka, Y., Ikematsu, M., Golnaraghi, A. R., Pourrahim, R., & Ohshima, K., (2009). Molecular characterization of *Turnip mosaic virus* isolates from *Brassicaceae* weeds. *Eur. J. Plant Pathol., 124*, 45–55.

Gamarra, H., Carhuapoma, P., Cumapa, L., González, G., Muñoz, J., Sporleder, M., & Kreuze, J., (2020). A temperature-driven model for *Potato yellow vein virus* transmission efficacy by *Trialeurodes vaporariorum* (*Hemiptera: Aleyrodidae*). *Virus Res., 289*, 198109.

Hasan, M., Sahi, G. M., Wakil, W., & Imanat, Y., (2003). Aphid transmission of *Sugarcane mosaic virus* (SCMV). *Pak. J. Agr. Sci., 40*(1, 2), 74–76.

Iutinska, O., Bysov, A., Petrenko, S., Budzanivska, I., Polischuk, V., & Shevchenko, O., (2012). Stability of plant virus genotype and risk assessment for arising of novel virus variants with elevated epidemic potential in anthropogenically transformed environment. *Bulletin of Taras Shevchenko National University of Kyiv-Biology, 3*, 5–8.

Jevremovic, D., Leposavić, A., & Paunovic, S., (2019). Genetic diversity of *Raspberry leaf blotch emaravirus* in red raspberries from Serbia. *Spanish J. Agricult. Res., 17*, 1, e1004.

Jones, R. A., (2006). Control of plant virus diseases. *Adv. in Virus Res., 67*, 205–244.

Karpicka-Ignatowska, K., Laska, A., Rector, B. G., Skoracka, A., & Kuczyński, L., (2021). Temperature-dependent development and survival of an invasive genotype of wheat curl mite, *Aceria tosichella*. *Exp. Appl. Acarol., 83*, 513–525.

Katis, N. I., Maliogka, V. I., & Dovas, C. I., (2012). Viruses of the genus *Allium* in the Mediterranean region. *Adv Virus Res., 84*, 163–208.

Kawakubo, Sh., Gao, F., Li, S., Tan, Z., Huang, Y. K., Adkar-Purushothama, C., Gurikar, C., et al., (2021). Genomic analysis of the brassica pathogen *Turnip mosaic potyvirus* reveals its spread along the former trade routes of the silk road. *PNAS, 118*(12), https://doi.org/10.1073/pnas.2021221118.

King, A. M. Q., Adams, M. J., Carstens, E. B., & Lefkowitz, E. J., (2012), *Virus Taxonomy: Classification and Nomenclature of Viruses: Ninth Report of the International Committee on Taxonomy of Viruses*. Elsevier Academic Press: San Diego.

Korkmaz, S., Tomitaka, Y., Onder, S., & Ohshima, K., (2008). Occurrence and molecular characterization of Turkish isolates of *Turnip mosaic virus*. *Plant Pathol., 57*, 1155–1162.

Kutsenko, O. V., Budzanivska, I. G., & Shevchenko, O. V., (2019). Genetic diversity of *Plum pox virus* in Ukraine. *Biopolym. Cell., 35*(6), 476–485.

Kyrychenko, A., Bohdan, M., Snihur, H., Shcherbatenko, I., & Antipov, I., (2021). First report of potato viruses infecting *Lamium purpureum* in Ukraine. *Mikrobiol. Z., 83*(6), 55–64.

Kyrychenko, A., Shcherbatenko, I., & Mishchenko, L., (2019). BCMV-UKR: Isolate of *Bean common mosaic virus* revealed in Ukraine. *Arch. Phytopathol. and Plant Prot., 52*(11, 12), 1005–1017.

Kyrychenko, A., Snihur, H., & Shevchenko, T., (2022). First report of *Cucumber mosaic virus* infecting *Alliaria petiolata* in Ukraine. *J. Plant Pathol.*, https://doi.org/10.1007/s42161-022-01107-0.

MacDiarmid, R., Rodoni, B., Melcher, U., Ochoa-Corona, F., & Roossinck, M., (2013). Biosecurity implications of new technology and discovery in plant virus research. *PLoS Pathog., 9*(8), e100333.

Majumder, S., Baranwal, V., & Joshi, S., (2008). Simultaneous detection of *Onion yellow dwarf virus* and *Garlic latent virus* in infected leaves and cloves of garlic by duplex RT-PCR. *J. Plant Pathol., 90*(2), 371–374.

Makkouk, K. M., & Kumari, S. G., (2009). Epidemiology and integrated management of persistently transmitted aphid-borne viruses of legume and cereal crops in West Asia and North Africa. *Virus Res., 141*(2), 209–218.

Maling, T., Diggle, A. J., Thackray, D. J., Siddique, K. H. M., & Jones, R. A. C., (2008). An epidemiological model for externally sourced vector-borne viruses applied to *Bean yellow mosaic virus* in lupin crops in a Mediterranean-type environment. *Phytopath., 98*(12), 1280–1290.

Massumi, H., Samei, A., Hosseini, P. A., Shaabanian, M., & Rahimian, H., (2007). Occurrence, distribution, and relative incidence of seven viruses infecting greenhouse-grown cucurbits in Iran. *Plant Dis., 91*, 159–163.

Mauck, K. E., De Moraes, C. M., & Mescher, M. C., (2014). Biochemical and physiological mechanisms underlying effects of *Cucumber mosaic virus* on host-plant traits that mediate transmission by aphid vectors. *Plant, Cell and Environment., 37*, 1427–1439.

Mavrič, P. I., Virščček, M. M., Lazarova, S., Peneva, V., Širca, S., & Urek, G., (2014). First detection of *Raspberry leaf blotch virus* in red raspberry in Bulgaria. *J. Plant Pathol., 96*(2), 431–439.

McGavin, W. J., Mitchell, C., Cock, P. J. A., Wright, K. M., & MacFarlane, S. A., (2012). Raspberry leaf blotch virus, a putative new member of the genus *Emaravirus*, encodes a novel genomic RNA. *J. Gen. Virol., 93*, 430–437.

McMechan, A. J., (2016). *Over-Summering Ecology of the Wheat Curl Mite (Aceria Tosichella Keifer).* A dissertation for the Doctorate of Philosophy, Major: Entomology, Lincoln, Nebraska.

Melcher, U., Verma, R., & Schneider, W. L., (2014). Metagenomic search strategies for interactions among plants and multiple microbes. *Front Plant Sci., 5*, 268.

Mikel, M. A., D'Arcy, C. J., & Ford, R. E., (1984). Seed transmission of *Maize dwarf mosaic virus* in sweet maize. *J. Phytopathol., 110*, 185–191.

Miller, Z., Menalled, F., Ito, D., Moffet, M., & Burrows, M., (2014). Impacts of crop variety and time of inoculation on the susceptibility and tolerance of winter wheat to *Wheat streak mosaic virus*. *Plant Dis., 98*, 1060–1065.

Mischenko, L. T., & Bojko, A. L., (2004). *Wheat Streak Mosaic Virus in Ukraine*. Reports of the National Academy of Sciences of Ukraine, 2, 176–180. In Ukrainian.

Mishchenko, L. T., Dunich, A. A., Skrypkina, I. Ya., & Kozub, N. O., (2019). Phylogenetic analysis of two Ukrainian isolates of *Wheat streak mosaic virus*. *Biopolym. Cell, 35*(1), 64–77.

Mishchenko, L., Dunich, A., Dashchenko, A., & Bondus, R., (2021b). Molecular characterization of *Cucumber mosaic virus* infecting *Echinacea purpurea* in Ukraine. *Arch. Phytopathol. and Plant Prot., 54*(15, 16), 1261–1276.

Mishchenko, L., Dunich, A., Molodchenkova, O., & Hlushchenko, L., (2021c). First report of *Cherry leaf roll virus* from *Sambucus nigra* in Ukraine. *J. Plant Pathol., 103*, 1077.

Mishchenko, L., Dunich, A., Smertenko, A., Dashchenko, A., Sovinska, R., & Korotieieva, H., (2020). First report of *Bean yellow mosaic virus* infecting *Gladiolus* sp. in Ukraine. *J. Plant Pathol., 102*, 923, 924.

Mishchenko, L., Nazarov, T., Dunich, A., Mishchenko, I., Ryshchakova, O., Motsnyi, I., Dashchenko, A., et al., (2021a). Impact of *Wheat streak mosaic virus* on peroxisome proliferation, redox reactions, and resistance responses in wheat. *Int. J. Mol. Sci., 22*(19), 10218.

Naumenko, L. A., (1973). Identification of maize mosaic virus. *Mikrobiol. Z., 35*(4), 468–472. In Ukrainian.

Overmyer, L. M., (2020). Factors influencing wheat curl mite *Aceria tosichella* keifer dispersal. *Dissertations and Student Research in Entomology, 65*.

Palukaitis, P., & Garcia-Arenal, F., (2003). Cucumoviruses. *Adv. Virus Res., 62*, 241–323.

Pavlova, S., Stakhurska, O., Budzanivska, I., Shevchenko, O., & Polischuk, V., (2017). Using geographical information system (GIS) technology to control the spread of *Plum pox virus* in Odessa region, Ukraine. *Acta Hort., 1163*, 147–152.

Pokorný, R., & Porubová, M., (2000). The occurrence of viral pathogens of the genus *Potyvirus* on maize (*Zea mays* L.) in the Czech Republic. *J. Plant Dis. and Prot., 107*(3), 329–336.

Polischuk, V. P., Shevchenko, O. V., Budzanivska, I. G., & Shevchenko, T. P., (2006). Abiotic environmental factors: Effects on epidemiology of plant virus infections. In: Cooper, I., Kuehne, T., & Polischuk, V., (eds.), *Virus Diseases and Crop Biosecurity (Nato Security through Science Series C)* (p. 160). Springer.

Polischuk, V., Budzanivska, I., Ryzhuk, S., Patyka, V., & Boyko, A., (2001). *Monitoring of Plant Virus Infections in Biocenoses of Ukraine*. Phytosociocenter: Kyiv, In Ukrainian.

Pozhylov, I., Snihur, H., & Budzanivska, I., (2021). Phylogenetic analysis of Ukrainian isolate of raspberry leaf blotch virus. *AGROFOR Int. J., 6*(1), 19–25.

Pozhylov, I., Snihur, H., Shevchenko, T., Budzanivska, I., Liu, W., Wang, X., & Shevchenko, O., (2022). Occurrence and characterization of *Wheat streak mosaic virus* found in mono- and mixed infection with high plains *Wheat mosaic virus* in winter wheat in Ukraine. *Viruses, 14*(6), 1220.

Pozhylov, I., Stakhurska, O., Shevchenko, O., & Budzanivska, I., (2018). *First Report of Raspberry Leaf Blotch Emaravirus in Raspberries in Ukraine* (p. 142). Books of Abstracts, The XXI Czech and Slovak Conference of Plant Protection, Brno.

Roossinck, M. J., (2002). Evolutionary history of *Cucumber mosaic virus* deduced by phylogenetic analyses. *J. Virol., 76*(7), 3382–3387.

Shepherd, R. J., & Holdeman, Q. L., (1965). Seed transmission of the johnsongrass strain of the *Sugarcane mosaic virus* in Maize. *Plant Dis., 49*, 468–469.

Sherevera, K., Shevchenko, T., Snihur, H., Budzanivska, I., & Shevchenko, O., (2019). First report of *Leek yellow stripe virus* on *Allium sativum* L. in Ukraine. *Agriculture and Forestry, 65*(4), 47–53.

Shevchenko, O. V., Budzanivska, I. G., Shevchenko, T. P., Polischuk, V. P., & Spaar, D., (2004). Plant virus infection development as affected by heavy metal stress. *Arch. Phytopathol. and Plant Prot., 37*(2), 139–146.

Shevchenko, O. V., Kamzel, O. A., Budzanivska, I. G., Shevchenko, T. P., Spaar, D., & Polischuk, V. P., (2007). Heavy metal soil contamination delays the appearance of virus-induced symptoms on potato but favors virus accumulation. *Arch. Phytopathol. and Plant Prot., 40*(6), 406–413.

Shevchenko, O., Shevchenko, T., Snihur, H., & Budzanivska, I., (2021). Epidemiological patterns, prevalence, and seasonal dynamics of different viruses in susceptible cucurbit crops. *Agriculture and Forestry., 67*(1), 73–82.

Shevchenko, O., Tymchyshyn, O., & Shevchenko, T., (2017). Traditional cultivation practices may efficiently prevent the virus spread in susceptible crops: A case study of *Turnip mosaic virus* in Ukraine. *AGROFOR Int. J., 2*(3), 84–91.

Shevchenko, O., Yasaka, R., Tymchyshyn, O., Shevchenko, T., & Ohshima, K., (2018). First evidence of the occurrence of *Turnip mosaic virus* in Ukraine and molecular characterization of its isolate. *J. Phytopathol., 166*, 429–437.

Shevchenko, T., Tymchyshyn, O., AlDalain, E., Bysov, A., Budzanivska, I., Shevchenko, O., & Polishchuk, V., (2015). The first evidence of subgroup IB isolates of *Cucumber mosaic virus* in Ukraine. *Biopolym. Cell, 31*(1), 57–62.

Shevchenko, T., Tymchyshyn, O., Kosenko, I., Budzanivska, I., Shevchenko, O., & Polishchuk, V., (2018). Molecular characterization and phylogenetic analysis of Ukrainian isolates of *Cucumber mosaic virus* based on the partial sequences of three genes. *Biopolym. Cell., 34*(1), 32–40.

Singh, K., Wegulo, S. N., Skoracka, A., & Kundu, J. K., (2018). Wheat streak mosaic virus: A century old virus with rising importance worldwide. *Mol Plant Pathol., 19*(9), 2193–2206.

Snihur, H. O., Bosiuk, V. J., Petrenko, S. M., Shevchenko, O. V., & Polischuk, V. P., (2009). Spread of cereal viruses in some agroecosystems of Ukraine. *Acta Physiologiae Plantarum., 31*(1), S-49.

Snihur, H., Kharina, A., Kaliuzhna, M., Chumak, V., & Budzanivska, I., (2021). First report of *Sugarcane mosaic virus* in *Zea mays* L. in Ukraine. *Mikrobiol. Z., 83*(5), 58–66.

Snihur, H., Petrenko, S., Kot, T., Shevchenko, O., & Polischuk, V., (2018). Widespread viral diseases endangering cereal crops in Ukraine. *Mikrobiol. Z., 80*(3), 103–114.

Snihur, H., Pozhylov, I., Budzanivska, I., & Shevchenko, O., (2020). First report of occurrence of high plains *Wheat mosaic virus* on different hosts in Ukraine. *J. Plant Pathol., 102*, 545, 546.

Snihur, H., Shevchenko, T., Sherevera, Kh., Budzanivska, I., & Shevchenko, O., (2019). First report of *Onion yellow dwarf virus* in Ukraine. *J. Plant Pathol., 101*, 1283.

Sovinska, R., Dunich, A., & Mishchenko, L., (2022). Co-infection of bean *Yellow mosaic virus* and *Cucumber mosaic virus* in *Gladiolus* sp. plants: Phylogenetic analysis of Ukrainian isolates. *Arch. Phytopathol. and Plant Prot., 55*(3), 303–330.

Taglienti, A., Tiberini, A., Ciampa, A., Piscopo, A., Zappia, A., Tomassoli, L., & Dell'Abate, M. T., (2020). Metabolites response to *Onion yellow dwarf virus* (OYDV) infection in 'Rossa di Tropea' onion during storage: A 1H HR-MAS NMR study. *J. Sci. Food and Agricult., 100*(8), 3418–3427.

Tatineni, S., & Hein, G. L., (2021). High plains *Wheat mosaic virus*: An enigmatic disease of wheat and corn causing the high plains disease. *Mol Plant Pathol., 22*(10), 1167–1179.

Thackray, D. J., Diggle, A. J., Berlandier, F. A., & Jones, R. A. C., (2004). Forecasting aphid outbreaks and epidemics of *Cucumber mosaic virus* in lupin crops in a Mediterranean-type environment. *Virus Res., 100*(1), 67–82.

Tóbiás, I., Shevchenko, O., Kiss, B., Bysov, A., Snihur, H., Polischuk, V., Salánki, K., & Palkovics, L., (2011). Comparison of the nucleotide sequences of *Wheat dwarf virus* (WDV) isolates from Hungary and Ukraine. *Pol. J. Microbiol., 60*(2), 125–131.

Trzmiel, K., (2009). First report of *Sugarcane mosaic virus* infecting maize in Poland. *Plant Dis., 93*(10), 1078.

Tsvigun, V. O., Rudneva, T. O., Shevchenko, T. P., Budzanivska, I. G., & Polishchuk, V. P., (2016). Strain attribution of Ukrainian isolates of *Zucchini yellow mosaic virus* and their occurrence in Ukraine. *Biopolym. Cell., 32*(3), 235–241.

Van, D. B. F., & Jeger, M. J., (2017). The basic reproduction number of vector-borne plant virus epidemics. *Virus Res., 241*, 196–202.

Van, D. P., (1994). Virus diseases of *Allium* species and prospects for their control. *Acta Hort., 358*, 299–306.

Walsh, J. A., & Jenner, C. E., (2002). *Turnip mosaic virus* and the quest for durable resistance. *Mol. Plant Pathol., 3*, 289–300.

Wosula, E. N., McMechan, A. J., Knoell, E., Tatineni, S., Wegulo, S. N., & Hein, G. L., (2018). Impact of timing and method of virus inoculation on the severity of wheat streak mosaic disease. *Plant Dis., 102*, 645–650.

Zindović, J., Virščëk, M. M., & Mavrič, P. I., (2015). First report of *Raspberry leaf blotch virus* (HTS/RT-PCR) in red raspberry in Montenegro. *J. Plant Pathol., 97*(2), 398.

# Index

## C

## N

## O



## P

## S



## T

## U



## V

## W

## Y



## Z